Theoretical Mechanics

THEORETICAL MECHANICS

E. Neal Moore
University of Nevada, Reno

John Wiley & Sons
New York · Chichester · Brisbane · Toronto · Singapore

Copyright © 1983, by John Wiley & Sons, Inc.

All rights reserved. Published simultaneously in Canada.

Reproduction or translation of any part of
this work beyond that permitted by Sections
107 and 108 of the 1976 United States Copyright
Act without the permission of the copyright
owner is unlawful. Requests for permission
or further information should be addressed to
the Permissions Department, John Wiley & Sons.

Library of Congress Cataloging in Publication Data:

Moore, E. Neal (Edwin Neal), 1934–
Theoretical mechanics.

Includes index.
1. Mechanics.   I. Title.

QA805.M74 1983   531   83-6841
ISBN 0-471-87488-4

Printed in the United States of America

10 9 8 7 6 5 4 3 2 1

This book is dedicated to
**Ruth,
Eric,
and
Julie**

Preface

This book has evolved from my teaching of a course in classical mechanics that is regularly given to graduate students at the University of Nevada, Reno. In the past 20 years, I have taught this one-semester course on several occasions, each time becoming more convinced of the need for a new textbook in this important area.

My overall selection of topics and illustrative examples differs somewhat from those found in existing textbooks, though naturally duplication of the standard material is impossible to avoid. I have included some topics that have been developed only in recent years and, on the other hand, have resurrected older material where it seemed appropriate. It is my contention that physicists should have a wide variety of material available to chose from, for there are those who prefer peach to chocolate or vanilla.

More topics have therefore been included than could normally be covered with any depth in a one-semester course. The presentation of topics has been structured so that no loss of continuity will occur if certain sections are omitted. The instructor can then slightly vary the content of the course from year to year. For example, the sections on Noether's theorem, the stress–strain tensors, nonlinear oscillations, and general relativity among others might be bypassed without detracting from the reader's comprehension of the remaining material.

Existing textbooks often place emphasis on the formal structure, the inherent beauty in the logical framework of mechanics. Some even go so far as to present the subject matter by a succession of theorems with rigorous proofs attached. Such an approach may be aesthetically pleasing to the instructor (who is largely familiar with the material). I have a strong pedagogical conviction, however, that students acquire their understanding more from experience in problem solving than from a study of formal proofs. As a consequence, I emphasize application of the methods discussed, giving illustrative examples immediately after most of the theorems. In general my approach often sacrifices rigor in the interests of clarity.

The book was intended primarily for students of physics at the introductory graduate level, but it might also be used successfully by persons in other disciplines, such as engineering, mathematics, and astronomy. Students with a good mathematical background could even employ this text at the senior level. It is assumed only that the reader has a familiarity with the basic undergraduate principles of mechanics and of calculus.

For the person who has background deficiencies or who is in need of review, Chapter 1 has been designed to provide a concise summary of the key undergraduate material in mechanics. In addition, the first four appendices supply essential formulas and give a brief overview of the mathematical areas involving vectors, Dirac delta functions, determinants, matrices, and complex variables.

Chapter 2 treats the methods of Lagrange and Hamilton, with the emphasis placed on several detailed examples of constrained motions. Here I introduce the Hamiltonian and the canonical equations at an earlier stage than in other books, for in my opinion the importance of the Hamiltonian (both in classical mechanics and in quantum mechanics) should not be underestimated.

Chapter 3 attempts to explain how variational principles are constructed and *used* in the solution of physical problems. Thus, the mere formulation of Hamilton's principle is not sufficient. An enterprising student will want to know how it is applied, and that is explicitly shown.

In Chapter 4 the two-body problem is discussed by means of time-honored procedures. Then I develop theorems for the many-body problem, employing the three-body case for specific examples. In this way an attempt has been made to broaden the student's understanding beyond the point reached with more elementary textbooks.

Chapters 5 and 6 develop more powerful mathematical tools, notably matrices, Cartesian tensors, and dyadics. Several different physical applications are made in the first of these chapters, as the mathematical machinery is being assembled; the emphasis, however, is on the rotation of rigid bodies. Although the division is not completely sharp, Chapter 5 is concerned in essence with the kinematics and Chapter 6 with the dynamics of rotation. In particular, the latter elucidates the distinction between finite and infinitesimal rotations, and gives a modern approach to the subject of the angular velocity about a time-dependent axis.

Chapter 7 begins with a review of simple oscillating system, then continues with more complicated situations involving transform methods and normal coordinates. In addition, introductions to the study of stability criteria and nonlinear phenomena are presented. Entire textbooks cannot do justice to these weighty subjects, but I feel there is a definite advantage in letting the student catch the flavor of the methods used in these areas.

Chapter 8 covers canonical transformations both from the standpoint of generating functions and from the currently fashionable symplectic approach. An analysis of the Hamilton–Jacobi method immediately follows. I also include a section on the classical time-development operator, a subject closely related to the discussion of Lie groups in Appendix F.

The final chapter contains a treatment of relativistic modifications to mechanics. Special relativity is developed by means of rotations in a four-dimensional space; a diagrammatic approach and a number of interesting physical applications are given. An introduction to general relativity is also presented, though rather briefly; however, the student should be able to understand the physical ideas and something of the mathematics required in this area.

Appendices E and F are actually an extension of the text into mathematical areas that some instructors will not want to pursue. But the general tensor analysis of Appendix E will have to be mastered by a student seriously interested in areas such as general relativity, while the Lie algebraic formalism of Appendix F is essential to current levels of research in classical mechanics as well as in particle physics. I have tried to make these appendices as painless as possible by giving several concrete examples and avoiding complexities in the formalism.

A few words on peripheral matters are in order at this point. When ever I make a brief comment of a historical or philosophical nature, I give a footnote on that page as a reference for an inquisitive reader. However, all the *primary* references are listed together at the back of the book; because many of my conceptual treatments and examples originated in one of these references, I recommend them highly.

Equations are numbered consecutively throughout each chapter, but only the more important ones (or those used later) are so designated. Occasionally an equation number is primed to indicate that it is a variation of the equation having the same but unprimed number, or is paired closely with the unprimed equation.

It is my hope that this book will help the student master the techniques required for the solution of complicated problems. Undoubtedly many of the physical concepts will be familiar, but the reader should gain experience with some new procedures that have proved fruitful in several fields of physics in addition to mechanics.

I thank W. T. Scott and P. L. Altick of the Physics Department and R. Macauley of the Mathematics Department at the University of Nevada, Reno. They have greatly encouraged me, and have made several detailed suggestions of value. I want also to express my appreciation to my ever faithful secretaries Mary Kaylor and Shirley Gallian for heroic hours spent in the typing of lengthy equations, as well as to Physics Editor Robert McConnin and the editorial staff and reviewers at Wiley for their splendid cooperation during all stages of the writing and production.

E. Neal Moore
Reno, Nevada
February, 1983

Contents

CHAPTER ONE Fundamental Principles

1.1	Introduction	1
1.2	Newton's Laws of Motion	2
1.3	Moments of Mass Distributions	7
1.4	Statics and Dynamics; Energy	11
1.5	Applications of the Basic Principles	19
1.6	The Gravitational Field and Potential	25
	Problems	31

CHAPTER TWO The Methods of Lagrange and Hamilton

2.1	Constraints and Generalized Coordinates	35
2.2	Applications of Lagrange's Equations	40
2.3	Generalized Momenta and System Invariants	47
2.4	The Hamiltonian and the Canonical Equations	51
2.5	Forces of Constraint	55
2.6	Noether's Theorem	64
	Problems	68

CHAPTER THREE Variational Principles

3.1	The Calculus of Variations	70
3.2	Hamilton's Principle	78
3.3	Other Variational Principles	83
	Problems	90

CHAPTER FOUR Central Forces

4.1	The Equations of Motion for Central Forces	92
4.2	The Inverse Square Law of Force	96
4.3	The Equivalent One-Dimensional Model and the Virial Theorem	106
4.4	Scattering and Cross Sections	111
4.5	The Many-Body Problem	115
	Problems	123

CHAPTER FIVE Matrices, Tensors, and Rotation Operators

5.1	Linear Transformations; Matrices	126
5.2	The Euler Angles and the Full Rotation Matrix	133
5.3	Cartesian Tensors and Dyadics	136
5.4	The Tensor of Inertia	141
5.5	Eigenvalue Equations; Principal Moments and Axes of the Inertia Tensor	144
5.6	Rotation Operators	151
5.7	The Stress–Strain Tensors	156
5.8	A Matrix Formulation of Linear Collision Theory	165
	Problems	169

CHAPTER SIX The Physics of Rotation

6.1	Euler's Theorem	173
6.2	Vectorial Representation of Rotations; Pseudovectors	176
6.3	Rotating Reference Systems; Angular Velocity	182
6.4	Time-Dependent Rotation Axes; Euler's Equations	189
6.5	Force-Free Motion of a Rigid Body; the Poinsot Construction	195
6.6	Miscellaneous Applications	202
	Problems	217

CHAPTER SEVEN Oscillations and Stability

7.1	Stability of Equilibrium	220
7.2	The One-Dimensional Damped Harmonic Oscillator; Vibration of Strings	221
7.3	Transform Methods for Oscillations	230
7.4	Normal Coordinates	237
7.5	Continuous Media	253
7.6	Stability of Motion; the Routh–Hurwitz Criterion	256

7.7	Oscillations in the Phase Plane	261
7.8	Nonlinear Equations; the Methods of Poincaré	267
	Problems	274

CHAPTER EIGHT Canonical Transformations

8.1	Hamilton's Canonical Equations and their Symmetry	277
8.2	Poisson Brackets and their Properties	279
8.3	Canonical Transformations	286
8.4	Generating Functions for Canonical Transformations	289
8.5	The Hamilton–Jacobi Equation	297
8.6	Action and Angle Variables	303
8.7	Infinitesimal Canonical Transformations and the Time Development Operator	309
8.8	Time-Dependent Perturbation Theory	311
	Problems	316

CHAPTER NINE The Theory of Relativity

9.1	Introduction	319
9.2	The Lorentz Transformation	321
9.3	The Brehme Diagram; Causality	327
9.4	Some Paradoxes of Relativity Theory	332
9.5	Four-Vectors	337
9.6	The General Theory of Relativity	346
9.7	The Riemann Curvature Tensor	352
9.8	The Schwarzschild Solution of the Field Equations	355
	Problems	363

APPENDIX A Vector Operations and Identities

A.1	Definitions and Basic Relationships	366
A.2	Vector Relationships in Orthogonal Curvilinear Coordinates	369
	Table of Useful Vector Relationships	373

APPENDIX B Dirac Delta Functions and Their Role in Physics

B.1	The Concept of the Delta Function	374
B.2	Delta Function Calculus	375
	Table of Useful Relationships Involving Dirac Delta Functions	377
B.3	Delta Functions in More than One Dimension	378

APPENDIX C Determinants and Matrices

C.1	Definition and Evaluation of Determinants	380
C.2	Properties of Determinants	382
C.3	Systems of n Linear Equations in n Unknowns	384
C.4	Properties of Matrices and Inverses	386
C.5	The Characteristic Equation of a Matrix	389

APPENDIX D Complex Variables and Contour Integration

D.1	Basic Concepts	392

APPENDIX E General Tensors and the Metric Tensor

D.1	Contravariance and Covariance	398
D.2	The Metric Tensor	402
D.3	Differentiation of Tensors	413
	Problems	418

APPENDIX F Groups, Lie Algebras, and the Formal Structure of Mechanics

F.1	The Theory of Groups	420
F.2	Lie Groups and Algebras	424
F.3	The Rotation Group in Three-Dimensional Space and the Group SU(2)	431
F.4	Other Groups of Interest in Physics	439
F.5	Lie Groups and Mechanics	441
	Problems	444
	References	446
	Index	451

A List of Significant Contributors to Mechanics

Name	Period of Life	Nationality	Partial Accomplishments
Newton, Sir Isaac	1643–1727	English	Laws of motion and gravitation with their implications.
(and chronologically, in order of birth)			
Galilei, Galileo	1564–1642	Italian	Aided development of laws of motion, stressed importance of observations.
Kepler, Johannes	1571–1630	German	Laws of planetary motion.
Bernoulli, Daniel	1700–1782	Swiss	Mathematics, fluids.
Euler, Leonhard	1707–1783	Swiss	Mathematics, Euler's equations for rotational motion.
D'Alembert, Jean le Rond	1717–1783	French	Mathematics, philosophy, D'Alembert's principle.
Lagrange, Joseph Louis	1736–1813	French	Mathematics, gravitational theory, Lagrange's equations.
Gauss, Karl Friedrich	1777–1855	German	Mathematics, field theory.
Poisson, Siméon Denis	1781–1840	French	Field theory, oscillations.
Hamilton, Sir William Rowan	1805–1865	English	Canonical equations, quaternions.
Riemann, Georg Friedrich Bernhard	1826–1866	German	Mathematics, Riemann spaces.
Lie, Marius Sophus	1842–1899	German	Mathematics, Lie groups and algebras.
Poincaré, Jules Henri	1854–1912	French	Mathematics, oscillations.
Einstein, Albert	1879–1955	German	Special and general relativity.

Theoretical Mechanics

Chapter One
FUNDAMENTAL PRINCIPLES

1.1 INTRODUCTION

Classical mechanics is a subject that appeals to a large and varied audience. Mathematicians, engineers, and scientists apply its principles daily in their work. In particular, the study of mechanics is essential to the training of a physicist; it is needed both for the development of mathematical skills and for the broadening of background knowledge.

The traditional approach to mechanics utilizes the notions of **particles** (point masses) and of extended **rigid bodies**. The particle is an obvious idealization, for every physical object must occupy a finite portion of space. Nevertheless, the concept is convenient because it allows the specification of position by a single vector, without regard for internal structure. In recent decades, the theory of relativity has taught us that the concept of a rigid body is also an idealization, since the size and shape of an object are greatly affected by its state of motion.

In addition, the role of the observer has largely been ignored in classical physics. By contrast, the interaction of the experimentalist with the system under investigation *has* generally been considered in quantum mechanics. Despite these and other limitations, we shall adopt the conventional particle–rigid body framework, because it has proved so useful.

The first chapter of this book is primarily a review of key undergraduate material. As a consequence, the reader is probably familiar with most of the topics covered. Many equations are written in vector form; a student whose background is deficient in that area should refer to Appendix A.

We frequently employ the **position vector** $\vec{r}$, which specifies the location of a particle relative to the origin; in rectangular coordinates (x, y, z), it has the form

$$\vec{r} = x\hat{i} + y\hat{j} + z\hat{k}, \tag{1.1}$$

where the unit vectors $\hat{i}, \hat{j}$, and $\hat{k}$ are constant vectors along the axes (Figure 1.1).

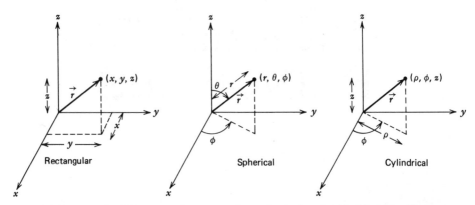

Figure 1.1 Position vector in rectangular, spherical, and cylindrical coordinate systems.

The time derivative of $\vec{r}$ will often be indicated by placing a dot above $\vec{r}$; it is the particle **velocity** $\vec{v}$ in the given coordinates. The time derivative of the velocity is the **acceleration** $\vec{a}$:

$$\vec{v} \equiv \frac{d\vec{r}}{dt} \equiv \dot{\vec{r}} = \dot{x}\hat{i} + \dot{y}\hat{j} + \dot{z}\hat{k}, \tag{1.2}$$

$$\vec{a} \equiv \frac{d\vec{v}}{dt} \equiv \dot{\vec{v}} \equiv \ddot{\vec{r}} = \ddot{x}\hat{i} + \ddot{y}\hat{j} + \ddot{z}\hat{k}. \tag{1.2'}$$

The position, velocity, and acceleration vectors are basic to the study of motion, also called **kinematics**.

1.2 NEWTON'S LAWS OF MOTION

The fundamental laws of classical mechanics were formulated by Sir Isaac Newton in the seventeenth century. While the concepts involved were not entirely original with him, he was the first who clearly stated and developed them. Much of this book is devoted to applications and extensions of the Newtonian ideas.

In earlier centuries, it had been thought that the external forces acting on a body were responsible for its motion, by pushing or pulling the object along. Friction experiments done by Galileo and others changed this point of view dramatically at the time of the Renaissance and resulted in the emergence of the concept of **inertia**—the resistance of an object to a *change* of its motion. Gradually, it came to be recognized that an unbalanced force $\vec{F}$ did not *cause* the motion, but instead *altered* it, producing acceleration. The mass m of a body is taken as the quantitative measure of its inertia; in principle, it can be found by observing the acceleration produced with a given force.

First Law (Law of Inertia)
Every body at rest tends to remain at rest, and every body in motion tends to maintain a constant velocity, unless acted on by an external unbalanced force.

Second Law (The Basic Equation of Motion)
An external unbalanced force $\vec{F}$ produces an acceleration $\vec{a}$ of a mass m according to the relation

$$\vec{F} = m\vec{a}. \tag{1.3}$$

Defining the **momentum** as

$$\vec{p} = m\vec{v}, \tag{1.4}$$

we can also write the second law in the form

$$\vec{F} = m\dot{\vec{v}} = \frac{d}{dt}(m\vec{v}) = \frac{d\vec{p}}{dt}, \tag{1.3'}$$

where the mass is assumed constant.

Third Law (Law of Action and Reaction)
For every force one body exerts on another body (the action), there is always a reaction force back on the first body; these paired forces are equal in magnitude and opposite in direction.

As an example of the third law, consider a parent who is administering a spanking to a child. The statement, "this hurts me as much as it does you," may or may not be true, depending on the emotional factors involved, but there is a reaction force on the parental hand which is certainly equal in magnitude to that of the blow delivered!

In many cases, the application of Newton's laws is not difficult. First, however, we must try to give rigorous definitions of **force** and **mass**, since these quantities are of paramount importance. It is at this point that some primary difficulties of interpretation have to be confronted.

The viewpoint of Mach[1] and many other writers is that the second law *is* the definition of force. Having accepted that idea, one can then apply the same force

---
[1] E. Mach, *The Science of Mechanics*, Open Court Publishing Co., La Salle, Illinois, 1902.

to two different bodies of masses, m_1 and m_2, obtaining from the second law

$$m_1 \vec{a}_1 = m_2 \vec{a}_2,$$

or

$$\frac{m_1}{m_2} = \frac{|\vec{a}_2|}{|\vec{a}_1|}.$$

Upon selection of a standard of mass for m_2, the acceleration ratio will determine the mass m_1 of any desired object. (The *ratio* will be the same for these two bodies, even if another force is employed with both.) In other words, the ratio of the measures of inertia is taken as inversely proportional to the acceleration ratio, hence the product $m\vec{a}$ is a convenient quantity for use in physics and is to be designated as the "force."

Strong objections can certainly be raised to this viewpoint in the sense that it takes a familiar physical phenomenon and elevates it to the status of a mathematical abstraction. If the second law is merely a definition, where is its connection to physical reality? Furthermore, a force defined in this way may be very useful in mechanics, but that does not guarantee that it will also prove helpful in electromagnetism.

The force can be defined in other ways, of course; one direct method is through the extension of a spring balance. The second law is then a truly observational postulate, yet this procedure still gives us an uneasy feeling. It is dependent on the details of the apparatus employed and therefore seems vaguely unsatisfactory.

Newton himself, in his famous work, the *Principia*, is unclear on his interpretation of these basic concepts that form the foundation of mechanics. The development of a theory must start from somewhere, however, even if the logical underpinnings are somewhat insecure.

In Newton's view, it was evidently easier to identify the case in which *no* force acts on a body; according to the first law, the object then remains at rest or moves at constant velocity. It is assumed that this motion is measured with respect to the particular frame of reference called an **inertial** (or **Newtonian**) **frame**. The first law is simply a special case of the second, with $\vec{F} = 0 = \vec{a}$; one suspects that Newton separated the laws with the intention of having the first law serve to define the inertial frame. The remaining laws are then understood to be valid only in such a reference system.

Once an inertial frame has been found, there are an infinite number of others moving uniformly relative to the first, and Newton's laws will be valid in all of these. Consider a second (primed) system moving along the x axis at constant speed v relative to the inertial (unprimed) frame:

$$x' = x - vt \quad \text{(and} \quad y' = y, \quad z' = z, \quad t' = t), \tag{1.5}$$

whereupon $\dot{x}' = \dot{x} - v$ and $\ddot{x}' = \ddot{x}$. The transformation from one such frame to another, called a **Galilean transformation**, preserves the value of acceleration so that the force law is unchanged.

Suppose, however, that a particle is at rest in the frame moving with velocity v and that v is no longer constant. The inertial (unprimed) observer sees an accelerated particle, which *must* be experiencing an actual physical force according to the second law; the "moving" observer, on the other hand, sees a particle at rest and concludes erroneously that *no* forces are acting. Newton's laws are therefore clearly invalid in a frame that is accelerated relative to an inertial frame (unless other "fictitious forces" are introduced, as will be discussed later).

The problem is that there is no good method of selecting even *one* inertial frame, either on conceptual or experimental grounds. Newton presupposed a background of absolute space and time in which an inertial frame was embedded, though he evidently found this position somewhat distasteful philosophically. As a practical matter, one merely specifies an approximate inertial frame in accordance with the needs of the problem under investigation. For elementary applications in the laboratory, a frame attached to the Earth usually suffices. For astronomical applications, however, one might have to choose a frame that is stationary relative to the background of distant fixed stars.

Alternative points of view are possible. In Chapter 9 we see that Einstein formulated the laws of physics for the general theory of relativity in a way that is independent of the coordinate system chosen. Almost a century ago, Ernst Mach outlined another scheme known today as **Mach's Principle**. It will be worthwhile to examine his ideas briefly, inasmuch as they are often mentioned in the literature.

Mach did not believe in the existence of an absolute space but felt that the inertial frame was determined locally through some type of interaction with all the matter in the universe. He did not attempt to specify the detailed nature of the interaction responsible for the inertia of a mass at a point, preferring to emphasize the significance of *relative* motion of all masses.

Thus, taking the Earth–moon system as an example, we can consider the moon in revolution about the Earth in the fixed framework of the distant stars. Or, alternatively, one might consider the shell of distant stars in revolution about the Earth at the center, with the moon fixed in place. Either view would be equally acceptable to Mach. But in the latter picture, the distant rotating shell of matter must be providing some force outward on the moon, for otherwise the Earth's gravitational pull would draw it inward. The same force of the distant matter is also responsible for the Earth's equatorial bulge, of course. By an extension of this concept, we should expect such a force to arise on a mass placed within any rotating shell of matter, a conclusion not in agreement with the Newtonian scheme and one which in principle could be tested experimentally.

In addition, a particle alone in the universe would show no effects of inertia (ignoring self-interaction). If other matter is gradually allowed to accumulate in the universe, then, according to Mach, the particle would (correspondingly) develop its own inertia through interaction with this matter. Now, inertial effects are never observed to depend on direction (the mass is the same regardless of the direction of motion). But since the matter in the neighborhood of any given location is generally not distributed isotropically, it would seem that the interaction responsible for the inertia must arise primarily from very distant matter.

The concepts of Mach are quite intriguing, but they do not solve the inertial frame problem. Instead, the emphasis is shifted to a mysterious interaction with masses far removed from our immediate vicinity, a phenomenon which cannot be directly investigated. It seems logically simpler therefore to follow the thinking of Newton and postulate the existence of an inertial frame (or at least some approximation to one) in whatever fashion is convenient for the problem at hand. With this understanding, we shall now proceed to a simple application that illustrates the basic approach in the Newtonian theory.

Example 1

A skydiver of mass m jumps from an airplane, experiencing a retarding force of air resistance which is proportional to the square of the velocity. Find the diver's velocity as a function of the time, the acceleration of gravity g, and the terminal velocity V.

Solution

The skydiver's weight mg downward and the force of air resistance upward are the only forces acting; so if we take the y-axis vertically upward, we have

$$F_y = -mg + kv^2.$$

The "terminal velocity" V is reached when these forces balance, so

$$F_y = 0 = -mg + kV^2,$$

or

$$V = \sqrt{mg/k}, \qquad k = mg/V^2.$$

In general, by the Second Law,

$$F_y = -mg + \frac{mgv^2}{V^2} = mg\left(\frac{v^2}{V^2} - 1\right) = m\frac{dv}{dt}.$$

Separating variables,

$$-\frac{g\,dt}{V^2} = \frac{dv}{V^2 - v^2},$$

which integrates at once to

$$-\frac{gt}{V^2} + C = \frac{1}{V}\tanh^{-1}(v/V).$$

The constant of integration $C = 0$ if $v = 0$ at $t = 0$, giving

$$\tanh^{-1}(v/V) = -gt/V,$$

or

$$v = V\tanh(-gt/V) = V\frac{e^{-gt/V} - e^{gt/V}}{e^{-gt/V} + e^{gt/V}} = -V\frac{1 - e^{-2gt/V}}{1 + e^{-2gt/V}}.$$

This is the desired answer. For times $t \gg V/2g$, the velocity is essentially terminal. Taking $V \simeq 50$ m/sec and $g = 9.8$ m/sec^2, we see that after about 15 sec of fall, terminal velocity is (approximately) attained; in actual practice, even less time may be required. The solution presented here is an exact one, except for the approximating of the force of air resistance by the term kv^2 in order to get a simple differential equation.

1.3 MOMENTS OF MASS DISTRIBUTIONS

In dealing with problems of equilibrium or rotational motion, it is useful to define the "moment of force," or **torque** $\vec{N}$ about a point, as

$$\vec{N} \equiv \vec{r} \times \vec{F} \tag{1.6}$$

and the "moment of momentum," or **angular momentum** $\vec{L}$, as

$$\vec{L} \equiv \vec{r} \times \vec{p}. \tag{1.7}$$

The presence of the vector $\vec{r}$ in these expressions classifies them as first "moments." Note that for a particle

$$\frac{d\vec{L}}{dt} = \frac{d}{dt}(\vec{r} \times \vec{p}) = \vec{v} \times \vec{p} + \vec{r} \times \dot{\vec{p}} = 0 + \vec{r} \times \vec{F} = \vec{N},$$

an equation which, when generalized to more complex systems, is the analogue of Newton's second law so far as rotational motion is concerned; that is, the vanishing of the net torque on a system requires that the angular momentum be a constant of the motion.

Another useful, related concept is that of the center of mass (often abbreviated CM), which is defined by averaging the position vector $\vec{r}$ over the mass distribution of a given body or configuration of particles. Thus, if $\vec{R}$ denotes the vector to the center of mass, M the total mass, and dV a volume element,

$$\vec{R} \equiv \frac{1}{M}\int \vec{r}\rho\, dV = \frac{\int \vec{r}\rho\, dV}{\int \rho\, dV} \quad \text{(solid body of density } \rho\text{)} \tag{1.8}$$

or

$$\vec{R} = \frac{1}{M}\sum_i m_i\vec{r}_i = \frac{\sum_i m_i\vec{r}_i}{\sum_i m_i} \quad \begin{pmatrix}\text{system of particles, the }i\text{th of which}\\ \text{has mass }m_i\text{ at position }\vec{r}_i\end{pmatrix}.$$

$$\tag{1.8'}$$

Example 2
Find the location of the center of mass for that part of a homogeneous sphere of radius a which lies between the z axis and the angle $\theta = \alpha$. (See Figure 1.2.)

Solution
From the symmetry, it is clear that we need calculate only the z coordinate of the mass center, which lies on the z axis. Spherical coordinates are naturally appropriate. Applying eq. (1.8) gives

$$\int z\rho\, dV = \rho\int_0^{2\pi}\int_0^{\alpha}\int_0^{a}(r\cos\theta)(r^2\sin\theta\, dr\, d\theta\, d\phi)$$
$$= \frac{2\pi\rho a^4}{4}\left[\frac{\cos^2\theta}{-2}\right]_0^{\alpha} = \frac{\pi\rho a^4}{4}(1 - \cos^2\alpha),$$

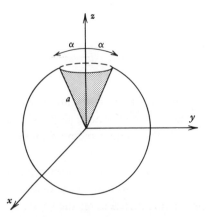

Figure 1.2 Cone-shaped portions of sphere.

and
$$\int \rho \, dV = \rho \int_0^{2\pi} \int_0^{\alpha} \int_0^{a} r^2 \sin\theta \, dr \, d\theta \, d\phi = \frac{2\pi\rho a^3}{3}(1 - \cos\alpha).$$
Hence, we have
$$z_{CM} = \frac{\int z\rho \, dV}{\int \rho \, dV} = \frac{3a}{8} \frac{(1 - \cos^2\alpha)}{(1 - \cos\alpha)} = \frac{3}{8}a(1 + \cos\alpha).$$

Example 3
An equilateral triangle of side S has a point mass $2m$ placed at one vertex and masses m at the other vertices. Find the CM. (See Figure 1.3.)

Solution
In this case, we can apply eq. (1.8') directly, giving
$$x_{CM} = \frac{2m \cdot \frac{S}{2} + m \cdot S + m \cdot 0}{4m} = \frac{S}{2},$$
$$y_{CM} = \frac{2m \cdot \frac{\sqrt{3}\,S}{2} + m \cdot 0 + m \cdot 0}{4m} = \frac{\sqrt{3}\,S}{4}.$$
Note, however, that eq. (1-8') is a special case of the more general formula (1.8); the latter reduces to the former when we express the density ρ for a system of point particles by means of the Dirac delta function. (The student will find a discussion of this function and its properties in Appendix B.) For this particular example, we would have
$$\rho = 2m\delta\left(x - \frac{S}{2}\right) \cdot \delta\left(y - \frac{\sqrt{3}\,S}{2}\right) \cdot \delta(z) + m\delta(x - S) \cdot \delta(y) \cdot \delta(z)$$
$$+ m\delta(x) \cdot \delta(y) \cdot \delta(z).$$

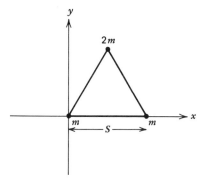

Figure 1.3 Triangle with point masses.

Integration over the triangle with this density expression will yield the above results at once.

For problems of rotational motion, a second moment of the mass distribution is often needed, and it will involve integration over a squared coordinate. Such a quantity is familiar from elementary physics as the **moment of inertia** I. Along with the **angular velocity** ω, the angle turned through per unit time for a rotating body, the moment of inertia (the equivalent of mass for rotational motion) enters into the analogues of eqs. (1.3) and (1.4) for simple rotational cases as

$$\vec{L} = I\vec{\omega} \quad \text{and} \quad \vec{N} = \frac{d\vec{L}}{dt} = \vec{\alpha},$$

with $\vec{\alpha}$ as the angular acceleration. (Later sections contain a much more elaborate discussion of these quantities.)

Example 4
A uniform wire of linear density τ is bent into the shape of a rectangle of sides a and b. Find the moment of inertia of this "skeleton figure" when rotated about an axis perpendicular to its plane and through the CM. (See Figure 1.4.)

Solution
For the region of the rectangle, we can write the density function as
$$\rho = \tau\delta(x)\delta(z) + \tau\delta(y - b)\delta(z) + \tau\delta(x - a)\delta(z) + \tau\delta(y)\delta(z).$$
The center of mass is easily seen to be at the location $x = a/2, y = b/2$, either by symmetry arguments or by means of eq. (1.8). Then, we have

$$I = \int \left[\left(x - \frac{a}{2}\right)^2 + \left(y - \frac{b}{2}\right)^2\right] \rho\, dV$$

$$= \tau \int_0^a dx \int_0^b \left[x^2 + y^2 - ax - by + \frac{a^2}{4} + \frac{b^2}{4}\right]$$

$$\times [\delta(x) + \delta(y - b) + \delta(x - a) + \delta(y)]\, dy$$

$$= \tau \left[\frac{a^3}{3} + a^2 b + \frac{a^3}{3} + \frac{b^3}{3} + b^2 a + \frac{b^3}{3}\right.$$

$$- a\left(\frac{a^2}{2} + ab + \frac{a^2}{2}\right) - b\left(\frac{b^2}{2} + ab + \frac{b^2}{2}\right)$$

$$\left. + \left(\frac{a^2}{4} + \frac{b^2}{4}\right)(b + a + b + a)\right]$$

$$= \tau\left[\frac{1}{6}a^3 + \frac{1}{6}b^3 + \frac{1}{2}a^2 b + \frac{1}{2}ab^2\right] = \frac{\tau}{6}(a + b)^3.$$

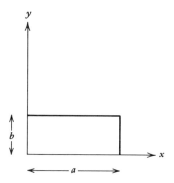

Figure 1.4 Skeleton rectangle.

(The integration limits must be extended a little beyond the values given, so there will be no problem in including the contributions of the delta functions.) Defining the *radius of gyration* k by $I \equiv Mk^2$, then with total mass $M = \tau 2(a + b)$, we have $k^2 = (a + b)^2/12$.

If instead of a skeleton figure we had considered a uniform plane sheet of the same total mass, the density would have been $\rho = (M/ab)\delta(z)$, leading to

$$I = \int_0^a dx \int_0^b dy \left[\left(x - \frac{a}{2} \right)^2 + \left(y - \frac{b}{2} \right)^2 \right] \left[\frac{M}{ab} \right]$$

$$= \frac{M}{ab}\left[\frac{a^3}{3}b + \frac{b^3}{3}a - a\frac{a^2}{2}b - ba\frac{b^2}{2} + \left(\frac{a^2}{4} + \frac{b^2}{4} \right)ab \right]$$

$$= \frac{M}{12}(a^2 + b^2),$$

and

$$k^2 = \frac{(a^2 + b^2)}{12}.$$

Thus, the spreading of the mass over the full area of the rectangle leads to a *smaller* moment of inertia than for the wire figure, a consequence of the fact that much of the mass is close to the axis of rotation and so contributes little to the integral.

1.4 STATICS AND DYNAMICS; ENERGY

The study of objects at rest, which are in equilibrium under the combined actions of several forces and/or torques, is known as **statics**; in such cases the vector sum of all forces and torques about any point must add to zero. This book is mainly concerned with problems of **dynamics**, which involve the motion of bodies. There are many dynamical situations, however, which can be treated by methods appropriate to statics, inasmuch as no accelerations are present (thus insuring that no net force or torque acts on the system).

As a simple illustration of this point, consider two flat bodies sliding along one another at a constant speed. A force is required to maintain the relative motion, for we know that microscopic irregularities on the two surfaces will lead to a frictional force opposing the motion, and the applied force must exactly counteract the friction.

The force of kinetic friction f is regarded as proportional to the normal force F_N with which the surfaces are pressed together:

$$f = \mu F_N. \tag{1.9}$$

The coefficient of kinetic friction μ is usually taken to be independent of the area of contact, the relative velocity, and the normal force, an approximation with a wide range of applicability. The limiting force of static friction between two objects about to slip relative to one another also satisfies an equation like (1.9), but with a larger value for μ.

As a further example, consider a paper chart you are pulling outward steadily with force F from the core of a reel of paper (see Figure 1.5). For small angle θ, the spool tends to move away from you, while for large θ, it tends to move toward you. At some intermediate value, the reel will undergo no translational motion. The conditions of equilibrium allow us to determine this particular value of θ as follows:

$$F \sin \theta = f = \mu F_N \quad \text{(net horizontal component of force} = 0\text{),}$$
$$F_N + F \cos \theta = W \quad \text{(net vertical component of force} = 0\text{),}$$
$$Fr = fR \quad \text{(net torque about center of spool} = 0\text{).}$$

Combining the first and last equations, we have

$$\sin \theta = \frac{f}{F} = \frac{r}{R},$$

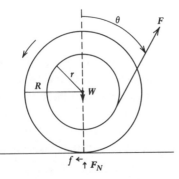

Figure 1.5 Spool of weight W unwinding.

Statics and Dynamics; Energy

so that the radii of the reel alone are sufficient to fix the critical angle; the coefficient of friction need not be known at all!

Now let us examine some more complicated situations in dynamics. Consider a system having a number of particles n, the ith of which has mass m_i and is located at a point specified by the vector $\vec{r}_i$ from the origin. Let $\vec{F}_i$ represent the net *external* force on the ith particle and $\vec{F}_{ij}$ the force on the ith particle due to the jth particle (a force which is *internal* to the system). We have from Newton's law

$$\dot{\vec{p}}_i = m_i \ddot{\vec{r}}_i = \vec{F}_i + \sum_{\substack{j=1 \\ (j \neq i)}}^{n} \vec{F}_{ij}.$$

Summing these equations of motion over all particles gives

$$\sum_{i=1}^{n} m_i \ddot{\vec{r}}_i = \sum_{i=1}^{n} \vec{F}_i + \sum_{\substack{i,j=1 \\ (j \neq i)}}^{n} \vec{F}_{ij}.$$

The double sum in the last term must vanish, however, because the internal forces cancel in pairs as a result of Newton's third law, which ensures that $\vec{F}_{kl} = -\vec{F}_{lk}$ for most forces commonly encountered. Calling the total external force $\vec{F} \equiv \sum_{i=1}^{n} \vec{F}_i$, one then has

$$\vec{F} = \sum_{i=1}^{n} m_i \ddot{\vec{r}}_i.$$

Writing eq. (1.8) in the form $M\vec{R} = \sum_i m_i \vec{r}_i$ and time differentiating twice yields

$$M\ddot{\vec{R}} = \sum_i m_i \ddot{\vec{r}}_i = \vec{F}. \tag{1.10}$$

Hence, the center of mass moves like a single particle of total mass M under the external force. If $\vec{F} = 0$, then the total linear momentum $M\dot{\vec{R}} = \vec{P} = \sum_i m_i \dot{\vec{r}}_i = \sum_i \vec{p}_i$ is a constant of the motion.

Next, consider the angular momentum of this system. Taking the cross product of $\vec{r}_i$ with the ith equation of motion produces

$$\vec{r}_i \times m_i \ddot{\vec{r}}_i = \frac{d}{dt}(\vec{r}_i \times m_i \dot{\vec{r}}_i) = \frac{d\vec{L}_i}{dt} = \vec{r}_i \times \vec{F}_i + \vec{r}_i \times \sum_{j(\neq i)} \vec{F}_{ij}.$$

Summing over i and letting $\vec{N} \equiv \sum_i \vec{r}_i \times \vec{F}_i$ be the total external torque on the system, we have

$$\dot{\vec{L}} \equiv \sum_i \frac{d\vec{L}_i}{dt} = \vec{N} + \sum_{\substack{i,j \\ (i \neq j)}} \vec{r}_i \times \vec{F}_{ij}.$$

Again, if pairs of internal torque terms are isolated, the third law can be used to give

$$\vec{r}_k \times \vec{F}_{kl} + \vec{r}_l \times \vec{F}_{lk} = (\vec{r}_k - \vec{r}_l) \times \vec{F}_{kl}.$$

But, $(\vec{r}_k - \vec{r}_l)$ is a vector lying along the line between the kth and lth particles; the force $\vec{F}_{kl}$ is said to be a **central** force if it also lies along this line, whereupon the cross product will vanish. Thus, for a system with internal central forces only,

$$\dot{\vec{L}} = \vec{N}. \tag{1.11}$$

As a further important step, let O be the origin of a Newtonian frame and O' be a point in arbitrary motion relative to O. (See Figure 1.6.) Then,

$$\vec{r}_i = \vec{r}_i' + \vec{r} \quad \text{and} \quad \dot{\vec{r}}_i = \dot{\vec{r}}_i' + \dot{\vec{r}},$$

or

$$\vec{v}_i = \vec{v}_i' + \vec{v}.$$

Since

$$\vec{L} = \sum_i \vec{r}_i \times \vec{p}_i = \sum_i \vec{r}_i \times m_i \vec{v}_i,$$

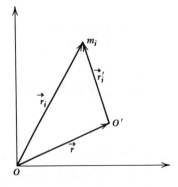

Figure 1.6 Position vectors for discussion of angular momentum.

Statics and Dynamics; Energy

we can substitute the vector relations for $\vec{r}_i$ and $\vec{v}_i$ to get

$$\vec{L} = \sum_i (\vec{r}_i' + \vec{r}) \times m_i(\vec{v}_i' + \vec{v})$$

$$= \sum_i (\vec{r}_i' \times m_i \vec{v}_i') + \left(\sum_i m_i \vec{r}_i'\right) \times \vec{v} + \vec{r} \times \left(\sum_i m_i \vec{v}_i'\right) + \left(\sum_i m_i\right) \vec{r} \times \vec{v}.$$

Next, choose O' to be the center of mass. In that case, we have

$$\sum_i m_i \vec{r}_i' = 0, \qquad \vec{r} = \vec{R}, \qquad \vec{v} = \dot{\vec{R}},$$

$$\vec{L} = \sum_i (\vec{r}_i' \times m_i \vec{v}_i') + \vec{R} \times M\dot{\vec{R}} \equiv \vec{L}' + \vec{R} \times M\dot{\vec{R}}. \tag{1.12}$$

The angular momentum arises as a sum of two parts, that about the center of mass $\vec{L}'$ and that about the origin from the entire mass as if concentrated at the CM and moving with it.

Now let us calculate the torque about O, observing that for this origin of a Newtonian frame, $\vec{F}_i = (d/dt)(m_i \dot{\vec{r}}_i)$. Proceeding as before,

$$\vec{N} = \sum_i \vec{r}_i \times \vec{F}_i = \sum_i (\vec{r}_i' + \vec{r}) \times \vec{F}_i = \vec{N}' + \vec{r} \times \vec{F},$$

where $\vec{N}' \equiv \sum_i \vec{r}_i' \times \vec{F}_i$ is the total torque about the arbitrary origin O'; note that while moments can be taken instantaneously about O' as well as O, the force on m_i cannot be written in terms of $m_i \ddot{\vec{r}}_i'$ alone, for in general O' is itself in accelerated motion relative to the Newtonian frame.

To investigate $\vec{N}'$ more fully, let us write

$$\vec{N}' = \sum_i \vec{r}_i' \times m_i \ddot{\vec{r}}_i = \sum_i \vec{r}_i' \times m_i (\ddot{\vec{r}}_i' + \ddot{\vec{r}})$$

$$= \frac{d}{dt}\left[\sum_i \vec{r}_i' \times m_i \dot{\vec{r}}_i'\right] + \left[\sum_i m_i \vec{r}_i'\right] \times \ddot{\vec{r}}$$

$$= \dot{\vec{L}}' + \left[\sum_i m_i \vec{r}_i'\right] \times \ddot{\vec{r}}. \tag{1.13}$$

The significance of this relation is that the familiar torque equation, (1.11), is augmented by the second term on the right when written for an arbitrarily moving origin O'. This term will vanish, leaving the usual form, under any of three conditions:

1. The term vanishes when O' is the center of mass, whereupon $\sum_i m_i \vec{r}_i'' \equiv 0$.
2. It vanishes when O' is not accelerated relative to O, so $\ddot{\vec{r}} \equiv 0$.
3. Finally, it vanishes when $\ddot{\vec{r}}$ is a vector directed along a line passing through the CM, for $\sum_i m_i \vec{r}_i''$ also lies along this line (it is just M times the vector from O' to the CM), and the cross product must be zero.

Before looking at some applications of these ideas, it is appropriate to introduce another familiar elementary concept. Suppose at a certain time a single particle of constant mass m is at a position specified by $\vec{r}_A$ and a short while later, by $\vec{r}_B$; $d\vec{r}$ is an element of path along the route between the positions of $\vec{r}_A$ and $\vec{r}_B$. (See Figure 1.7.) The **work** done by an external force $\vec{F}$ in going from A to B is defined as

$$W_{AB} \equiv \int_A^B \vec{F} \cdot d\vec{r}. \qquad (1.14)$$

From the second law of motion,

$$W_{AB} = \int_A^B \left(m \frac{d\vec{v}}{dt} \right) \cdot (\vec{v}\, dt) = m \int_A^B \frac{d}{dt} \left(\tfrac{1}{2} \vec{v} \cdot \vec{v} \right) dt = \tfrac{1}{2} m \left(v_B^2 - v_A^2 \right).$$

The work done is the *change* of the quantity $\tfrac{1}{2} m v^2$, which we call the **kinetic energy** T:

$$T \equiv \tfrac{1}{2} m v^2. \qquad (1.15)$$

Energy measures the capacity for doing work; kinetic energy is possessed by an object due to its motion. When the particle is at rest, $T = 0$, at least in the given Newtonian rest frame.

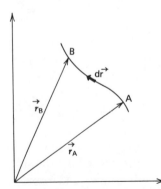

Figure 1.7 Arbitrary path for work calculation.

If the particle moves around a complete closed path, finally returning to the position specified by $\vec{r}_A$, and if the work done is zero around such a closed contour (indicated by the $\oint$ sign), then we say that the force is **conservative**. The force of gravity on Earth is a typical example; as an object falls under gravity, it gains kinetic energy by virtue of the work done by the field. If the velocity could suddenly be reversed, it would rise once more, losing kinetic energy as it did so. For a full round trip (ignoring air resistance), no net work would be done. A frictional force, however, is a good example of a nonconservative force; as friction always opposes the motion, $\vec{F} \cdot d\vec{r}$ is always negative and the traversal of a closed contour cannot give zero.

For a conservative force we note that from Stokes' theorem (see Appendix A)

$$\oint \vec{F} \cdot d\vec{r} = 0 = \oint \operatorname{curl} \vec{F} \cdot d\vec{S}.$$

The line integral of $\vec{F}$ vanishes around any closed path enclosing an arbitrary surface S. Then the curl of $\vec{F}$ must also vanish, and one can write

$$\vec{F} = -\nabla V \tag{1.16}$$

(the minus sign is conventional). V is the **potential energy**, which determines the force through a gradient operation; the addition of a constant to V therefore has no physical significance, and we are free to choose the zero of potential in any way we please. Thus, taking the force of gravity on Earth as constant over short distances, the potential energy will vary linearly with height, and its zero reference level can be taken at any convenient location for the problem of a mass moving in Earth's gravity.

Furthermore, we have

$$W_{AB} = \int_A^B \vec{F} \cdot d\vec{r} = -\int_A^B \nabla V \cdot d\vec{r} = -\int_A^B \left(\frac{\partial V}{\partial x} dx + \frac{\partial V}{\partial y} dy + \frac{\partial V}{\partial z} dz \right)$$

$$= -\int_A^B dV = V_A - V_B = -(V_B - V_A) = T_B - T_A$$

from our earlier relation. Then, one can write

$$T_A + V_A = T_B + V_B. \tag{1.17}$$

In the motion of a particle under the action of conservative forces, the total mechanical energy $E \equiv T + V$ is a constant. This theorem is the famous energy conservation theorem and may easily be generalized to more complex situations.

Let us return to the system of particles previously considered, taking O' in Figure 1.6 as the center of mass. For the ith particle, the kinetic energy is

$$T_i = \tfrac{1}{2} m_i (\dot{\vec{r}}_i \cdot \dot{\vec{r}}_i) = \tfrac{1}{2} m_i (\dot{\vec{r}}_i' + \dot{\vec{R}}) \cdot (\dot{\vec{r}}_i' + \dot{\vec{R}}).$$

Summing over all particles results in

$$T = \sum_i T_i = \tfrac{1}{2} \sum_i m_i (\dot{\vec{r}}_i')^2 + \left(\sum_i m_i \dot{\vec{r}}_i' \right) \cdot \dot{\vec{R}} + \tfrac{1}{2} \left(\sum_i m_i \right) (\dot{\vec{R}})^2.$$

Since $\sum_i m_i \vec{r}_i'' = 0$ when O' is the CM, the middle term drops out, leaving

$$T = \tfrac{1}{2} \sum_i m_i (\dot{\vec{r}}_i')^2 + \tfrac{1}{2} M (\dot{\vec{R}})^2. \tag{1.18}$$

Thus, the total kinetic energy is the sum of that relative to the CM and that of the total system mass moving with the velocity of the CM.

As a final comment on work-energy principles, we wish to consider the principle of **virtual work**, and/or **virtual displacements**. It is a useful technique for the study of objects in static equilibrium and was first stated clearly by Bernoulli in 1717. A virtual displacement $\delta \vec{r}$ refers to a displacement of some part of a physical system with time held fixed. Since any real physical motion requires an interval of time in which to take place, the virtual displacement is purely a conceptual one; but it must be compatible with the geometrical constraints imposed on the system. That virtual displacements are instantaneous means in effect that they cannot be altered even if the constraints are changing in time.

In general, virtual displacements are infinitesimal shifts about the configuration of static equilibrium. If the total force on the ith member of a set of particles is $(\vec{F}_T)_i$ (the index i can also refer to some part of a macroscopic body), the applied force is $\vec{F}_i$, and the force of a constraining surface is $(\vec{F}_c)_i$, then, in equilibrium,

$$(\vec{F}_T)_i = \vec{F}_i + (\vec{F}_c)_i = 0.$$

Taking the dot product with $\delta \vec{r}_i$, the virtual displacement, gives virtual work of zero. Summing over all members of the system produces

$$\sum_i (\vec{F}_T)_i \cdot \delta \vec{r}_i = \sum_i \vec{F}_i \cdot \delta \vec{r}_i + \sum_i (\vec{F}_c)_i \cdot \delta \vec{r}_i = 0.$$

Now, in fact, the work of the constraint forces $(\vec{F}_c)_i$ can usually be taken as zero. Assuming for example, that the surface along which a particle moves is frictionless, the normal force will be perpendicular to the displacement and the dot product will vanish. In addition, if it is assumed that the internal forces connecting two particles of a rigid body are along the line joining them, the work for the ith and jth particles is

$$(\vec{F}_c)_i \cdot \delta \vec{r}_i + (\vec{F}_c)_j \cdot \delta \vec{r}_j = (\vec{F}_c)_i \cdot \delta[\vec{r}_i - \vec{r}_j]$$

from Newton's third law. But $\vec{r}_i - \vec{r}_j$ is the vector between the particles, which is fixed in magnitude in a rigid body. Any change of this difference vector must then be perpendicular to the vector itself, hence to the force, so the dot product again vanishes. Assuming that the work done by the constraint forces is zero as in these typical cases, the **principle of virtual work** then states that

$$\sum_i \vec{F}_i \cdot \delta \vec{r}_i = 0 \qquad (1.19)$$

for the contribution of the applied forces alone. Now let us look at some detailed applications of the principles that have been outlined so far.

1.5 APPLICATIONS OF THE BASIC PRINCIPLES

As an initial example to demonstrate the use of the conservation of momentum and energy, we have selected a problem with far-ranging applications, from the collisions of billiard balls to those of elementary particles.

Example 5 Elastic Collision of Two Masses

Consider two masses m_1 and m_2 which collide with one another. The collision will be called **elastic**, by which we mean that internal energy states are unchanged and can therefore be ignored (for instance, electrons in atoms will not be excited to higher energy levels by the impact). Put more simply, the total kinetic energy of the two masses will be the same before and after the collision.

Let us take the initial velocities in the laboratory frame as $\vec{v}_1$ and $\vec{v}_2$, respectively. As only internal forces are involved, the CM (Figure 1.8) will continue to move at the rate

$$\dot{\vec{R}} = \frac{m_1 \vec{v}_1 + m_2 \vec{v}_2}{M} \qquad (M \equiv m_1 + m_2)$$

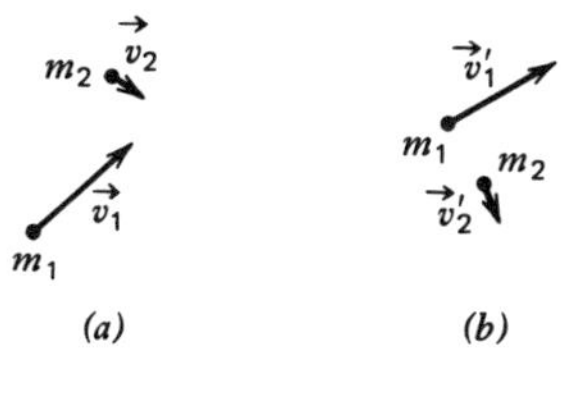

Figure 1.8 Two masses (*a*) before and (*b*) after collision.

throughout the interaction. We shall denote quantities by a subscript C when evaluated in the CM frame of reference, that coordinate system in which the CM is at rest. Making a Galilean transformation then produces

$$\vec{v}_{1C} = \vec{v}_1 - \dot{\vec{R}} = \frac{M\vec{v}_1 - (m_1\vec{v}_1 + m_2\vec{v}_2)}{M}$$

$$= \frac{m_2}{M}(\vec{v}_1 - \vec{v}_2) = -\frac{m_2}{M}\vec{v},$$

and, similarly,

$$\vec{v}_{2C} = \vec{v}_2 - \dot{\vec{R}} = \frac{m_1}{M}(\vec{v}_2 - \vec{v}_1) = \frac{m_1}{M}\vec{v}$$

where $\vec{v} \equiv \vec{v}_2 - \vec{v}_1$ is the initial *relative* velocity. Note that

$$m_1\vec{v}_{1C} = -m_2\vec{v}_{2C}.$$

The momenta are "equal and opposite," so the total momentum is zero in the CM frame.

Now let us indicate the velocities *after* the collision by primes. (If an interaction potential is present and produces forces in addition to those of the impact, the unprimed and primed velocities refer to those values before entering and after leaving the range of this potential.) The momentum vectors may be altered by the collision, but they will remain equal and opposite in the CM frame, so that

$$\vec{p}'_{1C} = -\vec{p}'_{2C} \quad \text{just as} \quad \vec{p}_{1C} = -\vec{p}_{2C}.$$

By conservation of kinetic energy, we can write

$$\tfrac{1}{2}m_1v_{1C}^2 + \tfrac{1}{2}m_2v_{2C}^2 = \frac{p_{1C}^2}{2m_1} + \frac{p_{1C}^2}{2m_2} = \frac{(p'_{1C})^2}{2}\left(\frac{1}{m_1} + \frac{1}{m_2}\right),$$

or

$$|p_{1C}| = |p'_{1C}|.$$

The momenta in the CM frame are unchanged in magnitude, so all that can happen is that the collision may rotate them in direction.

A few remarks might be in order at this point. The discussion above assumes that kinetic energy is conserved in the CM frame, as required by an elastic collision. But inasmuch as kinetic energy depends on the frame of reference for its value and is therefore not invariant under a Galilean transformation, one could

Applications of the Basic Principles

question whether it will be conserved as well in the laboratory or other reference frames. It turns out to be rather fortunate for the physicist: if a collision is elastic in one frame, it is also elastic in all others, provided only that momentum is conserved. (The proof is explored in the problem set.) The conservation of momentum is easily shown to be invariant under Galilean transformations, so there need be no worries in that regard.

Returning to the derivation, let $\hat{\varepsilon}$ be a unit vector along the final direction of $\vec{v}_{1C}''$ in the CM frame. We have

$$\vec{v}_{1C}'' = \frac{m_2 v}{M}\hat{\varepsilon}, \qquad \vec{v}_{2C}'' = -\frac{m_1 v}{M}\hat{\varepsilon},$$

or

$$\vec{p}_{1C}'' = mv\hat{\varepsilon} = -\vec{p}_{2C}'',$$

where $v \equiv |\vec{v}|$ is the magnitude of the relative velocity and $m \equiv m_1 m_2/M$ is the **reduced mass** of the system (to be encountered again in Chapter 4).

In the laboratory frame, we now find that

$$\vec{v}_1'' = \vec{v}_{1C}'' + \dot{\vec{R}} = \frac{m_2 v}{M}\hat{\varepsilon} + \frac{m_1 \vec{v}_1 + m_2 \vec{v}_2}{M}, \qquad (1.20)$$

$$\vec{v}_2'' = \vec{v}_{2C}'' + \dot{\vec{R}} = -\frac{m_1 v}{M}\hat{\varepsilon} + \frac{m_1 \vec{v}_1 + m_2 \vec{v}_2}{M}. \qquad (1.20')$$

Alternate expressions in terms of the momenta are

$$\vec{p}_1'' = mv\hat{\varepsilon} + \frac{m_1}{M}(\vec{p}_1 + \vec{p}_2)$$

and

$$\vec{p}_2'' = -mv\hat{\varepsilon} + \frac{m_2}{M}(\vec{p}_1 + \vec{p}_2).$$

The final momenta are now completely determined by knowledge of m_1, m_2, $\vec{v}_1$, and $\vec{v}_2$, except that the direction of the unit vector $\hat{\varepsilon}$ is unknown. To fix the orientation of $\hat{\varepsilon}$, we need to know more about the interaction: the nature of the potential, whether the collision was head-on or grazing, and so on.

There is a geometrical construction which may help in visualizing these results (Figure 1.9). Draw a circle of radius mv about a point O; then, $mv\hat{\varepsilon}$ will be a vector from O to some point on the circle. Choosing $\vec{R}$ horizontally toward the right, line segments of length $m_1 \vec{R}$ and $m_2 \vec{R}$ can be drawn on each side of O (these may be larger or smaller in length than the radius mv), giving the vector triangles of the $\vec{p}_1''$, $\vec{p}_2''$ relations, as is readily seen from the illustration.

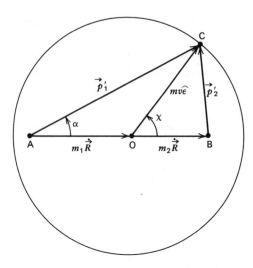

Figure 1.9 Geometrical representation of final momenta.

Now, suppose that m_2 is initially at rest ($\vec{v}_2 \equiv 0$), as would be the case when a beam of particles impinges on fixed target nuclei. Inasmuch as $\vec{v} = -\vec{v}_1$, $m_2\vec{R} = -m\vec{v}$ and point B lies on the circle. With $m_1\vec{R} = -m_1^2\vec{v}/M = (-m_1/m_2)m\vec{v}$, point A will be inside or outside the circle depending on the relative size of the masses. For $m_1 > m_2$ (A outside circle), there is a definite limit on the angle of deflection α of m_1; it can be no larger than in the situation when the momentum vector $\vec{p}_1'$ from A is tangent to the circle at C. Of course, there is no such restriction on α when $m_1 < m_2$ (A inside the circle). A careful study of the geometrical model makes these points quite clear; otherwise they are somewhat obscure.

A head-on collision in the case where $\vec{v}_2 = 0$ is of particular interest, since it is in effect a one-dimensional collision. In this situation, the angle χ of Figure 1.9 is π, $\vec{v} = -\vec{v}_1$, and $\hat{\epsilon} = \vec{v}/v$. From the eqs. (1.20), we then have

$$\vec{v}_1' = \frac{m_2}{M}\vec{v} - \frac{m_1}{M}\vec{v} = -\left(\frac{m_1 - m_2}{m_1 + m_2}\right)\vec{v},$$

$$\vec{v}_2' = -\frac{m_1}{M}\vec{v} - \frac{m_1}{M}\vec{v} = -\left(\frac{2m_1}{m_1 + m_2}\right)\vec{v}.$$

(It will be shown via the problem set that this is the largest magnitude of velocity which m_2 can acquire.) When $m_1 > m_2$ and point A is outside the circle, the momenta $\vec{p}_1'$ and $\vec{p}_2'$ will be in the same direction (that of $\vec{v}_1$), as is clear from the equations. On the other hand, for $m_1 < m_2$, the final momenta will be in opposite directions.

As a last example, consider the case of equal masses with $\vec{v}_2$ still taken as zero. Then $\vec{v} = -\vec{v}_1$, $m = \frac{1}{2}m_2$, and $\vec{R} = \frac{1}{2}\vec{v}_1$, with points A and B both on the circle,

Applications of the Basic Principles

so that eq. (1.20) leads to

$$\vec{p}_1' \cdot \vec{p}_2' = -m^2v^2 + m_1m_2\dot{\vec{R}}^2 + (m_2 - m_1)mv\hat{\varepsilon} \cdot \dot{\vec{R}}$$

$$= -\tfrac{1}{4}m_1^2v_1^2 + m_1^2\tfrac{1}{4}v_1^2 + 0 = 0.$$

The conclusion which follows is that the two masses must necessarily separate after the collision along paths that are at right angles, provided that

$$|\vec{p}_1'| \neq 0 \quad \text{and} \quad |\vec{p}_2'| \neq 0.$$

But clearly, $|\vec{p}_2'| \neq 0$, for some momentum must always be imparted to m_2 (assumed finite) if a collision is to occur. It *is* possible, however, to have a head-on collision in which *all* the energy is transferred to m_2, so that $\vec{v}_1' = 0$ and $\vec{v}_2' = \vec{v}_1$; otherwise, equal masses separate at right angles.

Example 6

A circular hoop of mass m and radius R is constrained to a vertical plane and projected down an inclined plane of angle α with an initial translational velocity v_0. It is also given an initial angular velocity ω_0 which tends to make it roll up the plane. If the coefficient of friction is μ and gravity is present, find ω_0 such that the sliding hoop just comes to rest at some point down the plane (Figure 1.10).

Solution

Take the x-axis along the plane. Then, eq. (1.3) becomes

$$m\ddot{x} = mg\sin\alpha - \mu mg\cos\alpha,$$

where eq. (1.9) has been used, with $mg\cos\alpha$ as the normal force. Canceling the mass, integrating, and inserting the initial condition, we have

$$\dot{x} = (\sin\alpha - \mu\cos\alpha)gt + v_0.$$

The hoop will come to rest at a time t_R when $\dot{x} = 0$, or

$$t_R = \frac{v_0}{g(\mu\cos\alpha - \sin\alpha)}.$$

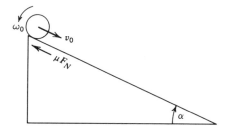

Figure 1.10 Hoop on inclined plane.

From the remarks following eq. (1.13), we know that we can apply the torque equation in the form $N = I\ddot{\theta}$ about the CM of the hoop; this gives

$$I\ddot{\theta} = mR^2\ddot{\theta} = -R\mu mg \cos \alpha,$$

$$\ddot{\theta} = -\frac{\mu g}{R} \cos \alpha,$$

the torque about the CM being entirely due to the frictional force (θ is an angle measured counterclockwise from the normal to the plane to some fixed diameter of the hoop). Integration yields

$$\dot{\theta} = -\frac{\mu g t}{R} \cos \alpha + \omega_0;$$

$$\dot{\theta} = 0 \text{ at time } t_R,$$

which yields

$$t_R = \frac{\omega_0 R}{\mu g \cos \alpha}.$$

Equating the two expressions for t_R gives

$$\frac{\omega_0 R}{\mu g \cos \alpha} = \frac{v_0}{g(\mu \cos \alpha - \sin \alpha)},$$

or

$$\omega_0 = \frac{\mu v_0 \cos \alpha}{R(\mu \cos \alpha - \sin \alpha)} = \frac{\mu v_0}{R(\mu - \tan \alpha)}.$$

In this problem, $\dot{\theta}$ decreases in time; but if the hoop had been spun initially in the opposite direction, the frictional torque would instead have increased $\dot{\theta}$. In that event, a state of pure rolling with $\dot{x} = R\dot{\theta}$ could have been obtained at some later time (the problem set explores such a possibility). Once the hoop is in a state of pure rolling, however, the point in contact with the plane is instantaneously at rest, so that the force acting is one of *static* friction and is no longer given by the expression (1.9).

Example 7 Virtual Work

A uniform ladder of length L and weight W is in equilibrium at an angle θ with the floor. If the wall and floor are smooth, the ladder must be held in place by a rope fastened at the bottom of the ladder and to a point of the wall at a height h above the floor. Find the tension T in the rope (Figure 1.11).

Solution
To find T by the method of virtual work, imagine a displacement δx of the foot of the ladder to the right, thus bringing the CM down on amount δy. From the

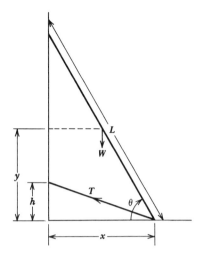

Figure 1.11 Ladder against wall.

geometry, we have

$$x = L\cos\theta, \quad \delta x = -L\sin\theta\,\delta\theta,$$
$$y = \frac{L}{2}\sin\theta, \quad \delta y = \frac{L}{2}\cos\theta\,\delta\theta.$$

The only relevant forces acting are T and W; so, from the principle of virtual work, we obtain

$$-T|\delta x|\frac{x}{\sqrt{h^2 + x^2}} + W|\delta y| = 0,$$

or

$$T = W\frac{\sqrt{h^2 + x^2}}{x}\frac{|\delta y|}{|\delta x|} = W\frac{\sqrt{h^2 + x^2}}{x}\frac{(L/2)\cos\theta}{L\sin\theta}$$
$$= \frac{W}{2}\frac{\sqrt{h^2 + x^2}}{x}\frac{x}{L\sin\theta} = \frac{\sqrt{h^2 + L^2\cos^2\theta}}{L\sin\theta}\frac{W}{2}.$$

A similar result is obtained by setting horizontal and vertical components of forces and moments about some point all separately equal to zero. In so doing, the smooth reaction forces at top and bottom of ladder must be included. For the virtual work calculation, they are not needed, as they are perpendicular to the chosen displacements.

1.6 THE GRAVITATIONAL FIELD AND POTENTIAL

In the process of developing his laws of motion, Newton also considered the force of attraction between two masses, such as the Earth and moon. He came to the conclusion that these forces should satisfy the law of action and reaction, and for two point masses m_1 and m_2 should lie along the line joining the masses, having a magnitude given by the famous inverse square **law of universal gravitation**:

$$|\vec{F}| = G\frac{m_1 m_2}{|\vec{r}_1 - \vec{r}_2|^2}. \qquad (1.21)$$

where G is a constant of proportionality which has been measured in the Cavendish torsion balance experiments on the forces between metallic spheres; its value is $6.67 \times 10^{-11} nt\text{-}m^2/kg^2$. Today, as in Newton's time, we can only shake our heads in bewilderment at the mysterious "action at a distance" concept embodied in this law. How the force between two objects not in contact can change instantly as the distance varies is difficult to comprehend and requires a study of field concepts.

The law as stated above strictly applies only to two point masses. Suppose we wish to use it for a computation of the force between a point mass m and an extended object of mass M (Figure 1.12). (For simplicity of calculation, it will be necessary for the extended object to have some regular geometrical shape.) We apply eq. (1.21) to m and an infinitesimal element of M, then integrate over the latter. Letting $\vec{r}$ be the position vector of m and $\vec{r}'$ that of an element of M, with $\rho(\vec{r}')$ the density function of the extended body, the generalization of (1.21) is

$$\vec{F} = Gm \int \frac{\rho(\vec{r}')dV'}{|\vec{r}' - \vec{r}|^3}(\vec{r}' - \vec{r}) \qquad (1.22)$$

for the force on m, the unit factor $(\vec{r}' - \vec{r})/|\vec{r}' - \vec{r}|$ having been inserted to allow for the direction of the force contribution from the element at $\vec{r}'$.

The **gravitational field intensity** $\vec{g}$ (or just the **field**) is defined as the force per unit mass at $\vec{r}$ (the mass should not be large enough to disturb the field appreciably in a measurement):

$$\vec{g}(\vec{r}) \equiv G \int \frac{\rho(\vec{r}')\,dV'}{|\vec{r}' - \vec{r}|^3}(\vec{r}' - \vec{r}). \qquad (1.23)$$

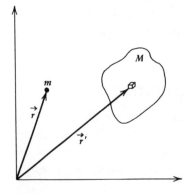

Figure 1.12 Vectors to point mass and extended object.

The Gravitational Field and Potential

Since the force can be found by taking the negative gradient of the potential energy, the gravitational potential $\Phi(\vec{r})$ is defined as the potential energy per unit mass for a particle at r, so that

$$\vec{g}(\vec{r}) = -\nabla \Phi(\vec{r}). \tag{1.24}$$

As in electrostatics, the potential is a convenient function on which to operate, inasmuch as it is a scalar rather than a vector like the field. The potential due to the extended object M is given by the expression

$$\Phi(\vec{r}) = -G \int \frac{\rho(\vec{r}')\, dV'}{|\vec{r}' - \vec{r}|}. \tag{1.25}$$

That this is true follows from taking the negative gradient of eq. (1.25) with respect to the components of the field point $\vec{r}$, an operation which is performed under the integral sign on the denominator alone. Noting that, for example, one has

$$\frac{\partial}{\partial x} \frac{1}{|\vec{r}' - \vec{r}|} = \frac{\partial}{\partial x} \frac{1}{|\vec{r} - \vec{r}'|} = \frac{\partial}{\partial x} \left[\frac{1}{\sqrt{(x-x')^2 + (y-y')^2 + (z-z')^2}} \right]$$

$$= \frac{-(x-x')}{[(x-x')^2 + (y-y')^2 + (z-z')^2]^{3/2}} = \frac{+(x'-x)}{|\vec{r}' - \vec{r}|^3},$$

there results

$$\nabla \left[\frac{1}{|\vec{r}' - \vec{r}|} \right] = \frac{\vec{r}' - \vec{r}}{|\vec{r}' - \vec{r}|^3}. \tag{1.26}$$

We can see that

$$-\nabla \Phi(\vec{r}) = G \int \frac{\rho(\vec{r}')\, dV'}{|\vec{r}' - \vec{r}|^3} (\vec{r}' - \vec{r}) = \vec{g}(\vec{r}),$$

as desired.

We note further that if we set $\vec{r} - \vec{r}' \equiv \vec{t}$, $t \equiv |\vec{t}|$, we find

$$\nabla \cdot \left[\frac{\vec{r} - \vec{r}'}{|\vec{r} - \vec{r}'|^3} \right] = \nabla \cdot \left(\frac{\vec{t}}{t^3} \right)$$

$$= \frac{[t^3 - t_x 3t^2(t_x/t)] + [t^3 - t_y 3t^2(t_y/t)] + [t^3 - t_z 3t^2(t_z/t)]}{t^6} = 0,$$

so long as $\vec{t} \neq 0$. But if we integrate this divergence over a volume in $\vec{t}$-space containing $\vec{t} = 0$, Gauss' theorem can be applied to yield

$$\int_V \nabla \cdot \left(\frac{\vec{t}}{t^3}\right) dV = \int_S \left(\frac{\vec{t}}{t^3}\right) \cdot d\vec{S} = \int_S \frac{t(t^2 \, d\Omega)}{t^3} = 4\pi,$$

where $d\Omega$ is an element of solid angle. In other words, the behavior at $\vec{t} = 0$ is that of a delta function:

$$\nabla \cdot \left(\frac{\vec{t}}{t^3}\right) = 4\pi \delta(\vec{t}),$$

hence

$$\nabla \cdot \left(\frac{\vec{r} - \vec{r}'}{|\vec{r} - \vec{r}'|^3}\right) = 4\pi \delta(\vec{r} - \vec{r}'). \tag{1.27}$$

Making use of eq. (1.23), we then have

$$\nabla \cdot \vec{g}(\vec{r}) = -G \int dV' \rho(\vec{r}') 4\pi \delta(\vec{r} - \vec{r}') = -4\pi G \rho(\vec{r}), \tag{1.28}$$

the basic field equation for the determination of $\vec{g}$ from the density ρ, which acts as a source function. In addition, from eq. (1.24) we can write

$$-\nabla \cdot \vec{g} = \nabla^2 \Phi = 4\pi G \rho, \tag{1.29}$$

which is Poisson's equation for the gravitational potential. With appropriate choice of boundary conditions, the field can be found from either of the last two equations; they *give the essence* of the Newtonian theory of gravitation.

As an example, consider a spherically symmetric distribution of mass within a volume V; the field lines must be radially inward by symmetry. Hence, integrating both sides of eq. (1.29) over a spherical volume of radius r and applying Gauss' law, we find that

$$\int_V \nabla \cdot \vec{g} \, dV = \int_S \vec{g} \cdot d\vec{S} = -|\vec{g}| 4\pi r^2 = -4\pi G \int \rho(\vec{r}) \, dV,$$

or

$$|\vec{g}| = \frac{GM(r)}{r^2}, \qquad \text{where } M(r) = \int_V \rho \, dV$$

is the total mass within the volume, whatever its radial distribution. Since the field

The Gravitational Field and Potential

and potential are the same as for a point mass M at the origin, it is easy to see why a spherical body attracts as if all the mass were concentrated at its center, a theory proved by direct integration in most elementary texts. Our proof has the advantage of being valid for any spherical distribution of mass, whether constant in density radially or not.

Let us now look at another example (Figure 1.13):

Example 8

A uniform sphere of mass M is embedded in a hole of radius R in an infinite thin plane having mass per unit area σ. Find the field at a distance d above the center of the sphere.

Solution

The field of the sphere, from our previous remarks, is just $-GM/d^2$, the origin being taken at the sphere center. For the plane, we take a ring of matter about the origin, with radius ρ:

$$g_z(\text{plane}) = -G \int_R^\infty \frac{\sigma 2\pi\rho \, d\rho}{(d^2 + \rho^2)} \frac{d}{\sqrt{d^2 + \rho^2}}$$

$$= -2\pi G\sigma d \left[\frac{-1}{\sqrt{d^2 + \rho^2}} \right]_R^\infty = -\frac{2\pi G\sigma d}{\sqrt{d^2 + R^2}},$$

the factor $d/\sqrt{d^2 + \rho^2}$ in the integrand giving the z component. The total field is then

$$g_z = -\frac{GM}{d^2} - \frac{2\pi G \, d\sigma}{\sqrt{d^2 + R^2}},$$

the other components being zero by symmetry.

One final remark is in order. There seems to be a common misconception that the force between a point mass and extended body can be found by use of

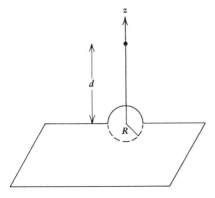

Figure 1.13 Sphere embedded in plane.

Newton's law, eq. (1.21), putting in the distance between the CM of the body and the point mass as the distance to be squared in the denominator. We trust it is clear from the above example that an integration is actually required (indeed, for *two* extended bodies, one must integrate over the elements of both); the CM is *not* intimately related to the force of gravity. However, such an idea offers a certain approximation to the truth when the point mass is at a distance large in comparison to the size of the extended object, as may be seen by the following argument.

Consider the vector diagram of Figure 1.14. $\vec{R}$ is directed from origin to CM, $\vec{d}$ from field point to CM, and $\vec{r}_c$ from CM to element of mass; $\vec{r}$ and $\vec{r}'$ are as before. From the vectors as shown, we see that

$$\vec{r}' - \vec{r} = (\vec{R} + \vec{r}_c) - \vec{r} = (\vec{R} - \vec{r}) + \vec{r}_c = \vec{d} + \vec{r}_c; \qquad \text{assume } d \equiv |\vec{d}| \gg |\vec{r}_c| \equiv r_c.$$

Upon expanding, one has

$$\frac{1}{|\vec{r}' - \vec{r}|} = \frac{1}{\sqrt{d^2 + r_c^2 + 2\vec{r}_c \cdot \vec{d}}} = \frac{1}{d\sqrt{1 + \left(\frac{2\vec{r}_c \cdot \vec{d}}{d^2} + \frac{r_c^2}{d^2}\right)}} \approx \frac{1}{d}\left(1 - \frac{\vec{r}_c \cdot \vec{d}}{d^2}\right).$$

The potential now becomes

$$\Phi(\vec{r}) \approx -\frac{G}{d} \int \left[1 - \frac{\vec{r}_c \cdot \vec{d}}{d^2}\right] \rho(\vec{r}') \, dV' = -\frac{G}{d} \int \rho(\vec{r}') \, dV' + \frac{G}{d^3} \vec{d} \cdot \int \vec{r}_c \rho(\vec{r}') \, dV'$$

$$= -\frac{GM}{d},$$

$$M \equiv \int \rho(\vec{r}') \, dV',$$

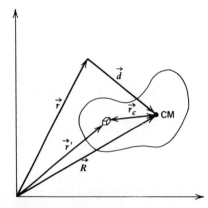

Figure 1.14 Vectors for potential calculation.

since the integral in the second term vanishes by definition of the CM and $\vec{r}_c$. Thus, we conclude that the common misconception *is* valid for large distances *d to the extent* that we ignore higher-order terms in the expansion of the radical, which naturally involve the moment of inertia and higher moments of the mass distribution for the extended object.

In concluding, we note that today the best test of the gravitational theory may be in the problem where it originated—the analysis of the moon's motions. The present scheme for treating the lunar motions (which have long been carefully recorded) was developed by Ernest William Brown[2] around the turn of the century. He considers the gravitational interactions of Sun, Earth, and Moon in an approximation procedure which is heavily dependent on the work of previous investigators (the three-body problem cannot be solved exactly). Brown's lengthy calculations verify the Newtonian predictions to an accuracy of six or more significant figures.

Astronomical applications of this kind do not normally require modifications to the Newtonian formalism from relativity or quantum mechanics. In addition, high-speed computers today make many calculations much more feasible than in earlier years. Nevertheless, a number of practical problems remain.

As a case in point, it has been estimated that tracking the motion of all the molecules in a gas container with reasonable precision (in position and velocity) for just one second would require that initial conditions be given to an accuracy of millions of significant figures! Otherwise the errors resulting from initial uncertainties in the data will accumulate rapidly and render the calculations meaningless. The modern computer has been a useful tool in the study of gravitation and of other theories, but it does not solve all our problems.

PROBLEMS

1.1 A skydiver falls with the velocity obtained in Section 1.2. Find the distance fallen as a function of the time t.

1.2 (*a*) Find the center of mass of the skeleton rectangle of figure 1.4, using the given density function.

(*b*) Find the moment of inertia of the skeleton rectangle when rotated about its diagonal.

1.3 A **couple** consists of a pair of forces which are equal in magnitude but oppositely directed, the lines of action of the forces being parallel yet not coincident. A couple tends to rotate (but not translate) a rigid body on

---

[2] E. W. Brown, "Theory of Motion of the Moon," *Mem. Roy. Ast. Soc.* **59**, 1 (1908).

which it acts. Show that any system of forces acting on the body is equivalent to a single force acting at an arbitrary point plus a couple.

1.4 Consider the spool of Figure 1.5 placed on an inclined plane of angle α. Show that rotation of the spool without translation can still be achieved, but that the appropriate angle θ now depends on the coefficient of friction as well as the radii of the spool and the angle α.

1.5 A shell is fired from a gun with V as the vertical component of its velocity. At the top of the trajectory, the shell explodes into two pieces of mass m_1 and m_2, giving additional kinetic energy K to the system (relative to the CM). Assuming that the pieces initially separate horizontally and later strike the ground a distance D apart, show that K is given by

$$\frac{g^2 D^2}{V^2} \frac{m_1 m_2}{2(m_1 + m_2)}.$$

1.6 A particle of mass m is accelerated along the positive x-direction by a constant force; it starts from rest at the origin of an inertial frame. A second reference frame moves with constant speed v_0 along the negative x-direction; initially, the two frames coincide.

(*a*) Find the velocity and position of the particle as a function of time in both reference frames.

(*b*) Find the work done by the force during a time interval t in both frames.

(*c*) Are the results of (*b*) different in the two cases? If so, are the laws of mechanics not different in the two inertial frames of reference? Explain your answer.

1.7 (*a*) Assume that momentum conservation for a system of particles is valid in some inertial frame. Prove that it is also valid in any other frame related to the first by a Galilean transformation.

(*b*) Assume that the total kinetic energy of a system is conserved in all Galilean frames of reference. Show *without* using Newton's laws that momentum will be conserved in all such frames.

(*c*) Assume that momentum is conserved in all frames related by Galilean transformation and kinetic energy is conserved in one frame. Show that kinetic energy is conserved in all such frames.

1.8 (*a*) Derive general expressions for the *magnitudes* of the velocities of the two masses undergoing a collision. That is, eliminate $\hat{\epsilon}$, $\vec{v}_1$, and $\vec{v}_2$ from the vector expressions in eq. (1.20), so that your results will be expressed entirely in terms of the masses, the angle χ, and the magnitude v of the relative velocity. (Assume that m_2 is at rest before the collision throughout this problem.)

Problems

(b) Calculate the maximum energy that mass 2 can acquire in the collision.

(c) Find the relationship between angles α and χ.

1.9 A ball of mass m and radius R rotates with an angular velocity ω_0 about a horizontal axis. If it is placed on a horizontal plane, the coefficient of friction being μ, how far can the ball travel before it is engaged in a pure rolling motion?

1.10 Two inclined planes intersect each other so as to form an equilateral triangle when viewed in vertical-plane cross section. A string runs over a smooth pulley fastened at the triangle apex to masses m_1 and m_2, one on each inclined plane. Use the method of virtual work to find the coefficient of static friction if the masses are on the verge of slipping in one direction. (Assume μ the same on both planes; neglect mass and stretching of string.)

1.11 Consider the force

$$F_x = 2axy + by^2z^2, \qquad F_y = ax^2 + 2bxyz^2, \qquad F_z = 2bxy^2z,$$

where a and b are constants.

(a) Find the potential energy associated with this force, taking the zero of potential at the origin.

(b) Prove that the force is conservative.

(c) Evaluate the line integral of the force around a square of unit area in the plane $z = 1$ with lower left-hand corner at the origin. Is your result expected?

1.12 A uniform rod of mass M lies on the z-axis between $z = 0$ and $z = L$.

(a) Find the gravitational force on a point mass m placed on the axis at distance z from the origin ($z > L$).

(b) If all the mass of the rod were concentrated at a point z_0 such that the force on m were unchanged, find the location of z_0.

(c) Show that z_0 is located almost at the center of mass when $z \gg L$.

(d) Show that the potential of the rod approaches that of a point mass at the origin as L goes to zero.

1.13 (a) A homogeneous right circular cylinder of radius R, height L, and mass M is placed along the z-axis between $z = 0$ and $z = L$. Find the gravitational potential of the cylinder on the axis at distance z from the origin ($z > L$).

(b) Show that your expression for the potential $\to 0$ as $z \to \infty$.

(c) Show that the potential approaches that of the rod of Problem 1.12 as $R \to 0$, L fixed. (If Problem 1.12 has not been assigned, you can work out the potential without much effort.)

(d) Show that the potential approaches that of a plane disk in the $z = 0$ plane as $L \to 0$, keeping R finite.
(e) Show that the disk expression of part (d) approaches that of a point mass at the origin as $R \to 0$.

1.14 The density of matter in a region of space is given by a spherically symmetrical distribution

$$\rho = \frac{C}{r(r^2 + b^2)^2},$$

where r is the radial distance from the origin, while b and C are constants.

(a) Find the gravitational field intensity for such a distribution of mass, using the field equation.
(b) Find the potential from either the field or Poisson's equation.
(c) Find the field also from the properties of a spherically symmetrical distribution of matter.

Chapter Two
THE METHODS OF LAGRANGE AND HAMILTON

2.1 CONSTRAINTS AND GENERALIZED COORDINATES

It has already been shown that the application of the laws of mechanics can be complicated considerably by the presence of constraints in a given physical problem. The constraints are frequently expressed as mathematical relationships, conditions that must be incorporated in any solution to the problem. In practice, this means that the components x_i of the particle position vectors are not all independent of one another but are related in such a way as to satisfy the constraints.

Furthermore, Newton's laws cannot be utilized until we know *all* the forces acting on a given body, including the forces of constraint. Inasmuch as the latter are often unknown, we need to reformulate the laws of motion so as to avoid constraining forces. In the following pages, we accomplish such a reformulation by means of the kinetic and potential energies rather than the forces, a method originally developed by Lagrange. Later in the chapter, we also examine a technique for finding the constraint forces.

Note first that constraints may be classified in a variety of ways. There are those which contain the time explicitly and are called **rheonomous** constraints (an example is a particle attached to an expanding balloon), while those which do not contain the time are known as **scleronomous**. These are formal definitions and not of much practical value.

It is most important, however, to fully understand the meaning of a **holonomic** constraint, which is one that is expressed by an equation of the general form

$$f(\vec{r}_1, \vec{r}_2, \ldots, t) = 0. \qquad (2.1)$$

This relationship connects the coordinates of one or more particles in a manner that is prescribed by the nature of the constraint; it is not necessary, however, that *all* the coordinates of each particle (and the time) actually appear in every equation of this kind. For example, the coordinates of a particle constrained to the surface of a sphere with radius a must satisfy the holonomic relation

$$x^2 + y^2 + z^2 - a^2 = 0.$$

If the particle were instead confined to a circular path of radius a in the xy plane, its coordinates would have to obey the two equations

$$x^2 + y^2 - a^2 = 0 \quad \text{and} \quad z = 0,$$

both of which express holonomic constraints.

Any constraint which is not capable of being expressed in the form (2.1) is said to be **nonholonomic**. Consider a particle inside a cylindrical container of radius R and height L. The constraint relations needed are of the type

$$\rho \leqslant R, \quad 0 \leqslant z \leqslant L,$$

which are in the form of inequalities and therefore nonholonomic.

A particular kind of nonholonomic constraint, called **anholonomic** or **Pfaffian**, is frequently of interest; it is one in which the equation of constraint is given in differential form. For a single particle, a velocity condition such as

$$A\,dx + B\,dy + C\,dz + D\,dt = 0$$

serves as an example. The coefficients A, B, C, and D are assumed to be known functions of the coordinates; even so, it may not be possible to integrate the relation so as to cast it into holonomic form. (That will be the case when the left side is not an exact differential, nor can any multiplicative factor be found to make it so.) The information required to perform the integration may be equivalent to a solution of the problem; therefore, as it stands, the differential equation does not provide a way of eliminating one of the coordinates as a dependent variable. By contrast, the holonomic relation (2.1) always permits a reduction to the minimum number of independent coordinates needed for a solution to a given problem; in essence, that is why this type of constraint is so important.

In the previous chapter, we have seen that if a system of particles is in equilibrium, the ith particle can undergo a virtual displacement $\delta \vec{r}_i$, and the net virtual work, when summed over all particles, will vanish:

$$\sum_i \vec{F}_i \cdot \delta \vec{r}_i + \sum_i (\vec{F}_c)_i \cdot \delta \vec{r}_i = 0, \qquad (2.2)$$

Constraints and Generalized Coordinates

where $\vec{F}_i$ is the applied force on the ith particle and $(\vec{F}_c)_i$ is the force of constraint on it. This relation follows from the fact that

$$\vec{F}_i + (\vec{F}_c)_i = 0.$$

Since the work of the constraint forces is normally zero (see Chapter 1), we have

$$\sum_i \vec{F}_i \cdot \delta \vec{r}_i = 0,$$

the latter certainly not implying that $\vec{F}_i = 0$! The coefficients of $\delta\vec{r}_i$ cannot be equated to zero because the $\delta\vec{r}_i$ are not all *independent*; they are related by the constraints on the system. We untangle this dependence by the introduction of new generalized coordinates not connected by constraint relations.

Before doing so, however, we need to generalize to the nonequilibrium situations of dynamics, making use of a technique devised by Bernoulli and D'Alembert and known as D'Alembert's principle. If the second law for the ith particle is cast in the form

$$\vec{F}_i + (\vec{F}_c)_i - \dot{\vec{p}}_i = 0,$$

this equation has exactly the form of the equilibrium condition, except for the addition of an inertial term $-\dot{\vec{p}}_i$ on the left side. With the inclusion of such a term, we have essentially reduced dynamics to statics. Equation (2.2) now becomes

$$\sum_i (\vec{F}_i - \dot{\vec{p}}_i) \cdot \delta \vec{r}_i = 0, \qquad (2.3)$$

where, as before, the work of constraint forces vanishes.

Now we proceed to express the $\vec{r}_i$ in terms of a new set of coordinates. Normally we would require $3N$ independent coordinates, or $3N$ **degrees of freedom**, to fully specify the motion of N particles in three-dimensional space. But if the system is constrained by j relations of the holonomic type, only $3N - j$ coordinates are really independent, so that the number of degrees of freedom has been reduced to $3N - j$. This reduction comes about because each constraint equation can be used to eliminate one of the $3N$ coordinates, expressing it in terms of the others.

We therefore define a new set of generalized coordinates q_k which will consist of $3N - j$ independent quantities which are convenient for the analysis of a particular problem. The position vectors $\vec{r}_i$ are related to the q_k through a set of transformation equations of the form

$$\vec{r}_i = \vec{r}_i(q_1, q_2, \ldots q_{3N-j}, t).$$

The reader is undoubtedly familiar with the use of specialized coordinate systems to facilitate the solution of many elementary problems having appropriate symmetries; that approach is generalized here to include constraints. Thus, a particle moving on the surface of a sphere would require the two polar angles θ and ϕ in order to specify its motion; the fact that the radius is some fixed constant is a constraint limiting the number of degrees of freedom to 2.

The choice of coordinates to be selected in a particular application is left to the individual, though it is dictated by the problem to a certain extent. With a little experience, several suitable choices might be devised. Furthermore, this approach does not limit us to the use of lengths and angles for the q's.

Let us take $n \equiv 3N - j$, writing

$$\vec{r}_i = \vec{r}_i(q_1, q_2, \ldots q_n, t) \equiv \vec{r}_i(q, t), \tag{2.4}$$

the last member being a short way of showing the functional dependence on all the q's. Differentiation of the transformation with respect to the time gives

$$\vec{v}_i = \sum_{j=1}^{n} \frac{\partial \vec{r}_i}{\partial q_j} \dot{q}_j + \frac{\partial \vec{r}_i}{\partial t},$$

while

$$\delta \vec{r}_i = \sum_{j=1}^{n} \frac{\partial \vec{r}_i}{\partial q_j} \delta q_j$$

(there is no time term in the last equation since $\delta t = 0$ for a virtual displacement). Note from the $\vec{v}_i$ expression that $\partial \vec{v}_i / \partial \dot{q}_k = \partial \vec{r}_i / \partial q_k$ ($\dot{q}_k$ appearing only once in the sum), and observe also that the $\dot{q}$'s are independent of q's, hence

$$\frac{\partial \vec{v}_i}{\partial q_k} = \sum_{j} \frac{\partial^2 \vec{r}_i}{\partial q_k \partial q_j} \dot{q}_j + \frac{\partial^2 \vec{r}_i}{\partial q_k \partial t} = \sum_{j} \frac{\partial}{\partial q_j} \left(\frac{\partial \vec{r}_i}{\partial q_k} \right) \dot{q}_j + \frac{\partial}{\partial t} \left(\frac{\partial \vec{r}_i}{\partial q_k} \right) = \frac{d}{dt} \left(\frac{\partial \vec{r}_i}{\partial q_k} \right).$$

Constraints and Generalized Coordinates

We then have for the inertial term of D'Alembert's principle

$$\sum_i \vec{\dot{p}}_i \cdot \delta \vec{r}_i = \sum_{i,j} \frac{d}{dt}(m_i \vec{v}_i) \cdot \frac{\partial \vec{r}_i}{\partial q_j} \delta q_j = \sum_{i,j} \left[\frac{d}{dt}\left(m_i \vec{v}_i \cdot \frac{\partial \vec{r}_i}{\partial q_j} \right) - m_i \vec{v}_i \cdot \frac{d}{dt} \frac{\partial \vec{r}_i}{\partial q_j} \right] \delta q_j$$

$$= \sum_{i,j} \left[\frac{d}{dt}\left(m_i \vec{v}_i \cdot \frac{\partial \vec{v}_i}{\partial \dot{q}_j} \right) - m_i \vec{v}_i \cdot \frac{\partial \vec{v}_i}{\partial q_j} \right] \delta q_j = \sum_j \left[\frac{d}{dt}\left(\frac{\partial T}{\partial \dot{q}_j} \right) - \frac{\partial T}{\partial q_j} \right] \delta q_j,$$

where we have used the kinetic energy in the form $T = \sum_i \frac{1}{2} m_i \vec{v}_i \cdot \vec{v}_i$.

One could also write that

$$\sum_i \vec{F}_i \cdot \delta \vec{r}_i = \sum_{i,j} \vec{F}_i \cdot \frac{\partial \vec{r}_i}{\partial q_j} \delta q_j \equiv \sum_j Q_j \delta q_j,$$

where

$$Q_j \equiv \sum_i \vec{F}_i \cdot \frac{\partial \vec{r}_i}{\partial q_j} \tag{2.5}$$

is the **generalized force**. (Q_j has the actual dimensions of force if, and only if, q_j is a length coordinate; if q_j is an angle, for example, then Q_j will be a torque.)

Combining all our results leads to D'Alembert's principle in the form

$$\sum_i (\vec{F}_i - \vec{\dot{p}}_i) \cdot \delta \vec{r}_i = 0 = \sum_j \left[Q_j - \frac{d}{dt}\left(\frac{\partial T}{\partial \dot{q}_j} \right) + \frac{\partial T}{\partial q_j} \right] \delta q_j.$$

When the constraints are given by holonomic relationships, the q's and δq's are all fully independent of one another, and the coefficient of each term in the sum must therefore vanish:

$$\frac{d}{dt}\left(\frac{\partial T}{\partial \dot{q}_j} \right) - \frac{\partial T}{\partial q_j} - Q_j = 0. \tag{2.6}$$

Suppose that the force is derivable from a potential energy V which does not depend explicitly on the time or on the generalized velocities $\dot{q}$ but only on position; these assumptions are usually valid. Then,

$$Q_j = \sum_i \vec{F}_i \cdot \frac{\partial \vec{r}_i}{\partial q_j} = -\sum_i \nabla_i V \cdot \frac{\partial \vec{r}_i}{\partial q_j} = -\sum_i \left(\frac{\partial x_i}{\partial q_j} \frac{\partial V}{\partial x_i} + \cdots \right) = -\frac{\partial V}{\partial q_j}$$

and eq. (2.6) becomes

$$\frac{d}{dt}\left(\frac{\partial T}{\partial \dot{q}_j}\right) - \frac{\partial T}{\partial q_j} + \frac{\partial V}{\partial q_j} = 0 = \frac{d}{dt}\left[\frac{\partial (T-V)}{\partial \dot{q}_j}\right] - \frac{\partial (T-V)}{\partial q_j}.$$

Defining the **Lagrangian** L as $L \equiv T - V$, we finally have the standard form of Lagrange's equations:

$$\frac{d}{dt}\left(\frac{\partial L}{\partial \dot{q}_j}\right) - \frac{\partial L}{\partial q_j} = 0, \qquad (2.7)$$

where

$$L \equiv T - V = L(q, \dot{q}, t). \qquad (2.8)$$

There is one such relation for each of the generalized coordinates; in the next section, we shall see how they are used in the solution of physical problems.

In a more general situation, suppose that Q_j is decomposed into a conservative part $-\partial V/\partial q_j$ and a nonconservative part $\mathscr{F}_j$ (either of which could be zero in a specific application):

$$Q_j = -\frac{\partial V}{\partial q_j} + \mathscr{F}_j,$$

where $\mathscr{F}_j$ might represent a force of friction or one of constraint; recall that $\mathscr{F}_j$ and hence Q_j need not even have the actual dimensions of force. Retracing the development following eq. (2.6), one finds now that

$$\frac{d}{dt}\left(\frac{\partial L}{\partial \dot{q}_j}\right) - \frac{\partial L}{\partial q_j} = \mathscr{F}_j. \qquad (2.7')$$

We shall see this relation again later in the chapter.

2.2 APPLICATIONS OF LAGRANGE'S EQUATIONS

Example 1   A Single Particle in Familiar Coordinate Systems

Let us describe the motion of a single particle of mass m in rectangular coordinates. The force acting on this mass will be denoted by $\vec{F}$. Whether a potential is specified or not, one can always write

$$Q_x = F_x \frac{\partial x}{\partial x} + F_y \frac{\partial y}{\partial x} + F_z \frac{\partial z}{\partial x} = F_x.$$

Applications of Lagrange's Equations

The kinetic energy is
$$T = \tfrac{1}{2}m(\dot{x}^2 + \dot{y}^2 + \dot{z}^2),$$

with
$$\frac{\partial T}{\partial \dot{x}} = m\dot{x}, \qquad \frac{\partial T}{\partial x} = 0,$$

so that eq. (2.6) produces
$$\frac{d}{dt}m\dot{x} - 0 - F_x = 0,$$

or
$$m\ddot{x} = F_x.$$

In a similar way, one has
$$m\ddot{y} = F_y,$$
$$m\ddot{z} = F_z.$$

These are the usual Newtonian equations of motion. (If the potential V had been given, (2.7) could have been employed directly after finding L.)

In view of the fact that no special rectangular symmetry was involved, cylindrical generalized coordinates could just as well have been used. The transformation (2.4) is then
$$x = \rho \cos\phi, \qquad y = \rho \sin\phi, \qquad z = z.$$

Differentiating, we find that
$$\dot{x} = \dot{\rho}\cos\phi - \rho\dot{\phi}\sin\phi, \qquad \dot{y} = \dot{\rho}\sin\phi + \rho\dot{\phi}\cos\phi, \qquad \dot{z} = \dot{z}.$$

Hence
$$T = \tfrac{1}{2}m(\dot{x}^2 + \dot{y}^2 + \dot{z}^2) = \tfrac{1}{2}m[\dot{\rho}^2 + \rho^2\dot{\phi}^2 + \dot{z}^2]$$

(where use has been made of the identity $\sin^2\alpha + \cos^2\alpha = 1$ for any angle α).

For the coordinate ρ, we have
$$\frac{\partial T}{\partial \dot{\rho}} = m\dot{\rho}, \qquad \frac{\partial T}{\partial \rho} = m\rho\dot{\phi}^2,$$

$$Q_\rho = F_x\frac{\partial x}{\partial \rho} + F_y\frac{\partial y}{\partial \rho} + F_z\frac{\partial z}{\partial \rho} = F_x\cos\phi + F_y\sin\phi = \vec{F}\cdot\left(\frac{\vec{\rho}}{\rho}\right) = F_\rho,$$

the component of force along the ρ direction, as expected. Plugging into eq. (2.6) gives

$$m\ddot{\rho} - m\rho\dot{\phi}^2 = F_\rho,$$

the radial equation of motion, with the centripetal acceleration term $-m\rho\dot{\phi}^2$ arising quite naturally in the calculation.

For the coordinate ϕ,

$$\frac{\partial T}{\partial \dot{\phi}} = m\rho^2\dot{\phi}, \quad \frac{\partial T}{\partial \phi} = 0,$$

$$Q_\phi = F_x(-\rho \sin\phi) + F(\rho \cos\phi) = \rho\vec{F}\cdot\hat{\varepsilon}_\phi = \rho F_\phi,$$

where $\hat{\varepsilon}_\phi = -\sin\phi\,\hat{i} + \cos\phi\,\hat{j}$ is a unit vector in the direction of increasing ϕ. The equation of motion is then, from (2.6),

$$\frac{d}{dt}(m\rho^2\dot{\phi}) = \rho F_\phi,$$

an angular momentum–torque relation of the usual type. (For the z coordinate, nothing is changed; we get $m\ddot{z} = F_z$ as before.)

Summarizing, the equations of motion for a single particle are

$$m\ddot{\rho} - m\rho\dot{\phi}^2 = F_\rho, \quad \frac{d}{dt}(m\rho^2\dot{\phi}) = \rho F_\phi, \quad m\ddot{z} = F_z \tag{2.9}$$

in cylindrical coordinates.

Let us now examine the situation in spherical coordinates, with the transformation $x = r\sin\theta\cos\phi$, $y = r\sin\theta\sin\phi$, $z = r\cos\theta$. Time differentiation produces

$$\dot{x} = \dot{r}\sin\theta\cos\phi + r\dot{\theta}\cos\theta\cos\phi - r\dot{\phi}\sin\theta\sin\phi,$$

$$\dot{y} = \dot{r}\cos\theta\sin\phi + r\dot{\theta}\cos\theta\sin\phi + r\dot{\phi}\sin\theta\cos\phi,$$

$$\dot{z} = \dot{r}\cos\theta - r\dot{\theta}\sin\theta,$$

whereupon

$$T = \tfrac{1}{2}m(\dot{x}^2 + \dot{y}^2 + \dot{z}^2) = \tfrac{1}{2}m(\dot{r}^2 + r^2\dot{\theta}^2 + r^2\sin^2\theta\,\dot{\phi}^2)$$

after algebraic simplification.

Applications of Lagrange's Equations

For the coordinate r,

$$\frac{\partial T}{\partial \dot{r}} = m\dot{r}, \qquad \frac{\partial T}{\partial r} = mr\dot{\theta}^2 + mr\sin^2\theta\,\dot{\phi}^2,$$

$$Q_r = F_x\frac{\partial x}{\partial r} + F_y\frac{\partial y}{\partial r} + F_z\frac{\partial z}{\partial r} = F_x\sin\theta\cos\phi + F_y\sin\theta\sin\phi + F_z\cos\theta$$

$$= \vec{F}\cdot\left(\frac{\vec{r}}{r}\right) = F_r.$$

With the aid of (2.6),

$$m(\ddot{r} - r\dot{\theta}^2 - r\sin^2\theta\,\dot{\phi}^2) = F_r,$$

the radial equation of motion.

Next, for the coordinate θ,

$$\frac{\partial T}{\partial \dot{\theta}} = mr^2\dot{\theta}, \qquad \frac{\partial T}{\partial \theta} = mr^2\sin\theta\cos\theta\,\dot{\phi}^2,$$

$$Q_\theta = F_x r\cos\theta\cos\phi + F_y r\cos\theta\sin\phi - F_z r\sin\theta = r\vec{F}\cdot\hat{\varepsilon}_\theta = rF_\theta,$$

with $\hat{\varepsilon}_\theta$ a unit vector in the direction of increasing θ. (See Appendix A for a further discussion of these unit vectors.) We have

$$\frac{d}{dt}(mr^2\dot{\theta}) - mr^2\sin\theta\cos\theta\,\dot{\phi}^2 = rF_\theta.$$

Finally, for the coordinate ϕ,

$$\frac{\partial T}{\partial \dot{\phi}} = mr^2\sin^2\theta\,\dot{\phi}, \qquad \frac{\partial T}{\partial \phi} = 0,$$

$$Q_\phi = -F_x r\sin\theta\sin\phi + F_y r\sin\theta\cos\phi + 0 = r\sin\theta\,\vec{F}\cdot\hat{\varepsilon}_\phi = r\sin\theta\,F_\phi,$$

and so

$$\frac{d}{dt}(mr^2\sin^2\theta\,\dot{\phi}) = r\sin\theta\,F_\phi.$$

In general, the spherical relations are somewhat complicated, but in special cases they readily reduce to more familiar equations. For example, if $\phi = $ constant,

$\dot{\phi} = 0 = F_\phi$, whereupon both F_θ and F_r simplify greatly. In summary, the spherical relations are

$$m(\ddot{r} - r\dot\theta^2 - r\sin^2\theta\,\dot\phi^2) = F_r,$$

$$\frac{d}{dt}(mr^2\dot\theta) - mr^2\sin\theta\cos\theta\,\dot\phi^2 = rF_\theta, \qquad (2.10)$$

$$\frac{d}{dt}(mr^2\sin^2\theta\,\dot\phi) = r\sin\theta\,F_\phi.$$

Example 2 Bead Moving on Helix

A smooth piece of wire is twisted into the shape of a helix with cylindrical equations $\rho = b$, $z = a\phi$, where a and b are constants. If the origin is a center of attractive force varying directly with the distance (constant k), find the equation of motion for a bead of mass m sliding on the wire.

Solution

In cylindrical coordinates, we have seen that

$$T = \tfrac{1}{2}m(\dot\rho^2 + \rho^2\dot\phi^2 + \dot z^2).$$

The potential energy is of the form

$$V = \tfrac{1}{2}k(\rho^2 + z^2),$$

so that

$$F_\rho = -\frac{\partial V}{\partial \rho} = -k\rho \quad \text{and} \quad F_z = -\frac{\partial V}{\partial z} = -kz.$$

The Lagrangian then becomes

$$L = T - V = \tfrac{1}{2}m(\dot\rho^2 + \rho^2\dot\phi^2 + \dot z^2) - \tfrac{1}{2}k(\rho^2 + z^2)$$
$$= \tfrac{1}{2}m(a^2 + b^2)\dot\phi^2 - \tfrac{1}{2}k(b^2 + a^2\phi^2)$$

when the two holonomic conditions $\rho = b$ and $z = a\phi$ are inserted in it. That is, when the constraints are taken into account, only one independent coordinate (ϕ) remains. Lagrange's equation is now

$$\frac{d}{dt}\left[m(a^2 + b^2)\dot\phi\right] + ka^2\phi = 0,$$

or

$$\ddot\phi = -\frac{ka^2}{m(a^2 + b^2)}\phi.$$

Alternatively, we could have used z as the independent coordinate and obtained

$$\ddot z = -\frac{ka^2}{m(a^2 + b^2)}z.$$

Or, finally, we could have chosen a coordinate S measured along the length of the helix, such that $S = b\phi$ and $z = (a/b)S$. Expressing L in terms of S leads at once to

$$\ddot{S} = -\frac{ka^2}{m(a^2 + b^2)} S.$$

No matter what coordinate is selected, the final result is a differential equation of simple harmonic type, showing that the bead will move along the wire in an oscillatory fashion about the origin.

Example 3 Coupled Springs in Oscillation

A mass M is suspended from the ceiling by a spring having constant K. A second mass m is suspended from the first by a spring of constant k. Neglecting spring masses, find the equations of motion. (Assume a uniform gravitational field of acceleration g.)

Solution

In solving a problem of this type, one's first inclination is to select vertical coordinates y_1' and y_2' measured downward from the ceiling. If that is done, the potential energy will contain terms of the form $\frac{1}{2}K(y_1' - y_{10}')^2$, where y_{10}' gives the equilibrium position of the uppermost spring. It will often be more convenient in this kind of problem to measure the generalized coordinates y_1 and y_2 *from* the equilibrium positions instead, thus eliminating constants such as y_{10}' from the expressions (Figure 2.1). (Naturally, such a choice will not affect the derivatives appearing in the kinetic energy.) The gravitational terms will also be taken as zero at the equilibrium position, giving for the energies

$$T = \tfrac{1}{2}M\dot{y}_1^2 + \tfrac{1}{2}m\dot{y}_2^2, \qquad V = \tfrac{1}{2}Ky_1^2 + \tfrac{1}{2}k(y_2 - y_1)^2 - Mgy_1 - mgy_2,$$

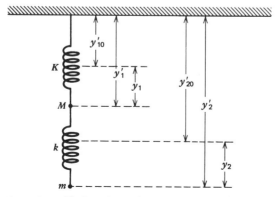

Figure 2.1 Choice of coordinates for coupled springs.

and for the Lagrangian

$$L = T - V = \tfrac{1}{2}M\dot{y}_1^2 + \tfrac{1}{2}m\dot{y}_2^2 - \tfrac{1}{2}Ky_1^2 - \tfrac{1}{2}k(y_2 - y_1)^2 + Mgy_1 + mgy_2.$$

The Lagrange equation for y_1, with the aid of

$$\frac{\partial L}{\partial \dot{y}_1} = M\dot{y}_1 \quad \text{and} \quad \frac{\partial L}{\partial y_1} = -Ky_1 + k(y_2 - y_1) + Mg,$$

is

$$M\ddot{y}_1 + Ky_1 - k(y_2 - y_1) - Mg = 0. \tag{2.11}$$

The Lagrange equation for y_2, using

$$\frac{\partial L}{\partial \dot{y}_2} = m\dot{y}_2 \quad \text{and} \quad \frac{\partial L}{\partial y_2} = -k(y_2 - y_1) + mg,$$

is

$$m\ddot{y}_2 + k(y_2 - y_1) - mg = 0. \tag{2.12}$$

The equations of motion (2.11) and (2.12) furnish a simple example of a *coupled* pair of differential equations, inasmuch as the variables y_1 and y_2 appear in both of them. It is possible to find linear combinations of y_1 and y_2 in terms of which the motion can be analyzed quite simply, but we shall postpone a discussion of the solutions of these equations until Chapter 7.

Example 4 Electrical Circuits

Consider a simple electrical circuit containing an inductance, a capacitance, and a switch. If the capacitor is charged and the switch closed, oscillations will be established in the circuit. Find the differential equation giving the charge on the capacitor as a function of time (Figure 2.2).

Solution

In this text, we have no intention of probing deeply into electromagnetism nor of pursuing the analogy between mechanical and electrical circuits. The Lagrangian scheme, however, is based on the concept of energy and therefore has applicabil-

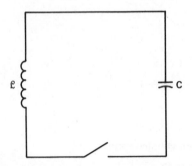

Figure 2.2 Oscillating electrical circuit.

ity that goes far beyond simple mechanical systems, and it seems well to illustrate the point with this simple example. Consider the following table:

Mechanical Concept	Electrical Analogue
Displacement (Position)	Charge Q
Velocity	Current
Force	Potential Difference
Mass	Inductance $\mathscr{L}$
Spring Constant	Reciprocal Capacitance $1/C$

We see from the table that the energy stored in the magnetic field of the inductance, well known from elementary principles as $\frac{1}{2}\mathscr{L}\dot{Q}^2$ (where Q is the instantaneous charge on the capacitor), is like a "kinetic" energy $\frac{1}{2}mv^2$. The energy stored in the capacitor is $\frac{1}{2}(1/C)Q^2$, much like $\frac{1}{2}kx^2$ for a spring. The Lagrangian then becomes

$$L = T - V = \tfrac{1}{2}\mathscr{L}\dot{Q}^2 - \tfrac{1}{2}(1/C)Q^2$$

and Lagrange's equation for the coordinate Q is

$$\frac{d}{dt}\left(\frac{\partial L}{\partial \dot{Q}}\right) - \frac{\partial L}{\partial Q} = 0 = \ddot{Q} + \frac{Q}{C},$$

or

$$\ddot{Q} = -\frac{1}{C}Q,$$

a familiar oscillator equation from electrical resonance theory.

2.3 GENERALIZED MOMENTA AND SYSTEM INVARIANTS

Suppose we have a system of particles with kinetic energy

$$T = \sum_i \tfrac{1}{2}m_i\left(\dot{x}_i^2 + \dot{y}_i^2 + \dot{z}_i^2\right)$$

and a potential energy that is independent of velocity. Then it is possible to write

$$\frac{\partial L}{\partial \dot{x}_j} = \frac{\partial T}{\partial \dot{x}_j} = m_j \dot{x}_j = (p_x)_j,$$

the x-component of momentum for the jth particle. Inasmuch as this kind of situation occurs frequently, it is convenient to define the **generalized momentum**, or **conjugate momentum** (conjugate to q_j) as

$$p_j \equiv \frac{\partial L}{\partial \dot{q}_j}. \tag{2.13}$$

Of course, the generalized coordinates q_j are not usually Cartesian and may not even have units of length, whereupon p_j will not have the customary dimensions of momentum. Note in particular that p_j has the units of angular momentum when q_j is an angle. The definition leads to familiar expressions for length or angle coordinates and has a broad range of applicability.

There are cases of importance, however, where eq. (2.13) does not give the expected result even in Cartesian systems. A frequently quoted example is that of a charged particle in an electromagnetic field. From the theory of electromagnetism, the Lagrangian is (in Gaussian units)

$$L = T - q\Phi + \frac{q}{c}\vec{A}\cdot\vec{v},$$

where Φ and $\vec{A}$ are the scalar and vector potentials of the field, q is the charge of the particle, and c is the speed of light. For the x coordinate, (2.13) yields

$$p_x = \frac{\partial L}{\partial \dot{x}} = \frac{\partial T}{\partial \dot{x}} + \frac{q}{c}A_x = m\dot{x} + \frac{q}{c}A_x,$$

so that the mechanical momentum in p_x must be augmented by another term involving A_x.

In terms of the generalized momentum, Lagrange's equations may be written as

$$\frac{d}{dt}(p_j) - \frac{\partial L}{\partial q_j} = 0,$$

or

$$\dot{p}_j = \frac{\partial L}{\partial q_j}. \tag{2.14}$$

The coordinate q_j is said to be **ignorable** or **cyclic** when it does not appear in the Lagrangian; in that event $\partial L/\partial q_j = 0$ and p_j is a constant of the motion. That is, the generalized momentum conjugate to an ignorable coordinate is conserved.

Even in simple problems, Lagrange's equations may be complicated in form and not readily amenable to exact solution. But such a solution may not be required; we may need or want only a rough description of the motion. The determination of constants of the motion such as p_j can be very informative, so we pursue this matter somewhat further.

Suppose we are given a generalized coordinate q_j for which the change dq_j corresponds to a translation of the entire physical system along some direction specified by a unit vector $\hat{\varepsilon}$. In such a situation, dq_j simply refers to an effective shift of the origin. If q_j, for example, represents the x-coordinate of the CM, then $\hat{\varepsilon} \equiv \hat{i}$, and if each particle shifts by the amount $dq_j = dx$,

$$\frac{\partial \vec{r}_i}{\partial q_j} = \hat{i},$$

so that

$$Q_j = \sum_i \vec{F}_i \cdot \frac{\partial \vec{r}_i}{\partial q_j} = \left(\sum_i \vec{F}_i\right) \cdot \hat{i} = \vec{F} \cdot \hat{i} = F_x,$$

which is the total force component along the direction of the translation. Inasmuch as there is nothing special about the x direction, then, for arbitrary $\hat{\varepsilon}$,

$$\frac{\partial \vec{r}_i}{\partial q_j} = \hat{\varepsilon}, \qquad Q_j = \vec{F} \cdot \hat{\varepsilon}.$$

With the potential energy assumed independent of velocity, we have

$$p_j = \frac{\partial L}{\partial \dot{q}_j} = \frac{\partial T}{\partial \dot{q}_j} = \frac{\partial}{\partial \dot{q}_j}\left(\frac{1}{2}\sum_i m_i \vec{v}_i \cdot \vec{v}_i\right) = \sum_i m_i \vec{v}_i \cdot \frac{\partial \vec{v}_i}{\partial \dot{q}_j}$$

$$= \sum_i m_i \vec{v}_i \cdot \frac{\partial \vec{r}_i}{\partial q_j} = \left(\sum_i m_i \vec{v}_i\right) \cdot \hat{\varepsilon},$$

the component of total momentum along the direction of translation. (We have made use here of $\partial \vec{v}_i / \partial \dot{q}_j = \partial \vec{r}_i / \partial q_j$ from Section 2.1.) Equation (2.14) now has the form

$$\dot{p}_j = \frac{\partial L}{\partial q_j} = -\frac{\partial V}{\partial q_j} \equiv Q_j = \vec{F} \cdot \hat{\varepsilon},$$

where $\partial T / \partial q_j = 0$ from the fact that a translation of the entire system cannot change the velocities or kinetic energies. Thus, the time rate of change of total linear momentum and the force are equal in their components along $\hat{\varepsilon}$.

If q_j is an ignorable coordinate,

$$\frac{\partial L}{\partial q_j} = -\frac{\partial V}{\partial q_j} = 0 = Q_j,$$

or
$$p_j = \text{constant}.$$

In other words, the total system momentum is *invariant* under such a translation. Conservation theorems of this kind are intimately related to the *symmetries* of the physical system, in this case the fact that the basic dynamical quantities are independent of q_j. Much of advanced physics is concerned with the connection between conservation theorems and physical symmetries.

A similar argument applies if q_j is instead chosen as an angular coordinate which refers to rotation of the system as a whole about an axis in the direction of some unit vector $\hat{n}$. The generalized force is now the component of the total torque about $\hat{n}$, and the momentum p_j is the corresponding component of the angular momentum. Again, it turns out that p_j is a constant of the motion whenever q_j is an ignorable coordinate, the invariance under rotation giving rise to the conservation of angular momentum. Thus, if any physical system is symmetrical about the z-axis such that the azimuthal angle ϕ does not appear in the energies (or the Lagrangian), the z-component of angular momentum will be conserved and that of the torque will vanish. This conclusion follows no matter how complex the system appears to be.

The constants of the motion require a little further consideration. There is a theorem which states that there are $2n$ such constants which are independent of one another for a system having n degrees of freedom. A formal proof of this theorem hardly seems necessary, for the basic idea behind it is intuitively evident. Thus, as a simple example, take a single particle constrained so as to move along the x axis under the action of some force which is a function of x alone. If the equation of motion can be integrated to yield a solution for the position x as a function of time, two constants must appear: the initial values of position and velocity or their equivalents. This conclusion is in accord with the theorem, for there is but one degree of freedom present.

In a more complicated situation, the initial values of coordinates and velocities (or momenta) are not always utilized, for they may not be the most convenient choice of constants for the analysis. Thus, the fixed angular momentum of a particle might be selected in preference to the initial value of its velocity.

In any event, the symmetries of a physical system as manifested in the Lagrangian *must* give rise to particular constants of the motion. The knowledge that certain functions of the coordinates and momenta actually remain constant during the motion can be a great help in simplifying the relevant equations and leading to their solution, in much the same way that a holonomic constraint makes possible the elimination of extraneous variables. We shall again discuss the invariances of the Lagrangian and related constants of the motion in the last section of this chapter, where we demonstrate the use of Noether's theorem.

2.4 THE HAMILTONIAN AND THE CANONICAL EQUATIONS

At this point, we introduce a very important quantity in physics called the **Hamiltonian** H and defined as

$$H \equiv \sum_j \dot{q}_j p_j - L. \qquad (2.15)$$

In order to understand its significance, consider a physical situation in which the constraint equations do not depend explicitly on the time, ensuring that the transformation (2.4) to the generalized coordinates will also have $\partial \vec{r}_i / \partial t = 0$, whereupon

$$\vec{v}_i = \frac{d\vec{r}_i}{dt} = \sum_k \frac{\partial \vec{r}_i}{\partial q_k} \dot{q}_k.$$

The kinetic energy is then a quadratic function of the velocities given by

$$T = \frac{1}{2} \sum_i m_i \vec{v}_i \cdot \vec{v}_i = \frac{1}{2} \sum_i m_i \left(\sum_k \frac{\partial \vec{r}_i}{\partial q_k} \dot{q}_k \right) \cdot \left(\sum_l \frac{\partial \vec{r}_i}{\partial q_l} \dot{q}_l \right)$$

$$= \frac{1}{2} \sum_i m_i \sum_{k,l} \left(\frac{\partial \vec{r}_i}{\partial q_k} \cdot \frac{\partial \vec{r}_i}{\partial q_l} \right) \dot{q}_k \dot{q}_l.$$

If it is assumed as usual that V is independent of velocities, then

$$\sum_j \dot{q}_j p_j = \sum_j \dot{q}_j \frac{\partial L}{\partial \dot{q}_j} = \sum_j \dot{q}_j \frac{\partial T}{\partial \dot{q}_j} = \sum_j \dot{q}_j \left[\frac{1}{2} \sum_i m_i \sum_{k,l} \left(\frac{\partial \vec{r}_i}{\partial q_k} \cdot \frac{\partial \vec{r}_i}{\partial q_l} \right) \frac{\partial}{\partial \dot{q}_j} (\dot{q}_k \dot{q}_l) \right].$$

But

$$\frac{\partial}{\partial \dot{q}_j} (\dot{q}_k \dot{q}_l) = \dot{q}_l \delta_{j,k} + \dot{q}_k \delta_{j,l},$$

which upon substitution leaves two identical terms in the j sum, producing $\sum_j \dot{q}_j p_j = 2T$. From eq. (2.15), one can now write

$$H = 2T - L = 2T - (T - V) = T + V, \qquad (2.16)$$

the total energy of the system. It is not surprising that the Hamiltonian is so important; under the conditions assumed, it is the energy! Note in particular that even if L and H both depend explicitly on the time through the potential energy V, the Hamiltonian will still be the total energy.

Much remains to be done. Suppose that all of the $\dot{q}$'s are eliminated from eq. (2.15) by expressing each $\dot{q}_j$ in terms of the corresponding p_j, making use of (2.13) in order to do so. Then the Hamiltonian will become a function $H(q, p, t)$, while the Lagrangian is still $L(q, \dot{q}, t)$. Differentiating the definition (2.15),

$$dH = \sum_j \dot{q}_j \, dp_j + \sum_j p_j \, d\dot{q}_j - \left(\sum_j \frac{\partial L}{\partial q_j} dq_j + \sum_j \frac{\partial L}{\partial \dot{q}_j} d\dot{q}_j + \frac{\partial L}{\partial t} dt \right).$$

The second and fourth sums cancel by virtue of the definition of p_j, leaving

$$dH = \sum_j \dot{q}_j \, dp_j - \sum_j \frac{\partial L}{\partial q_j} dq_j - \frac{\partial L}{\partial t} dt.$$

But regarding H as $H(q, p, t)$, we can also write

$$dH = \sum_j \frac{\partial H}{\partial q_j} dq_j + \sum_j \frac{\partial H}{\partial p_j} dp_j + \frac{\partial H}{\partial t} dt.$$

Equating these two expressions for dH, the coefficients of corresponding differentials can then be set equal to obtain

$$\dot{q}_j = \frac{\partial H}{\partial p_j}, \tag{2.17}$$

$$-\frac{\partial H}{\partial q_j} = \frac{\partial L}{\partial q_j} = \frac{d}{dt}\left(\frac{\partial L}{\partial \dot{q}_j}\right) = \dot{p}_j, \tag{2.17'}$$

$$\frac{\partial H}{\partial t} = -\frac{\partial L}{\partial t}. \tag{2.17''}$$

Furthermore, with the aid of eq. (2.17),

$$\frac{dH}{dt} = \sum_j \left(\frac{\partial H}{\partial q_j} \dot{q}_j + \frac{\partial H}{\partial p_j} \dot{p}_j \right) + \frac{\partial H}{\partial t} = \sum_j (-\dot{p}_j \dot{q}_j + \dot{q}_j \dot{p}_j) + \frac{\partial H}{\partial t} = \frac{\partial H}{\partial t}.$$

(2.18)

From (2.18) and the last member of the set (2.17), we see that the Hamiltonian is a constant of the motion unless both it and the Lagrangian depend explicitly on the time. (There are even cases where the transformation (2.4) depends on the time but H does not, whereupon it is still a constant but no longer the energy.) Equations (2.17) and (2.17'),

$$\dot{q}_j = \frac{\partial H}{\partial p_j}, \quad \dot{p}_j = -\frac{\partial H}{\partial q_j},$$

are known as **Hamilton's canonical equations**. They are a set of $2n$ first-order

The Hamiltonian and the Canonical Equations

equations which are completely equivalent to the n second-order equations of Lagrange. Even the symmetry aspects are still present, for an ignorable coordinate q_j (missing now in H) clearly leads to constant p_j. The first canonical equation usually reproduces eq. (2.13), giving nothing new. It is the second, or $\dot{p}_j$, relation that produces the equation of motion at once.

Before turning to examples, one point still needs to be discussed. The derivation of (2.17) has required a conservative force such that $Q_j = -\partial V/\partial q_j$, for we have used Lagrange's eqs. (2.7) which were based on that assumption. If a nonconservative force $\mathscr{F}_j$ is also present, we have instead the modified Lagrange eq. (2.7'):

$$\frac{d}{dt}\left(\frac{\partial L}{\partial \dot{q}_j}\right) - \frac{\partial L}{\partial q_j} = \mathscr{F}_j.$$

In this case, eq. (2.17') becomes

$$-\frac{\partial H}{\partial q_j} = \frac{\partial L}{\partial q_j} = \frac{d}{dt}\left(\frac{\partial L}{\partial \dot{q}_j}\right) - \mathscr{F}_j,$$

or

$$\dot{p}_j = -\frac{\partial H}{\partial q_j} + \mathscr{F}_j,$$

whereupon eq. (2.18) is modified to the form

$$\frac{dH}{dt} = \sum_j \mathscr{F}_j \dot{q}_j + \frac{\partial H}{\partial t}. \tag{2.18'}$$

The last expression shows that the energy will change in time at a rate determined by the nonconservative forces $\mathscr{F}_j$ (such as the force of friction), in addition to the effect of explicit time dependence.

Example 5 Particle Falling Under Gravity

A particle of mass m falls freely in a uniform gravitational field of acceleration g. Find the equation of motion by Hamilton's methods.

Solution

Taking the y-axis vertically upward and the zero of potential at the origin, we have

$$T = \tfrac{1}{2}m\dot{y}^2, \qquad V = mgy, \qquad L = \tfrac{1}{2}m\dot{y}^2 - mgy.$$

Lagrange's equation is given (for comparison of methodology) by

$$\frac{d}{dt}(m\dot{y}) + mg = 0,$$

or

$$\ddot{y} = -g,$$

as expected. The Hamiltonian is

$$H = \dot{y}p - L,$$

with

$$p = \frac{\partial L}{\partial \dot{y}} = m\dot{y}.$$

Then

$$\dot{y} = \frac{p}{m} \quad \text{and} \quad H = \frac{p^2}{2m} + mgy,$$

whereupon eq. (2.17) becomes

$$\dot{y} = \frac{\partial H}{\partial p} = \frac{p}{m}$$

and

$$-\dot{p} = \frac{\partial H}{\partial y} = mg = -m\ddot{y}, \quad \text{or} \quad \ddot{y} = -g.$$

The first of these relations was already known, while the second provides the equation of motion.

Example 6 Particle in Central Field Motion

Consider a particle of mass m undergoing central field motion in a plane with a potential function $V(r)$. Use Hamilton's equations to describe the motion in polar coordinates.

Solution

In the plane $z = 0$, we have $x = r\cos\theta$, $y = r\sin\theta$. These are the transformation equations to the generalized coordinates r, θ; they are time independent, so H is the total energy. Since

$$T = \tfrac{1}{2}m(\dot{r}^2 + r^2\dot{\theta}^2) \quad \text{and} \quad L = \tfrac{1}{2}m(\dot{r}^2 + r^2\dot{\theta}^2) - V(r),$$

$$p_r = \frac{\partial L}{\partial \dot{r}} = m\dot{r} \quad \text{and} \quad p_\theta = \frac{\partial L}{\partial \dot{\theta}} = mr^2\dot{\theta}.$$

In addition,

$$H = T + V = \frac{1}{2}m(\dot{r}^2 + r^2\dot{\theta}^2) + V(r) = \frac{p_r^2}{2m} + \frac{p_\theta^2}{2mr^2} + V(r).$$

Forces of Constraint

Equation (2.17) produces

$$\dot{r} = \frac{\partial H}{\partial p_r} = \frac{p_r}{m}, \qquad \dot{\theta} = \frac{\partial H}{\partial p_\theta} = \frac{p_\theta}{mr^2},$$

and

$$\dot{p}_r = -\frac{\partial H}{\partial r} = \frac{p_\theta^2}{mr^3} - \frac{\partial V(r)}{\partial r}$$

$$\dot{p}_\theta = -\frac{\partial H}{\partial \theta} = 0,$$

or

$$p_\theta = \text{constant} = \text{angular momentum}.$$

The radial equation of motion is

$$F_r = -\frac{\partial V}{\partial r} = \dot{p}_r - \frac{p_\theta^2}{mr^3} = m\ddot{r} - \frac{p_\theta^2}{mr^3} = m(\ddot{r} - r\dot{\theta}^2),$$

which has the $\ddot{r}$ term augmented by the familiar centripetal force. This relation is a differential equation for r as a function of time once the force law is given for F_r. (The expression for the constant angular momentum, of course, allows elimination of $\dot{\theta}$ in favor of r.) Example 1 of Section 2.2 obtained similar results through the use of cylindrical coordinates.

2.5 FORCES OF CONSTRAINT

Up to this point, we have considered the q_i's as independent generalized coordinates, all unnecessary coordinates having been eliminated through holonomic constraint relations. We now wish to establish a formalism which allows evaluation of forces of constraint. In so doing, it will be convenient to consider first a nonholonomic constraint wherein the constraint relations are in the differential form mentioned earlier. The coordinates q_i (no longer independent) for the moment will be any geometrically suitable set with differentials related by equations of the general type

$$\sum_{j=1}^{n} C_{kj}\, dq_j + C_{kt}\, dt = 0. \tag{2.19}$$

The index k labels the constraint relations. Let us assume in general that there are m constraint equations of this type, so that $k = 1, 2, \ldots m$.

As is customary, we now imagine virtual displacements δq_j with time held fixed, so that $\sum_j C_{kj}\delta q_j = 0$. The displacements δq_j are *connected* through these relations; that is, they are not independent of one another. It is for this reason that we now introduce **Lagrangian multipliers** λ_k (one for each constraint relation). The sole motivation for doing so lies in the fact that the λ_k's are essential to a scheme which renders the δq_j effectively independent.

It is clear that

$$\lambda_k \sum_j C_{kj} \delta q_j = 0$$

must hold; and if we sum on k, then

$$\sum_{k,j} \lambda_k C_{kj} \delta q_j = 0.$$

At present, the λ_k are undetermined multiplicative factors, each of which may be a function of the coordinates.

In the course of deriving eq. (2.6), we inserted the holonomic condition into the relation

$$\sum_j \left[Q_j - \frac{d}{dt}\frac{\partial T}{\partial \dot{q}_j} + \frac{\partial T}{\partial q_j} \right] \delta q_j = 0.$$

It was at that point that the independence of the δq_k's was required. Suppose the last two relations are now added, inserting the Lagrangian at the same time by use of the conservative force condition $Q_j = -\partial V/\partial q_j$ and (2.8):

$$\sum_{j=1}^n \left(\frac{\partial L}{\partial q_j} - \frac{d}{dt}\frac{\partial L}{\partial \dot{q}_j} + \sum_k \lambda_k C_{kj} \right) \delta q_j = 0.$$

The δq's are still dependent on one another through the relations (2.19); we now exploit that situation by *choosing* the m Lagrangian multipliers in such a manner as to make the coefficients of δq_j (for $j = 1$ to m) equal to zero. We then have

$$\sum_{j=m+1}^n \left(\frac{\partial L}{\partial q_j} - \frac{d}{dt}\frac{\partial L}{\partial \dot{q}_j} + \sum_k \lambda_k C_{kj} \right) \delta q_j = 0,$$

where the λ's are now known quantities. But the $(n - m)$ displacements δq_j that remain in the sum are, in effect, the independent ones, so that their coefficients must surely vanish. The upshot of all this is that

$$\frac{\partial L}{\partial q_j} - \frac{d}{dt}\frac{\partial L}{\partial \dot{q}_j} + \sum_k \lambda_k C_{kj} = 0$$

for *all* j's from 1 to n. These n equations along with the m constraint relations

Forces of Constraint 57

provide us sufficient information to fully determine the n coordinates q_j and the m multipliers λ_k.

Finally, to interpret the last equation, let us write it in the more common Lagrangian form

$$\frac{d}{dt}\frac{\partial L}{\partial \dot{q}_j} - \frac{\partial L}{\partial q_j} = \sum_k \lambda_k C_{kj}. \tag{2.20}$$

It is clear from the discussion following eq. (2.8) that the generalized force Q_j can be broken down into a part associated with a potential and a nonconservative part $\mathscr{F}_j$,

$$Q_j = -\frac{\partial V}{\partial q_j} + \mathscr{F}_j;$$

the result obtained is then eq. (2.7′):

$$\frac{d}{dt}\frac{\partial L}{\partial \dot{q}_j} - \frac{\partial L}{\partial q_j} = \mathscr{F}_j.$$

By comparison of (2.7′) with (2.20), we find $\mathscr{F}_j = \sum_k \lambda_k C_{kj}$. In other words, if a force $\mathscr{F}_j$ is supplied which is equivalent to the force of constraint (all constraints being removed), the system must be physically unaffected by the replacement. The form of the relations above means that we can identify such a force of constraint with $\sum_k \lambda_k C_{kj}$, so that it is found automatically from the Lagrangian multipliers when one uses this approach.

We have thus shown how to find the forces of constraint in a nonholonomic situation of differential type. But this case is surely of somewhat limited interest. Now we want to relate the derivation to holonomic constraints, which are much more common and appealing subjects for investigation. Note that the holonomic definition (2.1) can also be written as

$$\sum_j \frac{\partial f}{\partial q_j} dq_j + \frac{\partial f}{\partial t} dt = 0$$

which has exactly the same form as eq. (2.19) if the derivatives of f are identified with the coefficients C_{kj}.

At first sight, however, there is an apparent difficulty in the application of the formalism just derived to a holonomic situation. In the latter case, all coordinates q_j are understood to be independent; in the nonholonomic discussion, they were merely any appropriate set of coordinates, with additional constraining differential relations given. If we proceed similarly in the holonomic case, treating the coordinates q_j as a suitable set and *not* rendering them independent of one another by means of the relations (2.1), preferring instead to superimpose those relations in differential form (2.19), then there is an exact parallel to the nonholonomic case. We are then able to employ the same techniques for finding the holonomic forces of constraint, as the examples to follow demonstrate. Since the superfluous coordinates have not been eliminated with (2.1) from the beginning, there will be additional Lagrange equations to be written down, one for each constraint relation; these correspond to the "extra degrees of freedom" which in turn lead to the forces of constraint.

Example 7 Bead on Hoop

As a simple application of the ideas just considered, suppose we have a bead of mass m sliding down from the top of a frictionless hoop of radius a. What is the radial force of the hoop on the bead?

Solution

To solve this problem, note that there is only one independent variable involved; it is the angle θ which we shall measure from the vertical line through the center of the hoop. Taking the Lagrangian as

$$L = \tfrac{1}{2}ma^2\dot\theta^2 - mga\cos\theta,$$

Lagrange's eq. (2.7) becomes

$$\frac{d}{dt}\left(\frac{\partial L}{\partial \dot\theta}\right) - \frac{\partial L}{\partial \theta} = 0 = ma^2\ddot\theta - mga\sin\theta,$$

or

$$\ddot\theta = \frac{g}{a}\sin\theta.$$

Multiplying each side by $\dot\theta$ and integrating with respect to time gives

$$\frac{\dot\theta^2}{2} = -\frac{g}{a}\cos\theta + C.$$

The constant of integration C can be evaluated by setting $\dot\theta = 0$ for $\theta = 0$, whereupon

$$\dot\theta^2 = \frac{2g}{a}(1 - \cos\theta),$$

which means that the centripetal force on the bead must be

$$ma\dot\theta^2 = 2mg(1 - \cos\theta).$$

Forces of Constraint 59

The net reaction force λ of the hoop on the bead is merely the difference between the radial weight component and the centripetal force:

$$\lambda = mg\cos\theta - 2mg(1 - \cos\theta) = mg(3\cos\theta - 2).$$

At the top, this radial force is equal to the weight; at the angle whose cosine is $\frac{2}{3}$, the force is zero and the bead falls off the hoop.

We have demonstrated that the problem can be solved by elementary methods. Let us indicate how the solution is obtained by the Lagrangian multiplier formalism. We regard the position of the bead as specified by *two* variables r and θ, introducing the constraint of the hoop by the equation

$$r = a,$$

or

$$dr = 0$$

in differential form. From (2.19), it is easy to see that

$$C_{1r} = 1, \quad C_{1\theta} = C_{1t} = 0.$$

It is necessary to write two Lagrange equations of the form (2.20):

$$\frac{d}{dt}\left(\frac{\partial L}{\partial \dot{r}}\right) - \frac{\partial L}{\partial r} = \lambda_1 C_{1r} = \lambda_1$$

and

$$\frac{d}{dt}\left(\frac{\partial L}{\partial \dot{\theta}}\right) - \frac{\partial L}{\partial \theta} = 0,$$

where now we must write

$$L = \tfrac{1}{2}m(\dot{r}^2 + r^2\dot{\theta}^2) - mgr\cos\theta.$$

The solution of the θ equation proceeds as before and gives a result for $\dot{\theta}^2$ which can then be substituted into the radial equation (setting $r = a$, $\ddot{r} = 0$) to find λ_1. Of course, λ_1 is the constraint force and agrees with the answer obtained earlier.

Example 8 Atwood Machine as an Example of Constraint Forces

Consider two masses m_1 and m_2 at the ends of an inextensible string running over a smooth, small pulley. The masses of pulley and string are negligible. This is the familiar Atwood machine, analyzed so often by straightforward methods in elementary texts (Figure 2.3). Use the Lagrangian multiplier method to find the tension in the string as a force of constraint.

Solution

Measuring the y coordinates of the masses vertically downward as shown, we have

$$T = \tfrac{1}{2}m_1\dot{y}_1^2 + \tfrac{1}{2}m_2\dot{y}_2^2,$$
$$V = -m_1gy_1 - m_2gy_2,$$
$$L = \tfrac{1}{2}m_1\dot{y}_1^2 + \tfrac{1}{2}m_2\dot{y}_2^2 + m_1gy_1 + m_2gy_2.$$

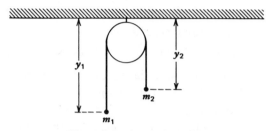

Figure 2.3 Atwood machine.

The two coordinates, however, are not independent; they are related by the constraint equation $y_1 + y_2 = l =$ length of string. Differentiating to put this in the standard form (2.19), we find $dy_1 + dy_2 = 0$. Then only one multiplier λ is needed, and the Lagrange equations, (2.20), become

$$m_1 \ddot{y}_1 - m_1 g = \lambda \cdot 1 = \lambda, \quad (y_1\text{-equation})$$
$$m_2 \ddot{y}_2 - m_2 g = \lambda \cdot 1 = \lambda. \quad (y_2\text{-equation})$$

The constraint relation tells us that $\ddot{y}_1 = -\ddot{y}_2$, which upon insertion in the y_2 equation yields

$$-m_2 \ddot{y}_1 - m_2 g = \lambda = m_1 \ddot{y}_1 - m_1 g,$$

or

$$\ddot{y}_1 = \frac{(m_1 - m_2)g}{(m_1 + m_2)}.$$

Putting this into the y_1-equation produces

$$\lambda = \frac{m_1(m_1 - m_2)g}{(m_1 + m_2)} - m_1 g = m_1 \left(\frac{-2 m_2 g}{m_1 + m_2} \right).$$

This is the sole force of constraint, which here is the string tension; the minus sign indicates that it acts upward on the masses. From the right side of the Lagrange equations, we see that the force on either mass has this same value λ, as would be expected.

Example 9 Bead on Helix

To illustrate in detail how the forces of constraint are found in a complex problem, consider again the bead on a helix discussed earlier in this chapter.

Solution

The constraint relations were $\rho = b$ and $z = a\phi$. In the differential form (2.19), these become

$$d\rho = 0, \quad dz - a \, d\phi = 0,$$

Forces of Constraint

leading to the identifications $C_{1\rho} = 1$, $C_{1z} = C_{1\phi} = 0$ from the first relation and $C_{2\rho} = 0$, $C_{2z} = 1$, $C_{2\phi} = -a$ from the second. The Lagrangian is

$$L = \tfrac{1}{2}m(\dot{\rho}^2 + \rho^2\dot{\phi}^2 + \dot{z}^2) - \tfrac{1}{2}k(\rho^2 + z^2),$$

giving the three Lagrange equations

$$m\ddot{\rho} - m\rho\dot{\phi}^2 + k\rho = \lambda_1 C_{1\rho} + \lambda_2 C_{2\rho} = \lambda_1, \quad (\rho\text{-equation})$$

$$\frac{d}{dt}(m\rho^2\dot{\phi}) = \lambda_1 C_{1\phi} + \lambda_2 C_{2\phi} = -a\lambda_2, \quad (\phi\text{-equation})$$

$$m\ddot{z} + kz = \lambda_1 C_{1z} + \lambda_2 C_{2z} = \lambda_2. \quad (z\text{-equation})$$

Inserting the constraint relations in the form $\rho = b$ and $\dot{z} = a\dot{\phi}$ into the ϕ-equation

$$\frac{mb^2}{a}\ddot{z} = -a\lambda_2,$$

or

$$\lambda_2 = -\frac{mb^2}{a^2}\ddot{z}.$$

Substitution of λ_2 into the z equation then yields

$$m\ddot{z} + kz = -\frac{mb^2}{a^2}\ddot{z},$$

or

$$\ddot{z} = -\frac{ka^2}{m(a^2 + b^2)}z.$$

This is an oscillator equation with solution

$$z = A \cos\left[\sqrt{\frac{ka^2}{m(a^2+b^2)}}\, t\right] + B \sin\left[\sqrt{\frac{ka^2}{m(a^2+b^2)}}\, t\right],$$

where A and B are constants dependent on the initial conditions. λ_2 is then a function of time given by

$$\lambda_2 = \frac{kb^2}{a^2 + b^2}z,$$

varying in the same sinusoidal fashion as z. From the ρ equation

$$\lambda_1 = kb - mb\dot{\phi}^2 = kb - \frac{mb}{a^2}\dot{z}^2,$$

which is readily calculated from differentiation of the solution for z above. λ_1, of course, is the radial force in the ρ-direction, while λ_2 is the z-component force on the bead; $-a\lambda_2$ is a torque which generates the ϕ-motion.

Example 10 A Time-Dependent Constraint

Consider a bead of mass m sliding on a straight wire without friction (Figure 2.4). If the wire is rotating at a constant angular speed ω in a vertical plane and gravity is present, we want to find the radial position of the bead as a function of time. (Assume $r = R$, $\dot{r} = V$ at time zero.) We also want to investigate the force of constraint of wire on bead, whether the Hamiltonian is a constant, and if it is equal to the energy.

Solution

Setting the angular coordinate $\theta = \omega t$, the equations of transformation are

$$x = r\cos\theta = r\cos\omega t, \quad y = r\sin\theta = r\sin\omega t,$$

whereupon the kinetic energy is

$$T = \tfrac{1}{2}m(\dot{r}^2 + r^2\dot{\theta}^2) = \tfrac{1}{2}m(\dot{r}^2 + \omega^2 r^2)$$

and the Lagrangian is

$$L = T - V = \frac{m}{2}(\dot{r}^2 + \omega^2 r^2) - mgr\sin\omega t.$$

The only independent coordinate present is the radial one, giving

$$\frac{d}{dt}\left(\frac{\partial L}{\partial \dot{r}}\right) - \frac{\partial L}{\partial r} = 0 = m\ddot{r} - m\omega^2 r + mg\sin\omega t$$

or

$$\ddot{r} - \omega^2 r = -g\sin\omega t$$

for the equation of motion. The solution which satisfies the initial conditions is then easily seen to be

$$r = R\cosh\omega t + \left(\frac{V}{\omega} - \frac{g}{2\omega^2}\right)\sinh\omega t + \frac{g}{2\omega^2}\sin\omega t.$$

The Hamiltonian is, from eq. (2.15),

$$H = p_r \dot{r} - L = \frac{p_r^2}{2m} - \tfrac{1}{2}m\omega^2 r^2 + mgr\sin\omega t.$$

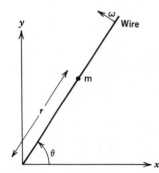

Figure 2.4 Bead sliding on wire in vertical plane.

Forces of Constraint

Since the transformation equations depend explicitly on the time as does H itself, the Hamiltonian is neither the energy nor a constant. To verify this, note that the energy is

$$E = T + V = \tfrac{1}{2}m(\dot{r}^2 + \omega^2 r^2) + mgr \sin \omega t \neq H,$$

and

$$\frac{dH}{dt} = \frac{p_r}{m}\dot{p}_r - m\omega^2 r\dot{r} + mg\dot{r}\sin\omega t + mg\omega r \cos\omega t = mg\omega r \cos\omega t \equiv \frac{\partial H}{\partial t}$$

upon substitution of $\dot{p}_r = -\partial H/\partial r = m\omega^2 r - mg \sin \omega t$ and $p_r = m\dot{r}$.

To find the force of constraint due to the wire, we have to allow for two "independent" coordinates in the Lagrangian, subject to the constraint

$$\theta = \omega t \quad \text{or} \quad d\theta - \omega\, dt = 0,$$

so that

$$C_r = 0,\ C_\theta = 1,\ C_t = -\omega.$$

Then,

$$L = \tfrac{1}{2}m(\dot{r}^2 + r^2\dot{\theta}^2) - mgr \sin \theta,$$

and

$$\frac{d}{dt}\left(\frac{\partial L}{\partial \dot{\theta}}\right) - \frac{\partial L}{\partial \theta} = \lambda = \frac{d}{dt}(mr^2\dot{\theta}) + mgr\cos\theta = 2m\omega r\dot{r} + mgr\cos\omega t$$

is the generalized force (or torque) of constraint.

From this last expression for L, the student might well be inclined to wonder why we could not calculate another Hamiltonian H' for the problem, considered as a situation with two degrees of freedom:

$$H' = p_r\dot{r} + p_\theta\dot{\theta} - L = H + p_\theta\dot{\theta} = H + \frac{p_\theta^2}{mr^2} = \frac{p_r^2}{2m} + \frac{p_\theta^2}{2mr^2} + mgr \sin \theta = E,$$

and

$$\frac{dH'}{dt} = \frac{p_r}{m}\dot{p}_r + \frac{p_\theta}{mr^2}\dot{p}_\theta - \frac{p_\theta^2}{mr^3}\dot{r} + mg\dot{r}\sin\theta + mgr\dot{\theta}\cos\theta = \frac{p_\theta}{mr^2}\dot{p}_\theta + mgr\dot{\theta}\cos\theta$$

$$= \dot{\theta}\left[\frac{d}{dt}(mr^2\dot{\theta}) + mgr\cos\theta\right] = \dot{\theta}\lambda,$$

after some obvious cancellation of terms. Viewing the problem as one having two variables and a constraint, the new Hamiltonian H' is now the energy, but it must change in time at a rate determined by the constraining torque, in accordance with the relation (2.18'). For a time-dependent problem, the meaning of the Hamiltonian depends on the way in which the problem is attacked.

The physical reason behind this conclusion is not difficult to see. The potential in the Lagrangian approach arises only from the applied and not the constraint forces; the latter do no work anyway for a constraining surface which is fixed in time, as with the helix just discussed. But where there is a time-dependent constraint (the effect of the moving wire in the present problem), the force of constraint may still be perpendicular to the instantaneous surface yet can now have a component along the direction of the actual displacement, making it possible for the constraining forces to do work. (In this case, the displacement of the bead is $d\vec{r} = dr\hat{e}_r + r\,d\theta\hat{e}_\theta$, and the element of work from the constraint force $\vec{F}_c$ is $\vec{F}_c \cdot d\vec{r} = \lambda\,d\theta$, with rate of change $\dot{\theta}\lambda$.) By treating the wire problem as one with two degrees of freedom, the modified Hamiltonian H' allows for the work actually done by the constraining force.

2.6 NOETHER'S THEOREM

Earlier in this chapter, it was noted that symmetries of the Lagrangian or Hamiltonian (usually dictated by ignorable coordinates) gave rise to constants of the motion; such parameters are of utmost importance in the analysis. The constants of the motion do not always come from the obvious symmetries of the Lagrangian, however, nor do they always have a simple form. Often they are expressed by complicated functions of coordinates and momenta which are invariant in time.

It would therefore seem desirable to develop a general approach which is not limited by the specific details of a given situation. Such a formalism was produced in 1918 by the noted mathematician Emmy Noether[1]. Her approach has been simplified and popularized by several writers in recent years. We shall give a condensed treatment at this point.

Consider a physical system described by generalized coordinates q_i and velocities $\dot{q}_i$. The Lagrangian $L(q, \dot{q}, t)$ is assumed to be known and will provide the correct equations of motion for the system when substituted into Lagrange's equations. Suppose that is possible to find a transformation to a new set of generalized coordinates

$$Q_i \equiv Q_i(q, \dot{q}, t, \varepsilon), \qquad T \equiv T(q, \dot{q}, t, \varepsilon) \qquad (2.21)$$

such that

$$(Q_i)_{\varepsilon=0} = q_i \quad \text{and} \quad T_{\varepsilon=0} = t, \qquad (2.22)$$

where ε is a continuous parameter independent of the coordinates or time. Note

---

[1] E. Noether, *Nachr. Gesell. Wissensch. Gottingen* **2**, 235 (1918).

that for all i's,

$$\frac{dQ}{dT} = \frac{dQ/dt}{dT/dt} = \frac{\dot{Q}}{\dot{T}}. \tag{2.23}$$

A new functional form will result in the Lagrangian if Q, dQ/dT, and T are inserted to replace q, $\dot{q}$, and t, respectively. Letting $\dot{F}$ be the total time derivative of some function $F(q, \dot{q}, t)$, Noether's theorem can be stated as follows:

If a transformation with the properties listed above can be found so that the equation

$$\left\{\frac{\partial}{\partial \varepsilon}\left[L\left(Q, \frac{dQ}{dT}, T\right)\dot{T}\right]\right\}_{\varepsilon=0} = \dot{F} \tag{2.24}$$

is satisfied, then the quantity

$$L(q, \dot{q}, t)\left(\frac{\partial T}{\partial \varepsilon}\right)_{\varepsilon=0} - F + \sum_i \frac{\partial L(q, \dot{q}, t)}{\partial \dot{q}_i}\left[\left(\frac{\partial Q_i}{\partial \varepsilon}\right)_{\varepsilon=0} - \dot{q}_i\left(\frac{\partial T}{\partial \varepsilon}\right)_{\varepsilon=0}\right]$$

must be a constant of the motion.

Let us prove the theorem before attempting to clarify its meaning with examples. To simplify the equations a little, we first set

$$A_i \equiv \left(\frac{\partial Q_i}{\partial \varepsilon}\right)_{\varepsilon=0} \quad \text{and} \quad B \equiv \left(\frac{\partial T}{\partial \varepsilon}\right)_{\varepsilon=0}. \tag{2.25}$$

We then want to prove that

$$LB - F + \sum_i \frac{\partial L}{\partial \dot{q}_i}[A_i - \dot{q}_i B] = \text{constant}. \tag{2.26}$$

Expanding Q_i and T in powers of ε and making use of eqs. (2.22) and (2.25), we have at once

$$Q_i = (Q_i)_{\varepsilon=0} + \left(\frac{\partial Q_i}{\partial \varepsilon}\right)_{\varepsilon=0}\varepsilon + \cdots = q_i + A_i\varepsilon + \cdots,$$

$$T = t + B\varepsilon + \cdots,$$

whereupon

$$\dot{Q}_i = \dot{q}_i + \dot{A}_i\varepsilon, \quad \dot{T} = 1 + \dot{B}\varepsilon.$$

From eq. (2.23), we conclude that

$$\frac{dQ_i}{dT} = \frac{\dot{Q}_i}{\dot{T}} = \frac{\dot{q}_i + \dot{A}_i \varepsilon}{1 + \dot{B}\varepsilon}.$$

Evaluation of these and related quantities at $\varepsilon = 0$ yields

$$(\dot{Q}_i)_{\varepsilon=0} = \dot{q}_i, \qquad \dot{T}_{\varepsilon=0} = 1, \qquad \left(\frac{dQ_i}{dT}\right)_{\varepsilon=0} = \dot{q}_i,$$

$$\left(\frac{\partial \dot{T}}{\partial \varepsilon}\right)_{\varepsilon=0} = \dot{B}, \qquad \left[\frac{\partial}{\partial \varepsilon}\left(\frac{dQ_i}{dT}\right)\right]_{\varepsilon=0} = \dot{A}_i - \dot{q}_i \dot{B}.$$

Observing that

$$\left[\frac{\partial L(Q, dQ/dT, T)}{\partial T}\right]_{\varepsilon=0} = \frac{\partial L(q, \dot{q}, t)}{\partial t} \equiv \frac{\partial L}{\partial t}, \text{ etc.},$$

we have by straightforward differentiation in eq (2.24):

$$\left\{\left[\sum_i \frac{\partial L}{\partial Q_i}\frac{\partial Q_i}{\partial \varepsilon} + \sum_i \frac{\partial L}{\partial(dQ_i/dT)}\frac{\partial(dQ_i/dT)}{\partial \varepsilon} + \frac{\partial L}{\partial T}\frac{\partial T}{\partial \varepsilon}\right]\dot{T} + L\frac{\partial \dot{T}}{\partial \varepsilon}\right\}_{\varepsilon=0} = \dot{F}.$$

Substitution of the various expressions given above for $\varepsilon = 0$ produces

$$\sum_i \frac{\partial L}{\partial q_i} A_i + \sum_i \frac{\partial L}{\partial \dot{q}_i}(\dot{A}_i - \dot{q}_i \dot{B}) + \frac{\partial L}{\partial t} B + L\dot{B} = \dot{F}. \tag{2.26'}$$

In all these terms, $L = L(q, \dot{q}, t)$; setting $\varepsilon = 0$ has removed the Q's, T, ε, etc. In the first sum on the left of eq. (2.26'), we now use Lagrange's equation

$$\frac{\partial L}{\partial q_i} = \frac{d}{dt}\left(\frac{\partial L}{\partial \dot{q}_i}\right)$$

and for $\partial L/\partial t$ substitute from the chain rule

$$\frac{\partial L}{\partial t} = \frac{dL}{dt} - \sum_i \frac{\partial L}{\partial q_i}\dot{q}_i - \sum_i \frac{\partial L}{\partial \dot{q}_i}\ddot{q}_i.$$

These relations are sufficient to turn eq. (2.26') into a perfect derivative:

$$\frac{d}{dt}\left[LB - F + \sum_i \frac{\partial L}{\partial \dot{q}_i}(A_i - \dot{q}_i B)\right] = 0.$$

Integration gives eq. (2.26) at once, concluding the proof of Noether's theorem.

Fortunately, the application of the theorem is often simpler algebraically than its proof. Consider as an example a single particle Lagrangian which is invariant under rotations about the z-axis. Taking the rotation angle as the parameter ε, the transformed coordinates are

$$X = x \cos \varepsilon - y \sin \varepsilon,$$

$$Y = x \sin \varepsilon + y \cos \varepsilon,$$

$$Z = z.$$

In addition, let $T \equiv t$. Equations (2.22) are then satisfied. Since the Lagrangian does not change under the rotation and $\dot{T} = 1$, the vital relation (2.24) can be satisfied by simply choosing F equal to a constant, which we take as zero for convenience. From (2.25),

$$A_x = \left(\frac{\partial X}{\partial \varepsilon}\right)_{\varepsilon=0} = -y, \quad A_y = x, \quad A_z = 0 = B,$$

whereupon eq. (2.26) ensures that

$$\frac{\partial L}{\partial \dot{x}} A_x + \frac{\partial L}{\partial \dot{y}} A_y = \frac{\partial L}{\partial \dot{y}} x - \frac{\partial L}{\partial \dot{x}} y = \text{constant}.$$

Using the conjugate momenta, this expression becomes

$$p_y x - p_x y = \text{constant} = (\vec{r} \times \vec{p})_z.$$

The z-component of angular momentum is constant, as expected. (The use of the momentum definition (2.13) implies that the potential energy must not be velocity dependent in an approach such as this one.)

As another important application of Noether's theorem, consider now a Lagrangian which is invariant under translations in time; that is, the Lagrangian will be explicitly independent of time. To satisfy eqs. (2.22) and (2.24), we choose

$$Q_i \equiv q_i, \quad T \equiv t + \varepsilon, \quad \frac{dQ_i}{dT} = \frac{\dot{Q}_i}{\dot{T}} \equiv \dot{q}_i, \quad F \equiv 0.$$

Equations (2.25) and (2.26) yield

$$A_i = 0, \quad B = 1,$$

and

$$L - \sum_i \frac{\partial L}{\partial \dot{q}_i} \dot{q}_i = \text{constant}.$$

From eq. (2.15), it is apparent that the Hamiltonian is a constant of the motion.

We have reached a conclusion in accord with eqs. (2.17) and (2.18), but now seen from a different perspective.

The above examples partially demonstrate the generality of Noether's theorem; many other situations could have been selected as well. With experience as a guide, the proper choice of transformation is often readily evident (this is the primary difficulty in the procedure). The form of the constant can then be determined in a straightforward fashion, sometimes leading to striking new insights for a given problem.

PROBLEMS

2.1 A particle of mass m is constrained to move on the surface of an inverted circular cone in a uniform gravitational field along the negative z-axis. If the cone has its tip at the origin and slopes upward around the z-axis at a rate a, set up Lagrange's equations for the motion. Then calculate the total energy as a function of the cylindrical coordinate ρ and the various constants involved.

2.2 A string passes over a fixed pulley with a mass m_1 attached to one end. At the other end, another pulley is fastened, with masses m_2 and m_3 attached to the ends of a string passing over it. By the use of Lagrangian techniques, find the acceleration of m_3 under gravity. (Neglect masses of pulleys and strings.) Show also that the acceleration of m_3 approaches g if $m_1 \ll m_2$, $m_1 \ll m_3$, or if $m_3 \gg m_1$, $m_3 \gg m_2$.

2.3 A straight piece of wire rotates with constant angular velocity ω in a horizontal plane about a vertical axis at one end. A bead of mass m slides smoothly on the wire, subject to an attractive force proportional to the distance (constant k) from the midpoint of the wire, which is of length $2R$. Solve Lagrange's equations to find the position of the bead as a function of time, assuming that at $t = 0$, $\rho = R$, $\dot{\rho} = \omega^2 R/\alpha$, where $\alpha \equiv \sqrt{(m\omega^2 - k)/m}$ is real. Then eliminate the time to find the path equation in polar coordinates.

2.4 A simple pendulum of length l and bob mass m is attached to a point of support which moves in time horizontally (along the x-axis) according to the function $X_0(t)$. If the pendulum is in a uniform gravitational field and constrained to a vertical plane, set up Lagrange's equations and describe the motion, taking $X_0(t) = A \sin \omega t$.

2.5 Two simple pendula of length l and bob mass m swing in a common vertical plane, being attached to two different support points. If the masses are connected by a spring of constant k, use the Lagrangian approach to formulate the equations of motion. (Assume small angles of oscillation.)

2.6 A double pendulum consists of two connected pendula swinging in a common vertical plane under gravity. The uppermost pendulum has length l_1 and bob of mass m_1 and is attached to a fixed point. The lower (length l_2, mass m_2) is attached to m_1. Determine the Lagrangian and the equations of motion.

2.7 Consider again the double pendulum of Problem 2.6. Find the generalized momenta. Express the Hamiltonian in terms of coordinates and momenta, and simply *indicate* how the equations of motion are obtained from H.

2.8 A *spherical* pendulum is one whose bob is not confined to a plane but instead is free to move to any point on a spherical surface of radius R about the point of suspension. Consider such a pendulum (bob mass m) in a uniform gravitational field of acceleration g. Find the conjugate momenta, and use them to derive Hamilton's canonical equations.

2.9 A bead of mass m slides on a smooth circular wire of radius a. The wire is fixed at one point and rotating in the xy-plane about that point with a constant angular velocity ω. Find the equation of motion through application of Hamilton's canonical equations. Then find the normal force on the bead exerted by the wire. (Find the force as a function of an appropriate angle and its time derivative.) Is the Hamiltonian the total energy? Is it constant in time?

2.10 A circular loop of wire is located in the xy-plane, with one point on it fixed at the origin and its center on the y-axis; the radius varies in time according to $r = a + bt^2$, where a and b are constants. Find the equations of motion for a bead of mass m sliding smoothly on the wire, and the normal force of wire on the bead (expressed as a function of an appropriate angular coordinate and its time derivative). Then calculate the modified Hamiltonian in two variables, showing that it is neither the kinetic energy nor a constant. Finally relate dH'/dt to the normal force on the bead.

2.11 Use Noether's theorem to prove that the total momentum component along the direction of the x-axis is constant whenever a system of particles has a Lagrangian that is invariant under translations in that direction.

2.12 Consider a free particle moving along the x-axis. The equation of motion is invariant under a Galilean transformation. Use this fact in connection with Noether's theorem to show that the velocity of the particle is constant. (Take ε as the relative velocity between two frames of reference.)

Chapter Three
VARIATIONAL PRINCIPLES

3.1 THE CALCULUS OF VARIATIONS

In many problems of physics or mathematics, we are interested in calculating the integral of some quantity and determining the conditions under which it takes on an extremum value (a maximum or minimum). Consider, for instance, a function f which is to be integrated over a path $y(x)$ connecting two fixed points (x_1, y_1) and (x_2, y_2) in the xy-plane (Figure 3.1). We shall make the simplifying assumption, appropriate for many situations, that the function f depends only on the independent variable x, the particular path selected $y(x)$, and the path derivative $y'(x) \equiv dy/dx$, so that we can write for the integral

$$I \equiv \int_{x_1}^{x_2} f(y, y', x)\, dx.$$

Clearly, the integral depends in general on the path chosen, for at a given value of x, the quantities y and y' and, hence, f can vary greatly from one path to another. Our problem, then, is to develop a procedure for obtaining the equation of the particular path $y(x)$ that makes the integral an extremum.

Let us begin by assigning a parameter α (which is real and of either sign) in some fashion to each of the possible paths between the two fixed points. The value $\alpha = 0$ would correspond by definition to the extremizing path $y(x, 0)$. Any particular path could then be labeled as

$$y(x, \alpha) = y(x, 0) + \alpha g(x),$$

where $g(x)$ is any function which vanishes at the two end points where all paths meet; this, of course, is only one of many possible ways of representing the path, but it is a convenient linear one. The integral I now depends on the parameter α

The Calculus of Variations

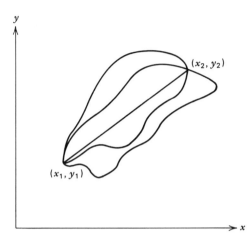

Figure 3.1 Possible paths connecting two points in the xy-plane.

through y and y' in f. The condition for an extremum is then

$$\left(\frac{\partial I}{\partial \alpha}\right)_{\alpha=0} = 0.$$

Differentiating under the integral sign (and recognizing that in this "one-dimensional" approach the variable x is independent of α), we obtain

$$\frac{\partial I}{\partial \alpha} = \int_{x_1}^{x_2} \left(\frac{\partial f}{\partial y}\frac{\partial y}{\partial \alpha} + \frac{\partial f}{\partial y'}\frac{\partial y'}{\partial \alpha}\right) dx.$$

Integrating by parts in the second term on the right produces

$$\int_{x_1}^{x_2} \frac{\partial f}{\partial y'}\frac{\partial^2 y}{\partial x\, \partial \alpha}\, dx = \frac{\partial f}{\partial y'}\frac{\partial y}{\partial \alpha}\bigg|_{x_1}^{x_2} - \int_{x_1}^{x_2} \frac{d}{dx}\left(\frac{\partial f}{\partial y'}\right)\frac{\partial y}{\partial \alpha}\, dx.$$

But $\partial y/\partial \alpha = g(x)$, which vanishes at the end points, so the integrated term drops out, and we have

$$\frac{\partial I}{\partial \alpha} = \int_{x_1}^{x_2} \left[\frac{\partial f}{\partial y} - \frac{d}{dx}\left(\frac{\partial f}{\partial y'}\right)\right]\frac{\partial y}{\partial \alpha}\, dx.$$

Now set $\alpha = 0$, multiply by $d\alpha$, let

$$\left(\frac{\partial I}{\partial \alpha}\right)_{\alpha=0} d\alpha \equiv \delta I \quad \text{and} \quad \left(\frac{\partial y}{\partial \alpha}\right)_{\alpha=0} d\alpha \equiv \delta y$$

to obtain

$$\delta I = \int_{x_1}^{x_2} \left[\frac{\partial f}{\partial y} - \frac{d}{dx}\left(\frac{\partial f}{\partial y'}\right) \right] \delta y \, dx.$$

δI is a variation of the integral brought about by a variation of the path $y(x)$, the latter caused by altering α an amount $d\alpha$ from its zero value.

If the bracketed expression in the integrand did not vanish, then $\delta y = g(x)\,d\alpha$ could be chosen of the same sign as this bracket, and δI could not be zero with an integrand positive for all x. But $\delta I = 0$ *is* the extremizing condition, which then can only be satisfied by requiring that

$$\frac{\partial f}{\partial y} - \frac{d}{dx}\left(\frac{\partial f}{\partial y'}\right) = 0. \qquad (3.1)$$

That is, I takes on an extremum value only when f satisfies eq. (3.1), which is called the **Euler–Lagrange equation**. But f is a function determined in advance by the nature of the problem under investigation, that is, it is a known function. The relation (3.1) is therefore in reality a differential equation which involves y and y'; where it can be solved, it leads to a solution $y(x)$, which is the correct path to follow to make I an extremum. Some examples to follow will clarify these points. (With slight change of notation, the reader will notice a striking identity between eq. (3.1) and the Lagrange equation of motion; the connection is discussed in the next section.)

Example 1 The Shortest Distance Between Two Points in a Plane
Find the curve in the xy-plane that gives the shortest distance from the origin to the point with coordinates (a, b).

Solution
Naturally, we know the solution to this problem before we begin. But this is a standard problem in the calculus of variations, and it is instructive to see how the answer can be obtained from that standpoint. The arc length along an arbitrary curve is given by

$$ds = \sqrt{dx^2 + dy^2} = \sqrt{1 + (y')^2}\, dx,$$

so we wish to minimize

$$I = \int_0^a \sqrt{1 + (y')^2}\, dx,$$

with $f \equiv \sqrt{1 + (y')^2}$. Since $\partial f/\partial y' = y'/[\sqrt{1 + (y')^2}]$, $\partial f/\partial y = 0$, eq. (3.1) gives

$$\frac{d}{dx}\left(\frac{\partial f}{\partial y'}\right) = 0 = \frac{d}{dx}\left(\frac{y'}{\sqrt{1 + (y')^2}}\right)$$

or

$$\frac{y'}{\sqrt{1 + (y')^2}} = C_1,$$

where C_1 is a constant. Solving for y',

$$y' = \frac{C_1}{\sqrt{1 - C_1^2}} = C_1',$$

another constant. Integration produces $y = C_1'x + C_2$, the equation of a straight line with slope C_1' and intercept constant C_2 to be determined from the two end points on the path. Since the curve must pass through the origin, $C_2 = 0$, and the condition of passing through the point (a, b) produces $C_1' = b/a$. Thus, the desired curve is the line $y = (b/a)x$, as expected.

It is clear by inspection that the line *minimizes* the value of I, though this has not been proved. The satisfying of eq. (3.1) merely means that the line should produce an extremum for I, and this will not always be a minimum.

Example 2 Particle Falling Freely Under Gravity

Next, as a simple dynamical application, consider the case of a mass m falling freely in a uniform gravitational field, the y-axis being taken as vertically upward.

Solution
We have $T = \frac{1}{2}m\dot{y}^2$, $V = mg(y - y_0)$, where y_0 is the height at time zero, chosen as the reference level for the potential energy. The Lagrangian is

$$L = T - V = \tfrac{1}{2}m\dot{y}^2 - mg(y - y_0).$$

Suppose we want to find the curve $y = y(t)$ that will extremize the time integral of the Lagrangian; the reason for doing so will soon be apparent. The variable x in the formalism is here replaced by the time t, the function f becomes the Lagrangian, and eq. (3.1) reads

$$\frac{\partial L}{\partial y} - \frac{d}{dt}\left(\frac{\partial L}{\partial \dot{y}}\right) = 0 = -mg - m\ddot{y},$$

or

$$\ddot{y} = -g.$$

Integration gives

$$\dot{y} = -gt + C_1,$$

$$y = -\frac{gt^2}{2} + C_1 t + C_2,$$

with C_1 and C_2 as constants. Clearly, $C_2 = y_0$, leading to the familiar function

$$y(t) = -\frac{gt^2}{2} + C_1 t + y_0.$$

To determine C_1, another value of y at a later time is needed.

Equation (3.1) is the condition that the time integral of L be an extremum; it has the form of Lagrange's equation and hence yields the equation of motion. This conclusion is not a result peculiar to this problem, but rather one of general validity, as we shall see later in the chapter.

Example 3 The Brachistochrone

Find the curve along which a particle of mass m must travel to get from the origin to point (x, y) in the least amount of time when falling (initial speed v_0) in a uniform gravitational field. The particle does not travel straight down, but instead has a frictionless constraint holding it along the curve.

This problem is another famous historical exercise in the calculus of variations. In fact, Bernoulli's interest in it is what led to the development of this branch of mathematical physics. (The name brachistochrone comes from the Greek for "shortest time.")

Solution

If we take the y axis vertically upward from the initial point, $T = \frac{1}{2}m(\dot{x}^2 + \dot{y}^2) = \frac{1}{2}mv^2$, $V = mgy$. By conservation of energy, $\frac{1}{2}mv^2 + mgy = \frac{1}{2}mv_0^2$ or $v = \sqrt{v_0^2 - 2gy}$. The time of descent to any point, taking ds as an element of path length, is

$$t = \int \frac{ds}{v} = \int \frac{\sqrt{1 + (y')^2}\, dx}{\sqrt{v_0^2 - 2gy}},$$

so that

$$f = \sqrt{\frac{1 + (y')^2}{v_0^2 - 2gy}}.$$

Using

$$\frac{\partial f}{\partial y'} = \frac{y'}{\sqrt{(v_0^2 - 2gy)\left[1 + (y')^2\right]}} \quad \text{and} \quad \frac{\partial f}{\partial y} = g\sqrt{\frac{1 + (y')^2}{(v_0^2 - 2gy)^3}},$$

then
$$\frac{\partial f}{\partial y} - \frac{d}{dx}\left(\frac{\partial f}{\partial y'}\right) = 0$$

becomes a formidable expression in terms of y'', y', and y. It is much easier to proceed if we employ a trick that is often useful with the calculus of variations, especially when x does not appear in the integrand f, as here. Simply write $ds = \sqrt{1 + (x')^2}\, dy$, $x' \equiv dx/dy$, thinking now of y as the independent and x the dependent variable, so $x = x(y)$; $f = f(x, x', y)$ and eq. (3.1) has the alternative form

$$\frac{\partial f}{\partial x} - \frac{d}{dy}\left(\frac{\partial f}{\partial x'}\right) = 0.$$

Here we have

$$t = \int \frac{\sqrt{1 + (x')^2}\, dy}{\sqrt{v_0^2 - 2gy}}, \qquad \frac{\partial f}{\partial x} = 0, \qquad \frac{\partial f}{\partial x'} = \frac{x'}{\sqrt{(v_0^2 - 2gy)[1 + (x')^2]}}.$$

Because $\partial f/\partial x$ vanishes, $\partial f/\partial x'$ is clearly a constant C in y:

$$\frac{x'}{\sqrt{(v_0^2 - 2gy)[1 + (x')^2]}} = C,$$

or

$$x' = \sqrt{\frac{(v_0^2 - 2gy)C^2}{1 - (v_0^2 - 2gy)C^2}} \equiv \sqrt{\frac{C_1 - y}{C_2 + y}},$$

where

$$C_1 \equiv \frac{v_0^2}{2g}, \qquad C_2 \equiv \frac{(1 - v_0^2 C^2)}{2gC^2}.$$

The last relation is integrated most conveniently by making a change of variable to θ, setting

$$y \equiv \left(\frac{C_1 - C_2}{2}\right) - \left(\frac{C_1 + C_2}{2}\right)\cos\theta, \qquad dy = \left(\frac{C_1 + C_2}{2}\right)\sin\theta\, d\theta,$$

whereupon

$$x' = \sqrt{\frac{(1 + \cos\theta)}{(1 - \cos\theta)}}.$$

We have

$$dx = x'\, dy = \left(\frac{C_1 + C_2}{2}\right)\sqrt{\frac{1 + \cos\theta}{1 - \cos\theta}}\sin\theta\, d\theta = \left(\frac{C_1 + C_2}{2}\right)d\theta(1 + \cos\theta),$$

which integrates to
$$x = \left(\frac{C_1 + C_2}{2}\right)(\theta + \sin\theta) + C'.$$
The parametric equations of the desired path are then
$$x = C' + \left(\frac{C_1 + C_2}{2}\right)(\theta + \sin\theta),$$
$$y = \left(\frac{C_1 - C_2}{2}\right) - \left(\frac{C_1 + C_2}{2}\right)\cos\theta = -C_2 + \left(\frac{C_1 + C_2}{2}\right)(1 - \cos\theta).$$
These are the equations of a cycloid, the path traced by a fixed point on the circumference of a wheel of radius $(C_1 + C_2)/2$ which rolls along a straight line, as is readily shown. Again, we have only indicated that the time of descent integral should have a stationary value for the cycloid path, not that it is a minimum, but that is indeed the case.

Example 4 Isoperimetric Problems

Suppose we are given a variational problem in which an auxiliary condition is imposed in the form of an integral which must have a certain definite value. That is, we wish to find $y(x)$ such that the integral $I = \int_{x_1}^{x_2} f(y, y', x)\, dx$ has its usual extremum, but in addition the integral $\int_{x_1}^{x_2} \sigma(y, y', x)\, dx$ must give the constant C_0 for this same path $y(x)$. The name "isoperimetric problem" was originally coined for the situation when the integral I referred to the maximum area that could be enclosed by a curve of fixed perimeter C_0.

It is possible to solve such problems by the construction of a new function
$$F \equiv f + \lambda\sigma,$$
where λ is an unknown Lagrangian multiplier that is independent of x. To proceed, form the integral
$$I' \equiv \int_{x_1}^{x_2} F\, dx = \int_{x_1}^{x_2} (f + \lambda\sigma)\, dx$$
and note that the procedure outlined earlier in this section leads to
$$\delta I' = \int_{x_1}^{x_2} \left[\frac{\partial F}{\partial y} - \frac{d}{dx}\left(\frac{\partial F}{\partial y'}\right)\right]\delta y\, dx.$$

At first, it appears that the variation of I' would give zero just as did the variation of I, for the $\lambda\sigma$ term in F integrates to the constant λC_0 and therefore contributes nothing to the variation. Consequently, the Euler–Lagrange equation should hold

The Calculus of Variations

for the function F:

$$\frac{\partial F}{\partial y} - \frac{d}{dx}\left(\frac{\partial F}{\partial y'}\right) = 0. \tag{3.1'}$$

While this conclusion is correct, a little more needs to be said. The variation δy occurring in $\delta I'$ is not completely arbitrary here; only those variations of path are allowed in which the constraining integral over σ will have the prescribed value C_0. By analogy with the Lagrangian multiplier discussion of nonholonomic constraints in the previous chapter, we *choose* λ so that eq. (3.1') will be true.

The solution of eq. (3.1') for $y(x)$ will involve the unknown λ as well as the ubiquitous integration constants, but these can all be determined through application of the auxiliary condition $C_0 = \int_{x_1}^{x_2} \sigma \, dx$ along with insertion of the y values at x_1 and x_2.

To illustrate this method, consider a fence of length L constructed in such a manner as to connect two points of a wall that are a distance $2a$ apart (Figure 3.2). What must be the shape of the fence if it is to enclose the maximum area consistent with its fixed length?

The area enclosed by the fence is $\int_{-a}^{a} y \, dx$, while the length of fence is $\int_{-a}^{a} \sqrt{1 + (y')^2} \, dx = L$. Thus, we have $f = y$, $\sigma = \sqrt{1 + (y')^2}$, $F = y + \lambda \sqrt{1 + (y')^2}$. The Euler–Lagrange equation becomes

$$1 - \frac{d}{dx}\left[\frac{\lambda y'}{\sqrt{1 + (y')^2}}\right] = 0.$$

Integrating directly,

$$\frac{\lambda y'}{\sqrt{1 + (y')^2}} = x + C_1.$$

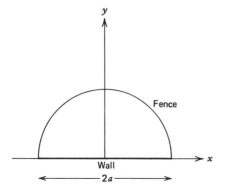

Figure 3.2 Fence enclosing maximum area.

Solving for y', we find

$$y' = \frac{-(x + C_1)}{\sqrt{\lambda^2 - (x + C_1)^2}};$$

the negative sign was chosen upon extracting the square root for y' so that the slope will be negative for positive x. Integrating again gives

$$y = \sqrt{\lambda^2 - (x + C_1)^2} + C_2.$$

The condition that $y = 0$ at $x = a$ yields $C_2 = -\sqrt{\lambda^2 - (C_1 + a)^2}$, while the fact that $y = 0$ at $x = -a$ produces $C_1 = 0$, leaving

$$y = \sqrt{\lambda^2 - x^2} - \sqrt{\lambda^2 - a^2}.$$

It remains to determine λ from the length integral:

$$L = \int_{-a}^{a} \sqrt{1 + (y')^2}\, dx = \int_{-a}^{a} \sqrt{1 + \left(\frac{x^2}{\lambda^2 - x^2}\right)}\, dx$$

$$= \lambda \int_{-a}^{a} \frac{dx}{\sqrt{\lambda^2 - x^2}} = 2\lambda \arcsin\left(\frac{a}{\lambda}\right),$$

or

$$\frac{\lambda}{a} = \frac{1}{\sin(L/2\lambda)}.$$

We see from the last equation or from y that $\lambda \geq a$, but this transcendental relation is not readily solved for λ. To make one possible solution explicit, let us choose L equal to πa, whereupon $\lambda = a$ and $y = \sqrt{a^2 - x^2}$. For this choice, we find that $y(x)$ is a semicircle of radius a centered on the origin; such a conclusion is not unexpected intuitively. In general, the result is a portion of a circle of radius λ with center displaced along the y axis.

3.2 HAMILTON'S PRINCIPLE

Let us next consider a physical situation closely related to the mathematical treatment of the previous section. Suppose we have a set of N particles described by the $3N$ coordinates x_1 through z_N, each of which is a function of time (and can be determined by solving the equations of motion when the forces are given). Let

Hamilton's Principle

the initial configuration be defined by specifying values of all $3N$ coordinates at an initial instant of time t_i, and let a final configuration be defined similarly at a later time t_f. One can even speak of a $3N$-dimensional "configuration space" in which a single point fixes the position of all N particles completely. (Often, this configuration space, however, is defined in terms of the generalized coordinates q_i rather than the rectangular coordinates x_i.) In such a space, the entire motion of the physical system takes place along some curve joining the initial and final points; this curve is a multidimensional entity which normally bears no resemblance to the real physical path followed by any of the particles in the actual three-dimensional space. The independent variable, time, plays the role of a parameter that marks the position of the system as it moves along the curve between the two end points.

Now imagine that the system moves instead along a neighboring path in configuration space; any constraints imposed on the system will be maintained, and the time interval $t_f - t_i$ along the curve as well as the location of the end points will also be unaltered.

For example, a different path is readily obtained by changing the way in which one or more of the coordinates depends on the time. Now the Newtonian laws have definite solutions consistent with the initial and final conditions so that, if the original curve satisfied these equations of motion, the modified path certainly cannot do so. To analyze this matter further, let us call δx_i the "variation" of x_i and mean by it the difference between the x_i values on the neighboring path and the "true" path at an arbitrary time t. Recalling from Chapter 1 that work done by the forces acting on a system is the negative of the change in potential energy, we have for the variation of potential between the two paths

$$-\delta V = \sum_{i=1}^{N} \vec{F}_i \cdot \delta \vec{r}_i = \sum_i m_i \ddot{\vec{r}}_i \cdot \delta \vec{r}_i,$$

where the force $\vec{F}_i$ on the ith particle is expressed by means of the second law of Newton. But the right side can be written as

$$\sum_i m_i \left[\frac{d}{dt}(\dot{\vec{r}}_i \cdot \delta \vec{r}_i) - \dot{\vec{r}}_i \cdot \delta \dot{\vec{r}}_i \right],$$

provided that $(d/dt)\delta \vec{r}_i = \delta \dot{\vec{r}}_i$, which will be substantiated shortly (and indeed would probably never be questioned by the reader were it not for the modern student's great familiarity with the noncommuting operators of quantum mechanics). Since the variation of the kinetic energy is

$$\delta T = \sum_i m_i \dot{\vec{r}}_i \cdot \delta \dot{\vec{r}}_i,$$

we find that

$$\delta T - \delta V = \delta L = \sum_i m_i \frac{d}{dt}(\dot{\vec{r}}_i \cdot \delta \vec{r}_i).$$

Integrating with respect to time, note that the right side must vanish, since it is an exact differential with a $\delta \vec{r}_i$ factor that gives zero at the limits (because of the requirement that all varied paths begin and end on the same two points). Then we have

$$\int_{t_i}^{t_f} \delta L\, dt = 0 = \delta \int_{t_i}^{t_f} L\, dt. \tag{3.2}$$

The variational principle embodied in eq. (3.2) is known as **Hamilton's principle**; it follows directly from Newton's laws and provides an alternative method for deriving the equations of motion, a formulation which is independent of the coordinates used. (It is not any help, however, in solving these equations.) the condition (3.2) states that the time integral of the Lagrangian must be an extremum for the true path of the motion; in practice, the integral is usually a minimum.

To complete the proof satisfactorily and to understand how the principle can be applied, the commuting of the time derivative and the variation symbol must be explored more fully. Consider any two of the $3N$ coordinates, denoted by x and y, and examine the projection of the motion in the xy-plane. Let AB be a segment of the true path, and CD a segment of some varied path nearby (Figure 3.3). Points A and C are corresponding points on the two paths in the sense that they refer to the same value of time; similarly for B and D. If the coordinates of A are chosen as (x, y), then those of C are $(x + \delta x, y + \delta y)$, while B is at $(x + dx, y + dy)$. In going from A to B to D, we could write the coordinates of D as

$$[x + dx + \delta(x + dx), y + dy + \delta(y + dy)],$$

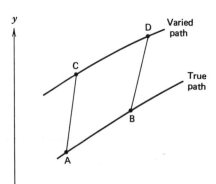

Figure 3.3 Path analysis for Hamilton's principle.

Hamilton's Principle

whereas in going from A to C to D we would instead write

$$[x + \delta x + d(x + \delta x), y + \delta y + d(y + \delta y)]$$

for the coordinates of D. Equating the two expressions for D, we find that

$$\delta(dx) = d(\delta x), \delta(dy) = d(\delta y),$$

so the variation symbol will commute with any derivative operator. This fact allows us to write relations such as

$$\delta(\dot{x}^2) = 2\dot{x}\,\delta(\dot{x}) = 2\dot{x}\frac{d}{dt}(\delta x),$$

which make it possible to express the variation of more complicated quantities in terms of just δx and thus allows the use of Hamilton's principle in a simple way, as shown below.

There are some persons who might reach the above conclusion by a strictly algebraic procedure. For this purpose, note that

$$\delta y \equiv y(x, \alpha) - y(x, 0) = \alpha g(x), \qquad \frac{d}{dx}(\delta y) = \alpha \frac{dg}{dx}$$

(since it is zero on the true path, α is equivalent to the earlier $d\alpha$), while for the variation of the derivative,

$$\delta\left(\frac{dy}{dx}\right) = \frac{dy(x,\alpha)}{dx} - \frac{dy(x,0)}{dx} = \frac{dy(x,0)}{dx} + \alpha\frac{dg}{dx} - \frac{dy(x,0)}{dx} = \alpha\frac{dg}{dx}.$$

Then, $d(\delta y) = \delta(dy)$ follows at once.

Example 5

A particle of mass m falls freely in a uniform gravitational field. Find the equation of motion by Hamilton's principle.

Solution

Taking the y-axis vertically upward, the Lagrangian is as usual of the form

$$L = \tfrac{1}{2}m\dot{y}^2 - mgy,$$

whereupon

$$\delta\int_{t_i}^{t_f}L\,dt = 0 = \int_{t_i}^{t_f}\delta L\,dt = m\int[\dot{y}\,\delta(\dot{y}) - g\,\delta y]\,dt$$

$$= m\int\left[\dot{y}\frac{d}{dt}(\delta y) - g\,\delta y\right]dt.$$

Integrating by parts in the first term, we have

$$m[\dot{y}\,\delta y]_{t_i}^{t_f} - m\int \delta y \cdot \ddot{y}\,dt - m\int g\,\delta y\,dt = -m\int (\ddot{y}+g)\,\delta y\,dt,$$

the integrated term vanishing with $\delta y = 0$ at both limits. Finally, we obtain

$$\ddot{y} + g = 0$$

as the equation of motion, a condition forced on us by the vanishing of the last integral inasmuch as the δy still left in the integrand is completely arbitrary. Other problems are attacked similarly.

The approach employed here is quite different from that of Example 2 in the previous section, where the techniques of the calculus of variations led immediately to Lagrange's equation for the falling particle. In general, the method above, using Hamilton's principle directly, is straightforward but laborious. It is not recommended as a daily procedure for problem solving, but is being pursued here to enhance our understanding of mechanics, and because such variational principles are often used as a starting point in other branches of physics.

One matter remains to be investigated. Suppose that the Lagrangian L depends on more than one coordinate, as is frequently the case. Let us call one of these variables y_i and its derivative $y_i' \equiv dy_i/dx$, where x is the "independent" or parametric variable (which is most commonly the time in physical applications). The derivation of Section 3.1 is easily generalized to include this situation; the end result is that the Euler–Lagrange equation holds true for each of the variables y_i:

$$\frac{\partial f}{\partial y_i} - \frac{d}{dx}\left(\frac{\partial f}{\partial y_i'}\right) = 0, \qquad (3.1'')$$

where the index i runs through the number of y_i present. If the integral to be varied is that of Hamilton's principle, that is,

$$I = \int L(q_i, \dot{q}_i, t)\,dt,$$

then we have at once from the Euler–Lagrange equation

$$\frac{d}{dt}\left(\frac{\partial L}{\partial \dot{q}_i}\right) - \frac{\partial L}{\partial q_i} = 0,$$

the familiar set of Lagrange's equations. We have shown that Newton's laws lead to Hamilton's principle, which in turn leads to Lagrange's equations. There are other ways of proceeding, but all methods are essentially equivalent in that they eventually produce the same equations of motion for a given physical problem. In

Other Variational Principles

the next section, we consider some alternative procedures of historical importance; they are a bit more complicated to apply but illustrate the wide variety of formalisms available in this area.

3.3 OTHER VARIATIONAL PRINCIPLES

The Principle of Least Action

In discussing Hamilton's principle, we used a δ-variation between corresponding points of two adjacent paths, these points being identified by equivalent values of the time parameter. The δ-variation, in other words, was really our old friend the virtual displacement, in which time is held fixed. But such displacements are not always physically realizable in the motion; the varied path may not be a possible trajectory, even though the constraints imposed on the system are not violated. Such is the case when the Hamiltonian (in most examples a constant equal to the total energy) is not conserved in the variation process. The principle of least action requires a new kind of variation between adjacent paths, called a **Δ-variation**, that is designed to keep the Hamiltonian constant. To do so, the system representative point in configuration space may have to move faster or slower than in the previous discussion. This means that, even though the two end points of the motion are the same in space for all paths selected, the time of traversal between those points is no longer constant, but must vary from one path to another in such a way as to keep the Hamiltonian fixed. With this understanding, one can define the **action S** as follows:

$$S \equiv \sum_i \int_{t_0}^{t} p_i \dot{q}_i \, dt = \sum_i \int_{t_0}^{t} p_i \, dq_i. \tag{3.3}$$

The principle of least action requires that

$$\Delta S = 0, \tag{3.4}$$

the Δ-variation being as described above. Despite the name, the action is not always minimized, but simply extremized. (The principle was first stated rather badly by Maupertuis[1] and later developed by Euler and Lagrange.)

The proof of this principle can be sketched rather quickly. For any coordinate q, it is clear from the discussion of the two types of variation that

$$\Delta q = \delta q + \dot{q} \, \Delta t,$$

where Δt represents the change in time produced as we move from one path to

---

[1]R. Lindsay and H. Margenau, *Foundations of Physics* (John Wiley, New York 1936), p. 133.

another. Then, for an arbitrary function $f(q, t)$,

$$\Delta f = \sum_i \frac{\partial f}{\partial q_i} \Delta q_i + \frac{\partial f}{\partial t} \Delta t = \sum_i \frac{\partial f}{\partial q_i} \delta q_i + \left(\sum_i \frac{\partial f}{\partial q_i} \dot{q}_i + \frac{\partial f}{\partial t} \right) \Delta t = \delta f + \dot{f} \Delta t,$$

a relation just like that for the coordinate q. From the definition, eq. (2.15), of the Hamiltonian,

$$\Delta S = \Delta \int_{t_0}^{t} \left(\sum_i p_i \dot{q}_i \right) dt = \Delta \int_{t_0}^{t} (L + H) \, dt = \Delta \int_{t_0}^{t} L \, dt + \Delta [H(t - t_0)]$$

$$= \Delta \int_{t_0}^{t} L \, dt + H \Delta t,$$

where all trajectories begin at a common time t_0 but end at some variable time t. Taking f as $\int_{t_0}^{t} L \, dt$ and applying the Δ variation rule yields

$$\Delta \int_{t_0}^{t} L \, dt = \delta \int_{t_0}^{t} L \, dt + L \Delta t,$$

or

$$\Delta S = \delta \int_{t_0}^{t} L \, dt + (L + H) \Delta t,$$

the first term not vanishing as usual because of the variable upper limit. But from the δ commutation property, we can write (time t is varied by Δ but not δ)

$$\delta \int_{t_0}^{t} L \, dt = \int_{t_0}^{t} \sum_i \left(\frac{\partial L}{\partial q_i} \delta q_i + \frac{\partial L}{\partial \dot{q}_i} \delta \dot{q}_i \right) dt$$

$$= \int_{t_0}^{t} \sum_i \left(\frac{d}{dt} \left(\frac{\partial L}{\partial \dot{q}_i} \right) \delta q_i + \frac{\partial L}{\partial \dot{q}_i} \frac{d}{dt} \delta q_i \right) dt = \int_{t_0}^{t} \sum_i \frac{d}{dt} \left(\frac{\partial L}{\partial \dot{q}_i} \delta q_i \right) dt$$

$$= \sum_i \left(\frac{\partial L}{\partial \dot{q}_i} \delta q_i \right)_{t_0}^{t} = \sum_i \left[\frac{\partial L}{\partial \dot{q}_i} (\Delta q_i - \dot{q}_i \Delta t) \right]_{t_0}^{t} = - \left(\sum_i p_i \dot{q}_i \right) \Delta t,$$

since Δq_i vanishes at the end points of the motion for all paths and $\Delta t = 0$ at the initial time t_0. Finally, we obtain

$$\Delta S = \left(- \sum_i p_i \dot{q}_i + L + H \right) \Delta t = 0$$

by the definition of the Hamiltonian.

Other Variational Principles 85

Owing to the complexity of the Δ-variation, the least action principle is seldom used today. As a simple application, however, consider a particle of mass m with no applied forces acting:

$$T = \tfrac{1}{2}m\dot{r}^2 = H - V = \text{constant (since } V = \text{constant)}$$

or

$$|d\vec{r}| = dt\sqrt{\frac{2(H-V)}{m}} \equiv dq.$$

The action is

$$S = \int_{t_0}^{t} p\, dq = \int m\sqrt{\frac{2(H-V)}{m}} \left[\sqrt{\frac{2(H-V)}{m}}\right] dt = 2(H-V)\int_{t_0}^{t} dt$$

$$= 2(H-V)(t - t_0),$$

$$\Delta S = 0 \sim \Delta(t - t_0).$$

That is, the particle moves between two fixed points so as to minimize the time of travel. Alternatively, we can write

$$S = \sqrt{2m(H-V)}\int dq,$$

$$\Delta S = 0 \sim \Delta \int dq;$$

the particle minimizes the length of travel between the end points, or moves along a straight line.

Gauss' Principle of Least Constraint

Variational principles exist in abundance in the history of mechanics. Most of them have fallen into disfavor as physicists evidently preferred the other methods emphasized in this book, which are simpler to apply. It seems appropriate to mention here as typical one such idea, developed by Gauss[2] in 1829, which does not require the path variation scheme of the previous approaches.

Suppose that the ith member of a system of particles undergoes the vector displacement $\vec{r}_i$ during the time increment Δt, when considered as free to move

---

[2] R. Lindsay, *Physical Mechanics*, 2nd edition (Van Nostrand, New York 1950), p. 253.

under the influence of the applied forces acting. In addition, there are usually constraint forces acting as well; when their effect is included, the mass m_i will instead be displaced along the vector $\vec{r}_i'$ (Figure 3.4). Let $\Delta \vec{r}_i$ be the vector difference $\vec{r}_i' - \vec{r}_i$, which in a practical sense measures the way in which the constraints affect the ith particle. Gauss' principle states that the quantity $\sum_i m_i (\Delta \vec{r}_i)^2$, when summed over all particles, must be a *minimum* for the actual motion as compared with any other variation of the system motion which does not violate the geometrical constraints. This sum was designated by Gauss as the "total constraint," and so the principle got its name accordingly.

In order to construct a proof of this principle, let us note first that displacements from rest in a fixed time interval Δt are proportional to accelerations or forces. Under the influence of the applied forces, the displacement for one particle is

$$|\vec{r}| = \bar{v}\Delta t = \frac{v+0}{2}\Delta t = \frac{v\Delta t}{2},$$

where v is the final speed and $\bar{v}$ is the average speed during the interval Δt. (The acceleration is taken as constant for this brief period.) Since the magnitude of the acceleration is

$$a = \frac{v-0}{\Delta t} = \frac{F}{m},$$

then

$$v = \frac{F}{m}\Delta t$$

in terms of the net applied force F and mass m. We therefore have

$$|\vec{r}| = \frac{F}{2m}(\Delta t)^2$$

for the displacement. In the same way, we can write

$$|\Delta \vec{r}| = \frac{F_c}{2m}(\Delta t)^2.$$

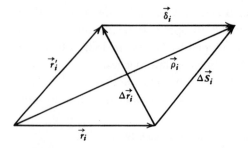

Figure 3.4 Vector triangles for Gauss' principle.

Inasmuch as $|\Delta \vec{r}|$ directly measures the effect of the constraints, F_c can be identified with the net force of constraint.

Now suppose that the ith particle of our system undergoes a displacement $\vec{\rho}_i$ (rather than $\vec{r}_i'$) through the action of both the applied and the constraint forces; correspondingly, $\Delta \vec{r}_i$ will be replaced by $\Delta \vec{S}_i$ (see Figure 3.4). The proof of the principle consists of showing that for *any* hypothetical $\vec{\rho}_i$, the total constraint of Gauss must be greater than for the vector $\vec{r}_i'$ found in the actual motion.

From the geometry of the figure, we have

$$\Delta \vec{S}_i = \Delta \vec{r}_i + \vec{\delta}_i \quad \text{and} \quad (\Delta \vec{S}_i)^2 = (\Delta \vec{r}_i)^2 + \delta_i^2 + 2\Delta \vec{r}_i \cdot \vec{\delta}_i.$$

In accordance with our discussion above, the quantity $\Delta \vec{r}_i$ is proportional to $(\vec{F}_c)_i/m_i$; and therefore, if the $(\Delta \vec{S}_i)^2$ expression is multiplied by m_i and summed over i, we obtain

$$\sum_i m_i (\Delta \vec{S}_i)^2 = \sum_i m_i (\Delta \vec{r}_i)^2 + \sum_i m_i \delta_i^2 + \left[\sum_i (\vec{F}_c)_i \cdot \vec{\delta}_i \right] (\Delta t)^2.$$

The last term on the right must vanish, since $\vec{\delta}_i$ can be viewed as a possible virtual displacement and we assume that forces of constraint can do no work with such a displacement. The first term on the right is the total constraint of Gauss, and the second term (δ_i^2) is always positive; hence, the total constraint with any $\Delta \vec{S}_i$ (on the left) must always be greater than that for the $\Delta \vec{r}_i$ of the actual motion, thus establishing the principle.

This "geometrical proof" is of historical interest but is somewhat lacking in clarity. An analytical proof more directly related to the method of application seems preferable, so let us proceed with it before turning to an example. Note from the relations which follow eq. (2.4) in Section 2.1, Chapter 2, that

$$\vec{v}_i = \sum_j \frac{\partial \vec{r}_i}{\partial q_j} \dot{q}_j + \frac{\partial \vec{r}_i}{\partial t},$$

whereupon

$$\ddot{\vec{r}}_i = \dot{\vec{v}}_i = \sum_j \frac{d}{dt}\left(\frac{\partial \vec{r}_i}{\partial q_j} \right) \dot{q}_j + \sum_j \frac{\partial \vec{r}_i}{\partial q_j} \ddot{q}_j + \sum_j \frac{\partial^2 \vec{r}_i}{\partial q_j \partial t} \dot{q}_j + \frac{\partial^2 \vec{r}_i}{\partial t^2}$$

and therefore $\partial \ddot{\vec{r}}_i / \partial \ddot{q}_k = \partial \vec{r}_i / \partial q_k$ by inspection. Making use of D'Alembert's principle, eq. (2.3), in the form

$$\sum_i (m_i \ddot{\vec{r}}_i - \vec{F}_i) \cdot \delta \vec{r}_i = 0 = \sum_i (m_i \ddot{\vec{r}}_i - \vec{F}_i) \cdot \left(\sum_k \frac{\partial \vec{r}_i}{\partial q_k} \delta q_k \right),$$

we have

$$\sum_{i,k}\left[m_i\ddot{\vec{r}}_i\cdot\frac{\partial\ddot{\vec{r}}_i}{\partial\ddot{q}_k}-\vec{F}_i\cdot\frac{\partial\ddot{\vec{r}}_i}{\partial\ddot{q}_k}\right]\delta q_k = \sum_k\left[\sum_i m_i\frac{\partial}{\partial\ddot{q}_k}\left(\frac{1}{2}\ddot{\vec{r}}_i^2\right)-Q_k\right]\delta q_k.$$

Assuming that the displacements δq_k are independent, we can then write

$$\frac{\partial}{\partial\ddot{q}_k}\left(\frac{1}{2}\sum_i m_i\ddot{\vec{r}}_i^2\right)-Q_k = 0,$$

or

$$\frac{\partial}{\partial\ddot{q}_k}\left[\sum_i\frac{1}{2m_i}(m_i\ddot{\vec{r}}_i-\vec{F}_i)^2\right]=0, \quad (3.5)$$

since the last expression is just

$$\frac{\partial}{\partial\ddot{q}_k}\left\{\sum_i\frac{1}{2m_i}[m_i^2\ddot{\vec{r}}_i^2+\vec{F}_i^2-2\vec{F}_i\cdot(m_i\ddot{\vec{r}}_i)]\right\} = \frac{\partial}{\partial\ddot{q}_k}\left(\frac{1}{2}\sum_i m_i\ddot{\vec{r}}_i^2\right)-\sum_i\vec{F}_i\cdot\frac{\partial\ddot{\vec{r}}_i}{\partial\ddot{q}_k}$$

$$= \frac{\partial}{\partial\ddot{q}_k}\left(\frac{1}{2}\sum_i m_i\ddot{\vec{r}}_i^2\right)-Q_k,$$

as obtained above.

Differentiating this expression once more, we have

$$\frac{\partial^2}{\partial\ddot{q}_k^2}\left[\sum_i\frac{1}{2m_i}(m_i\ddot{\vec{r}}_i-\vec{F}_i)^2\right] = \frac{\partial^2}{\partial\ddot{q}_k^2}\left(\frac{1}{2}\sum_i m_i\ddot{\vec{r}}_i^2\right) = \frac{\partial}{\partial\ddot{q}_k}\left(\sum_i m_i\ddot{\vec{r}}_i\cdot\frac{\partial\vec{r}_i}{\partial q_k}\right)$$

$$= \sum_i m_i\left(\frac{\partial\vec{r}_i}{\partial q_k}\right)^2 > 0.$$

Inasmuch as the second derivative is positive, the function

$$\sum_i\frac{1}{2m_i}(m_i\ddot{\vec{r}}_i-\vec{F}_i)^2$$

has been minimized with respect to the acceleration $\ddot{q}_k$. This quantity is in fact proportional to the total constraint of Gauss.

The preceding proof gives no clear indication of the method of application, yet a standard procedure is usually followed. As an example, consider a simple pendulum of length l in a uniform gravitational field; at some instant t, the

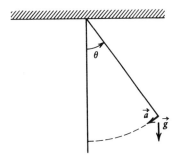

Figure 3.5 Analysis of pendulum motion.

pendulum makes an angle θ with the vertical (Figure 3.5). If the bob could move freely at this point, it would fall straight down a distance $\frac{1}{2}g(\Delta t)^2$ in time Δt. (At least, it would do so if started from the instantaneous rest position when θ is a maximum.) The bob is actually constrained so as to move along a circular arc. Assuming that the arc is essentially a straight line for a short interval Δt, the bob will move a distance $\frac{1}{2}a(\Delta t)^2$, where $a = -l\ddot{\theta}$ is the acceleration along the arc. The total constraint is given by the expression

$$m(\Delta \vec{r})^2 = m(\vec{r}' - \vec{r})^2 = \frac{m}{4}(\Delta t)^4[\vec{a} - \vec{g}]^2 = \frac{m}{4}(\Delta t)^4(a^2 + g^2 - 2\vec{a} \cdot \vec{g})$$

$$= \frac{m}{4}(\Delta t)^4(a^2 + g^2 - 2ga\cos(90° - \theta)) = \frac{m}{4}(\Delta t)^4(a^2 + g^2 - 2ga\sin\theta)$$

from the geometry, with $\vec{a}$ along the arc and $\vec{g}$ directed downward. Minimizing this result by differentiating with respect to the acceleration and setting the derivative equal to zero gives

$$\frac{m}{4}(\Delta t)^4[2a - 2g\sin\theta] = 0, \qquad a = g\sin\theta,$$

or finally,

$$\ddot{\theta} = \frac{-g}{l}\sin\theta,$$

the usual equation of motion.

The above is typical of the procedure usually followed. The displacements involved are proportional to accelerations, and the minimizing is then done with respect to the latter, which leads at once to the equation of motion. Other examples are given in the problem set.

Gauss' principle has more than mere historical significance. Late in the nineteenth century, the physicist Gibbs and the mathematician Appell independently discovered a dynamical formulation[3] utilizing what is today called the

---

[3] See E. Desloge, *Classical Mechanics*, Vol. II (Wiley–Interscience, New York 1982), Chapter 69.

Gibbs–Appell function S:

$$S \equiv \tfrac{1}{2}\sum_i m_i \ddot{x}_i^2.$$

The procedure for application of this function involves differentiations with respect to appropriate accelerations, and it is reminiscent of the Gaussian technique. The method is useful in the treatment of anholonomic constraints, where the Gibbs–Appell function can be employed in much the same manner as is done with the kinetic energy—Lagrangian approach to holonomic constraints.

PROBLEMS

3.1 The shortest distance between two points on a surface lies along a curve called the **geodesic** of that surface. Show that the geodesics of a sphere are great circles of that surface, using the calculus of variations.

3.2 Find the curve (in the xy-plane) joining the origin to the point $x = 1, y = 1$ and for which the integral of the function $2(y')^2 + \tfrac{1}{2}(y/x)^2 - (yy'/x)$ has an extremum. Show that this integral has a larger value for the parabolas $y = x^2$ and $x = y^2$ than for the curve giving the extremum.

3.3 The ends of a uniform inextensible string of length l are connected to two fixed points at the same level, a distance $2a$ apart. Find the curve along which the string must hang if it is to have its center of mass as low as possible. (This problem is of considerable historical interest, for it indicates the approach of Bernoulli to the determination of the form of the freely hanging string or chain.)

3.4 As simple mathematical exercises in the area of the calculus of variations, consider the following:

(*a*) Suppose one has a family of curves of the form

$$y = kx + \alpha \sin nx,$$

where k is a positive constant, n is an integer, and α is a small constant. Prove that the integral of $(y')^2$ between $x = 0$ and 2π has its minimum value for this family when $\alpha = 0$; explicitly evaluate the integral to do so.

(*b*) Find $y = y(x)$ such that the variation of the integral of

$$f = y'^2 + z'^2 - ay$$

is zero, where $z = ay$ and a is constant. (The primes indicate x differentiation.)

3.5 A one-dimensional simple harmonic oscillator has a Lagrangian given by
$$L = \tfrac{1}{2}m\dot{x}^2 - \tfrac{1}{2}kx^2$$
where m is the mass and k is the force constant. Assume that the oscillator differential equation has a solution of the form
$$x = A\cos\sqrt{k/m}\,t + B\sin\sqrt{k/m}\,t,$$
with $x = x_0$ at time zero and $x = D$ at time t'. Show that the constants are
$$A = x_0 = D\sec\sqrt{k/m}\,t', \quad B = 0$$
through the use of Hamilton's principle. (Do *not* employ the equation of motion for the oscillator.)

3.6 The one-dimensional simple harmonic oscillator has a potential function
$$V = \tfrac{1}{2}kx^2.$$
Use Hamilton's principle as in the text to find the equation of motion for the oscillator.

3.7 A simple pendulum in one plane swings in a uniform gravitational field. Find the equation of motion through the application of Hamilton's principle.

3.8 A particle of mass m is constrained to the surface of a sphere of radius a. If this surface is smooth so that no tangential forces act, show that the action S is proportional to the length of path traveled.

3.9 The Atwood machine is a familiar device consisting of an inextensible string passing smoothly over a pulley, with two masses m_1 and m_2 connected to the ends of the string. Find the acceleration of the masses in a uniform gravitational field through use of Gauss' principle of least constraint.

3.10 Two particles of masses m_1 and m_2 are connected to the ends of an inextensible string which runs smoothly over a pulley at the peak of two intersecting smooth inclined planes (see Figure 3.6). Find the acceleration of the masses under gravity by Gauss' principle of least constraint.

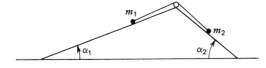

Figure 3.6 Configuration for Problem 3.10.

Chapter Four
CENTRAL FORCES

4.1 THE EQUATIONS OF MOTION FOR CENTRAL FORCES

Consider a system of two particles with masses m_1 and m_2 and position vectors $\vec{r}_1, \vec{r}_2$, acted on by internal forces alone. Clearly this system has six degrees of freedom in general, corresponding to the six independent components of the position vectors. But instead of choosing these components as the generalized coordinates, let us select the three components of the center of mass vector $\vec{R}$ defined in Chapter 1, and the three components of the relative vector $\vec{r}$, defined as

$$\vec{r} \equiv \vec{r}_2 - \vec{r}_1, \qquad |\vec{r}| \equiv r. \tag{4.1}$$

Using eq. (1.8),

$$\vec{R} = \frac{m_1 \vec{r}_1 + m_2 \vec{r}_2}{m_1 + m_2},$$

we find that

$$\vec{r}_1 = \vec{R} - \left(\frac{m_2}{m_1 + m_2}\right)\vec{r}, \qquad \vec{r}_2 = \vec{R} + \left(\frac{m_1}{m_1 + m_2}\right)\vec{r}, \tag{4.2}$$

linear relations which are readily verified. The kinetic energy can now be expressed as

$$T = \frac{1}{2}m_1 \dot{\vec{r}}_1^2 + \frac{1}{2}m_2 \dot{\vec{r}}_2^2 = \frac{1}{2}(m_1 + m_2)\dot{\vec{R}}^2 + \frac{1}{2}\left(\frac{m_1 m_2}{m_1 + m_2}\right)\dot{\vec{r}}^2.$$

Normally, the potential energy is a function of r alone, so that the Lagrangian $L = T - V(r)$ does not contain $\vec{R}$. The components of the CM motion are then

The Equations of Motion for Central Forces

constants, as is always the case for an ignorable coordinate. The first term of T or L is constant and so can be disregarded in later discussions; the remainder of L is what one would expect to get if a *single particle* of "reduced mass"

$$m \equiv \frac{m_1 m_2}{m_1 + m_2} \qquad (4.3)$$

moved at distance r from a fixed center of force at the origin. In effect, we are now dealing with a fictitious one-body problem, the effective mass m being less than either m_1 or m_2, hence the designation of "reduced" mass. If the mass m_2 becomes very large relative to m_1, then m is nearly the same as m_1.

Inasmuch as $\vec{F}(r) = -(\partial V/\partial r)\hat{e}_r$, where $\hat{e}_r = \vec{r}/r$, any central force lies along the radial direction (either inward or outward) and depends on r alone. Letting $|\vec{F}(r)| \equiv F(r)$ and noting that

$$\frac{\partial F_x}{\partial y} = \frac{\partial}{\partial y}\left[\frac{x}{r}F(r)\right] = x\frac{\partial}{\partial r}\left[\frac{F(r)}{r}\right]\frac{\partial r}{\partial y} = \frac{xy}{r}\frac{\partial}{\partial r}\left[\frac{F(r)}{r}\right] = \frac{\partial F_y}{\partial x},$$

we see that the z component of curl $\vec{F}$ vanishes; from the symmetry,

$$\text{curl } \vec{F}(r) = 0. \qquad (4.4)$$

Any central force is then a conservative force, as is already clear from the fact that the force is the negative gradient of a scalar potential function.

From the arguments presented in Chapter 2, it will be recognized that the spherically symmetry of the potential energy ensures that the angular momentum $\vec{L} = \vec{r} \times \vec{p}$ is a constant of the motion. (The angular coordinate for a rotation about any axis is ignorable.) Thus, $\vec{L}$ has a fixed direction in space as determined by the initial conditions, and both $\vec{r}$ and $\vec{p}$ must lie in the plane perpendicular to $\vec{L}$ so as to satisfy the cross product properties. Since any central field orbit lies in a plane, we can use plane-polar coordinates r and θ to describe the motion; the full machinery of spherical coordinates is not really needed. Therefore, the central field problem has been reduced from six degrees of freedom to only two.

The Lagrangian can now be written as

$$L = \tfrac{1}{2}m(\dot{r}^2 + r^2\dot{\theta}^2) - V(r),$$

since

$$\dot{\vec{r}}^2 = v^2 = \dot{r}^2 + r^2\dot{\theta}^2,$$

and the two equations of motion are consequently

$$\frac{d}{dt}\left(\frac{\partial L}{\partial \dot{r}}\right) - \frac{\partial L}{\partial r} = 0 = m\ddot{r} - mr\dot{\theta}^2 + \frac{\partial V}{\partial r}, \tag{4.5}$$

$$\frac{d}{dt}\left(\frac{\partial L}{\partial \dot{\theta}}\right) - \frac{\partial L}{\partial \theta} = 0 = \frac{d}{dt}(mr^2\dot{\theta}) = \dot{p}_\theta, \qquad p_\theta = \text{constant}. \tag{4.6}$$

The constant value of p_θ is simply $\mathscr{L}$, the magnitude of the angular momentum, a vector that is perpendicular to the orbit plane. Furthermore, it then follows that the quantity $\frac{1}{2}r^2\dot{\theta}$ is a constant also, and since $(\frac{1}{2}r)r\,d\theta = \frac{1}{2}r^2\,d\theta$ is the element of area dA in polar coordinates, we see that dA/dt has a fixed value during the motion. That the radius vector sweeps out equal areas in equal times has been known since 1609, even before Newton! This **law of areas** is the second of Kepler's planetary laws, and it is a general property of a central force field.

We now have

$$p_\theta = mr^2\dot{\theta} = \mathscr{L} \tag{4.7}$$

which upon substitution in eq. (4.5) produces

$$m\ddot{r} - mr\dot{\theta}^2 = m\ddot{r} - \frac{\mathscr{L}^2}{mr^3} = -\frac{\partial V}{\partial r} = F(r), \tag{4.8}$$

the fundamental radial equation of motion.

Since the central field is conservative, the total energy E is another constant of the motion. It is given by

$$E = T + V = \frac{1}{2}m(\dot{r}^2 + r^2\dot{\theta}^2) + V(r) = \frac{1}{2}m\dot{r}^2 + \frac{1}{2}\frac{\mathscr{L}^2}{mr^2} + V(r). \tag{4.9}$$

The complete specification of this plane motion requires that we determine four constants of integration, chosen in many situations as the initial values of position and velocity (r, θ, $\dot{r}$, and $\dot{\theta}$) at time zero. It is traditional here, however, to select instead the energy and angular momentum as two of these constants. The information given is really equivalent, for the initial values of $\dot{r}$ and $\dot{\theta}$ could easily be expressed in terms of E and $\mathscr{L}$. Furthermore, the transition to quantum mechanics (where energy and angular momentum are the significant quantities) is simplified by such a choice.

The formal solution of the equations of motion can now be completed by solving eq. (4.9) for $\dot{r}$,

$$\dot{r} \equiv \frac{dr}{dt} = \sqrt{\frac{2}{m}\left[E - V(r) - \frac{\mathscr{L}^2}{2mr^2}\right]},$$

The Equations of Motion for Central Forces

or upon solving for dt and integrating both sides,

$$t = \int_{r(t=0)}^{r(t)} \frac{dr}{\sqrt{\frac{2}{m}\left[E - V(r) - \frac{\mathscr{L}^2}{2mr^2}\right]}}. \tag{4.9'}$$

Once the potential is given, the right side of eq. (4.9') can be integrated *in principle* and solved for r as a function of time. From eq. (4.7), one then obtains

$$\dot{\theta} \equiv \frac{d\theta}{dt} = \frac{\mathscr{L}}{mr^2},$$

hence

$$\theta(t) - \theta_{(t=0)} = \frac{\mathscr{L}}{m}\int_0^t \frac{dt}{r^2(t)},$$

which may be integrated to yield $\theta(t)$. Elimination of the parameter t between $r(t)$ and $\theta(t)$ will finally provide the equation of the orbit.

The above procedure provides a complete solution to the equations of motion, but the integrals involved cannot always be done simply if indeed at all. Furthermore, it seems a roundabout way of arriving at the thing of greatest interest, which is usually the orbit equation for a given law of force. Fortunately, it turns out that there is a simpler, more direct way of deriving the orbit equation. If eq. (4.7) is rearranged somewhat, one obtains

$$\frac{d}{dt} = \frac{\mathscr{L}}{mr^2}\frac{d}{d\theta} \tag{4.7'}$$

as a differential operator relationship. Applying this identity twice produces

$$\frac{d^2}{dt^2} = \frac{\mathscr{L}}{mr^2}\frac{d}{d\theta}\left(\frac{\mathscr{L}}{mr^2}\frac{d}{d\theta}\right),$$

whereupon (4.8) becomes

$$\frac{\mathscr{L}}{r^2}\frac{d}{d\theta}\left(\frac{\mathscr{L}}{mr^2}\frac{dr}{d\theta}\right) - \frac{\mathscr{L}^2}{mr^3} = F(r).$$

Applying the substitution $\mu \equiv 1/r$, which is rather standard in such problems, we find that $d\mu/d\theta = -(1/r^2)(dr/d\theta)$, and

$$-\mathscr{L}\mu^2\frac{d}{d\theta}\left(\frac{\mathscr{L}}{m}\frac{d\mu}{d\theta}\right) - \left(\frac{\mathscr{L}^2}{m}\mu^3\right) = F\left(\frac{1}{\mu}\right),$$

or

$$\frac{\mathcal{L}^2\mu^2}{m}\left(\frac{d^2\mu}{d\theta^2} + \mu\right) = -F\left(\frac{1}{\mu}\right). \tag{4.10}$$

This is the basic differential equation for the orbit. Once the law of force is supplied, it can be solved for $\mu(\theta)$ and the reciprocal taken for $r(\theta)$. As an example, the important case of the inverse square force (such as the laws of gravity or electrostatics) is treated in the next section, after which we shall return to the general properties of central fields.

4.2 THE INVERSE SQUARE LAW OF FORCE

Consider the attractive inverse square force, of the form $F(r) = -k/r^2$, where k is a positive constant. Then, $F(1/\mu) = -k\mu^2$; eq. (4.10) gives

$$\frac{d^2\mu}{d\theta^2} + \mu = \frac{mk}{\mathcal{L}^2},$$

a relation familiar in form from elementary study of oscillations and having an obvious general solution as follows:

$$\mu = \frac{mk}{\mathcal{L}^2}[1 + \varepsilon \cos(\theta - \theta')],$$

or

$$r = \frac{(\mathcal{L}^2/mk)}{[1 + \varepsilon \cos(\theta - \theta')]}, \tag{4.11}$$

where ε and θ' are arbitrary constants of integration, the factor $mk/\mathcal{L}^2$ being taken in front of the μ solution for later convenience. The physical meaning of θ' is clear at once upon differentiation of r:

$$\frac{dr}{d\theta} = \frac{\varepsilon\mathcal{L}^2}{mk}\frac{\sin(\theta - \theta')}{[1 + \varepsilon\cos(\theta - \theta')]^2}.$$

When $\theta = \theta'$, $dr/d\theta = 0$ (provided $\varepsilon \neq -1$, which we will see is true), and the vanishing of this derivative marks the turning points on the orbit, called **apsidal points**, where r reaches its maximum or minimum values. Orienting the x axis to pass through such a turning point, we can choose $\theta' = 0$ with no significant loss of generality; the angle $\theta = \pi$ would then give another turning point if that angle

The Inverse Square Law of Force

is actually possible on the orbit. In fact, it is easy to show (by substitution of $-\theta$ for θ in the differential equation) that the orbit is always symmetrical about the apsidal points, whether the force is inverse square in nature or not. Next, we turn our attention to the remaining constant ε; it can be related simply to the energy. From the operator identity (4.7′), we have

$$\dot{r} = \frac{\mathscr{L}}{mr^2}\frac{dr}{d\theta} = \frac{\mathscr{L}}{m}\left(\frac{mk}{\mathscr{L}^2}\right)^2\left(\frac{\varepsilon\mathscr{L}^2}{mk}\right)\sin(\theta - \theta') = \frac{\varepsilon k}{\mathscr{L}}\sin(\theta - \theta'),$$

whereupon, from (4.9),

$$E = \frac{1}{2}m\dot{r}^2 + \frac{\mathscr{L}^2}{2mr^2} - \frac{k}{r}$$

$$= \frac{1}{2}m\left[\frac{\varepsilon k}{\mathscr{L}}\sin(\theta - \theta')\right]^2 + \frac{\mathscr{L}^2}{2m}\left\{\frac{mk}{\mathscr{L}^2}[1 + \varepsilon\cos(\theta - \theta')]\right\}^2$$

$$- k\left\{\frac{mk}{\mathscr{L}^2}[1 + \varepsilon\cos(\theta - \theta')]\right\}$$

$$= (\varepsilon^2 - 1)\frac{mk^2}{2\mathscr{L}^2},$$

and therefore

$$\varepsilon = \sqrt{1 + \frac{2\mathscr{L}^2 E}{mk^2}}. \tag{4.12}$$

The constant ε is then fully determined by E and $\mathscr{L}$, in other words, by the initial conditions for the motion.

In order to see the geometrical meaning of ε, called the **eccentricity** of the orbit, we must review the properties of conic sections in analytical geometry (Figure 4.1). A conic curve is defined as the locus of a point moving so that the ratio of its distance from a fixed point, the **focus**, to its distance from a fixed line, the **directrix**, is a constant ε. From Figure 4.1, we have

$$\varepsilon = \frac{r}{AB},$$

or

$$r = \varepsilon \cdot AB = \varepsilon \cdot CD = \varepsilon(OD - r\cos\theta),$$

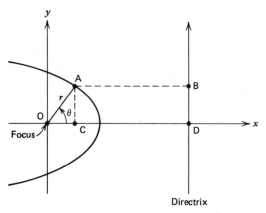

Figure 4.1 Geometry of a conic curve.

which upon solving for r produces

$$r(1 + \varepsilon \cos \theta) = \varepsilon \cdot OD = \text{constant } C,$$

and

$$r = \frac{C}{1 + \varepsilon \cos \theta},$$

a relation agreeing precisely with eq. (4.11) where $\theta' = 0$. Note that the focus is at the origin, the location of the force center.

The various types of conics are obtained with each in a particular range of the eccentricity ε, which by eq. (4.12) is directly related to the energy. Thus, we can write the following table from (4.12) and the properties of the eccentricity in analytical geometry:

Eccentricity	Energy E	Type of Curve
> 1	> 0	Hyperbola
1	0	Parabola
< 1 (but > 0)	< 0	Ellipse
0	$- mk^2/2\mathscr{L}^2$	Circle

The more familiar forms of the conic equations as expressed in rectangular coordinates can be related easily to the polar form (4.11), with appropriate values of ε. The basic point to be stressed here is that an inverse square central field orbit is always a conic curve, the particular type being determined by the energy.

It turns out that there is a sophisticated vectorial way of arriving at the orbit equation, (4.11), which makes use of the eccentricity in *vector* form. To see how this goes, consider the application of Newton's second law to the particle of reduced mass m. The equation of motion is

$$\frac{d\vec{p}}{dt} = -\frac{k}{r^2}\left(\frac{\vec{r}}{r}\right).$$

Taking the cross product of this equation with the constant angular momentum vector $\vec{L}$ gives

$$\vec{L}\times\dot{\vec{p}} = \frac{d}{dt}(\vec{L}\times\vec{p}) = -\frac{k}{r^3}(\vec{L}\times\vec{r}).$$

But

$$\frac{1}{r^3}(\vec{L}\times\vec{r}) = \left(\frac{m}{r^3}\right)(\vec{r}\times\dot{\vec{r}})\times\vec{r} = \left(\frac{m}{r^3}\right)[r^2\dot{\vec{r}} - (\vec{r}\cdot\dot{\vec{r}})\vec{r}] = m\frac{d}{dt}\left(\frac{\vec{r}}{r}\right),$$

using the definition of $\vec{L}$ and the fact that

$$\frac{dr}{dt} = \dot{x}\frac{\partial r}{\partial x} + \dot{y}\frac{\partial r}{\partial y} + \dot{z}\frac{\partial r}{\partial z} = \frac{\vec{r}\cdot\dot{\vec{r}}}{r}$$

to simplify the triple cross product expression. We then have

$$\frac{d}{dt}\left(\vec{L}\times\vec{p} + mk\frac{\vec{r}}{r}\right) = 0,$$

which means that the vector

$$\vec{\Sigma} \equiv \frac{\vec{L}\times\vec{p}}{mk} + \frac{\vec{r}}{r}$$

is a constant of the motion, called the **eccentricity** (or **Runge–Lenz**) vector.

In order to make contact with the prior theory of conic sections, note that the magnitude squared of $\vec{\Sigma}$ is given by

$$\Sigma^2 = \frac{1}{m^2k^2}(\vec{L}\times\vec{p})\cdot(\vec{L}\times\vec{p}) + 1 + \frac{2}{mkr}(\vec{L}\times\vec{p})\cdot\vec{r}$$

$$= \frac{1}{m^2k^2}[\mathscr{L}^2p^2 - (\vec{L}\cdot\vec{p})^2] + 1 + \frac{2}{mkr}[\vec{L}\cdot(\vec{p}\times\vec{r})]$$

$$= \frac{\mathscr{L}^2p^2}{m^2k^2} - \frac{2\mathscr{L}^2}{mkr} + 1 = \frac{2\mathscr{L}^2}{mk^2}\left(\frac{p^2}{2m} - \frac{k}{r}\right) + 1 = \frac{2\mathscr{L}^2E}{mk^2} + 1$$

($\vec{L}\cdot\vec{p} = 0$ since orbit plane is perpendicular to $\vec{L}$), which agrees precisely with eq.

(4.12), showing that the eccentricity vector has the magnitude required to make its name appropriate. Since $\vec{\Sigma}$ clearly lies in the plane of the orbit, let us designate the angle between $\vec{\Sigma}$ and $\vec{r}$ as $\pi - \theta$. Then, dotting $\vec{\Sigma}$ with $\vec{r}$,

$$\vec{\Sigma}\cdot\vec{r} = r\Sigma(-\cos\theta) = \frac{(\vec{L}\times\vec{p})\cdot\vec{r}}{mk} + r = \frac{\vec{L}\cdot(\vec{p}\times\vec{r})}{mk} + r = -\frac{\mathscr{L}^2}{mk} + r,$$

or

$$r = \frac{(\mathscr{L}^2/mk)}{1 + \Sigma\cos\theta},$$

which is eq. (4.11) with $\theta' = 0$. The direction of $\vec{\Sigma}$ must lie along the conic symmetry axis as a consequence, inasmuch as the assumed angle between $\vec{\Sigma}$ and $\vec{r}$ leads to this functional form in θ. It should be understood that nothing new has been obtained through this approach; it is simply an interesting alternative way of reaching the orbit equation.

In celestial mechanics, the orbit of greatest importance is the ellipse. The terminology and techniques used in that discipline differ somewhat from the usual physics approach, and we should like to discuss them briefly. Consider the ellipse shown in Figure 4.2, inscribed within a circle of radius a = RC. Through the point P at which the mass is instantaneously located on the orbit, construct a perpendicular RD to the major axis AB. The polar angle θ at the focus F is called in celestial mechanics the **true anomaly**, while the angle BCR, called Φ, is known as the **eccentric anomaly**; they change in value as the orbit is traversed. We see from the figure that

$$FC = CD - FD = a\cos\Phi - r\cos\theta,$$

and

$$FC = a - BF = a - r_{\min} = a - \frac{\mathscr{L}^2}{mk(1 + \varepsilon)}.$$

Here $r_{\min}$ is the smallest value of r possible on the orbit, while $r_{\max} = \mathscr{L}^2/[mk(1 - \varepsilon)]$ is the largest such value. Adding,

$$r_{\min} + r_{\max} = \frac{\mathscr{L}^2}{mk}\left[\frac{1}{1+\varepsilon} + \frac{1}{1-\varepsilon}\right] = \frac{2\mathscr{L}^2}{mk(1-\varepsilon^2)} = 2a,$$

hence,

$$\frac{\mathscr{L}^2}{mk} = a(1 - \varepsilon^2) \quad \text{and} \quad r_{\min} = a(1 - \varepsilon),$$

The Inverse Square Law of Force

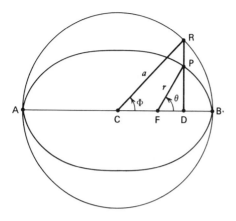

Figure 4.2 Ellipse and auxiliary circle.

and we find that

$$FC = a - a(1 - \varepsilon) = a\varepsilon = a\cos\Phi - r\cos\theta.$$

Substituting for $r\cos\theta$ from eq. (4.11), we have

$$FC = a\varepsilon = a\cos\Phi - \frac{(\mathscr{L}^2/mk) - r}{\varepsilon} = a\cos\Phi - \frac{a(1-\varepsilon^2)}{\varepsilon} + \frac{r}{\varepsilon}$$

or

$$r = a(1 - \varepsilon\cos\Phi). \qquad (4.13)$$

This equation relates the radial distance to the eccentric anomaly; the angular coordinate θ can be found from eq. (4.11) once Φ and thus r are determined, so we turn our attention to Φ itself.

In order to relate Φ to the time, we proceed to solve eq. (4.9) for $\dot{r}$ as before:

$$\dot{r} = \sqrt{\frac{2}{m}\left(E + \frac{k}{r} - \frac{\mathscr{L}^2}{2mr^2}\right)},$$

or

$$\frac{dr}{\sqrt{\frac{2}{m}\left(E + \frac{k}{r} - \frac{\mathscr{L}^2}{2mr^2}\right)}} = dt.$$

We wish to express the constant energy in terms of the orbital parameters a and ε. To do so, note that since the energy is constant, it may be evaluated at any point on the orbit, such as at $r = r_{min}$:

$$E = \frac{1}{2}mr_{min}^2 \dot{\theta}^2 - \frac{k}{r_{min}} = \frac{1}{2}\frac{\mathcal{L}^2}{mr_{min}^2} - \frac{k}{r_{min}}$$

$$= \frac{1}{2}\frac{\mathcal{L}^2}{m\left[\frac{\mathcal{L}^2}{mk(1+\varepsilon)}\right]^2} - \frac{k}{\left[\frac{\mathcal{L}^2}{mk(1+\varepsilon)}\right]} = \frac{mk^2}{\mathcal{L}^2}\left[\frac{(1+\varepsilon)^2}{2} - (1+\varepsilon)\right]$$

$$= \frac{mk^2}{2\mathcal{L}^2}(\varepsilon^2 - 1) = -\frac{k}{2a},$$

using the $\mathcal{L}^2$ relation obtained above. Upon substitution of this result into the dt expression,

$$dt = \frac{r\,dr}{\sqrt{\frac{2}{m}\left(-\frac{k}{2a}r^2 + kr - \frac{\mathcal{L}^2}{2m}\right)}} = \frac{r\,dr}{\sqrt{\frac{2}{m}\left[-\frac{k}{2a}r^2 + kr - \frac{ak}{2}(1-\varepsilon^2)\right]}}$$

$$= \frac{\sqrt{\frac{ma}{k}}\,r\,dr}{\sqrt{\varepsilon^2 a^2 - (r-a)^2}}.$$

Differentiating (4.13) and inserting here for $r\,dr$, we obtain after simplifying

$$dt = \sqrt{\frac{ma}{k}}\,a(1 - \varepsilon\cos\Phi)\,d\Phi.$$

To cast this relation into the form usually seen, we observe that Kepler's law of areas applied to the ellipse as a whole yields

$$\frac{dA}{dt} = \frac{1}{2}r^2\dot{\theta} = \frac{\mathcal{L}}{2m} = \frac{\pi ab}{T},$$

where T is the period of the motion and πab is the area of the ellipse. Making use of the relation $b = a\sqrt{1-\varepsilon^2}$, well known from analytical geometry, and solving for T,

$$T = \frac{2\pi ab m}{\mathcal{L}} = 2\pi a^2\sqrt{1-\varepsilon^2}\,\frac{m}{\mathcal{L}} = 2\pi\sqrt{\frac{ma^3}{k}}, \qquad (4.14)$$

where we have used $\mathcal{L}^2/mk = a(1-\varepsilon^2)$. Equation (4.14) is one form of Kepler's

The Inverse Square Law of Force

third (harmonic) law. It was established on the basis of observational data early in the seventeenth century. Later, Newton was able to provide a theoretical foundation for the law which depends for its validity on the inverse square law of gravitational force. Substituting (4.14) into the relation for dt produces

$$\frac{2\pi \, dt}{T} = (1 - \varepsilon \cos \Phi) \, d\Phi.$$

Integrating with the condition that $\Phi = 0$ at $t = 0$, one obtains

$$\frac{2\pi t}{T} = \Phi - \varepsilon \sin \Phi.$$

In time t, the radius vector would have moved through an angle $2\pi t/T$ if it were moving at a uniform angular velocity (which it isn't, of course). This angle is called the **mean anomaly** M in celestial mechanics:

$$M = \Phi - \varepsilon \sin \Phi. \tag{4.15}$$

Equation (4.15) is **Kepler's equation**, which is transcendental and can only be solved by approximation methods. (A pocket calculator could be useful in this regard.) In order to use eq. (4.15), one selects a time t, determines M, solves the equation for Φ as accurately as necessary, and finally uses the earlier relations (4.13) and (4.11) to obtain r and θ.

As an interesting practical application of the inverse square theory, consider spacecraft missions to the outermost planets. The energy requirements would be far beyond the capabilities of modern rocket technology if it were not possible to utilize gravity in a judicious fashion along the way.

In order for a spacecraft of mass m to reach the outermost fringes of the solar system, it must leave the Earth (a distance R_E from the mass M_S of the sun) with a velocity of escape v_{es}, which can be found from

$$E = 0 = \tfrac{1}{2} m v_{es}^2 - \frac{G m M_S}{R_E},$$

v_{es} being the minimum required so that the spacecraft will have practically all its kinetic energy converted to potential by the time it gets far away from the sun (Figure 4.3). Solving, we find

$$v_{es} = \sqrt{\frac{2 G M_S}{R_E}} = 42 \text{ km/sec}.$$

But the Earth's orbital velocity in its almost circular orbit around the sun, or

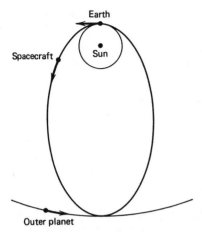

Figure 4.3 Spacecraft launched from moving Earth.

$v_E = 2\pi R_E/\text{year}$, is determined upon equating centripetal and gravitational force expressions:

$$\frac{GmM_S}{R_E^2} = mv_E^2/R_E,$$

giving

$$v_E = \sqrt{\frac{GM_S}{R_E}} = \frac{1}{\sqrt{2}} v_{es} = 30 \text{ km/sec}.$$

The fact that v_E is of the same order as v_{es} means that the required initial velocity can be reduced to only 12 km/sec ($= 42 - 30$) if we launch in the direction of Earth's orbital motion. We assume additional energy is available for escape from *Earth's* gravity, and that the launch is timed so that the spacecraft and planet will meet at aphelion on the spacecraft orbit. In effect, the vehicle has been inserted into an elliptical orbit much like a typical planet and is moving under the sun's gravity toward its ultimate rendezvous. The orbital parameters are readily calculated from the known radial distances at aphelion and perihelion.

The time of flight for the mission described above is on the order of several years, but it can be shortened dramatically (with no greater initial velocity requirements) by means of an energy boost en route from the planet Jupiter. Suppose the spacecraft is launched so as to make a close passage by Jupiter, being deflected in the process into a new elliptical orbit and acquiring significantly greater kinetic energy, so as to hasten its progress toward one of the outer planets (Figure 4.4).

The Inverse Square Law of Force

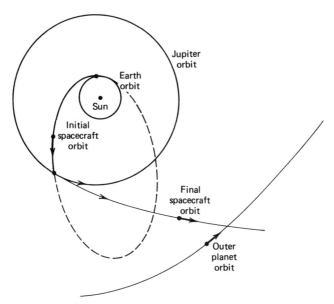

Figure 4.4 Orbits illustrating Jupiter swingby effect.

Let $\vec{p}$ and $\vec{p}'$ represent the momenta of the spacecraft before and after the Jupiter passage; these are measured in a sun-centered "inertial" frame in which Jupiter is moving with velocity $\vec{v}_J$. Let $\vec{p}_J$ and $\vec{p}'_J$ represent these same momenta in a frame in which Jupiter is at rest. Making a Galilean transformation, we know that

$$\vec{p} = \vec{p}_J + m\vec{v}_J, \qquad \vec{p}' = \vec{p}'_J + m\vec{v}_J.$$

Then, upon subtraction, the change of momentum in the "collision" with Jupiter is the same in either frame:

$$\Delta \vec{p} \equiv \vec{p}' - \vec{p} = \vec{p}'_J - \vec{p}_J \equiv \Delta \vec{p}_J.$$

But the change of kinetic energy is

$$\Delta T = \frac{p'^2}{2m} - \frac{p^2}{2m} = \frac{(\vec{p}'_J + m\vec{v}_J)^2}{2m} - \frac{(\vec{p}_J + m\vec{v}_J)^2}{2m}$$

$$= \left(\frac{p'^2_J}{2m} - \frac{p^2_J}{2m}\right) + \vec{v}_J \cdot (\vec{p}'_J - \vec{p}_J)$$

$$= \Delta T_J + \vec{v}_J \cdot \Delta \vec{p}_J.$$

Since the collision is elastic in the Jupiter frame, $\Delta T_J = 0$, leaving $\Delta T = \vec{v}_J \cdot \Delta \vec{p}$

as the change in the inertial frame. Thus, the kinetic energy gained by the spacecraft at the expense of Jupiter is $\Delta T = \vec{v}_J \cdot (\vec{p}' - \vec{p})$; since the initial orbit fixes the value of $\vec{v}_J \cdot \vec{p}$, ΔT will be maximized when $\vec{p}'$ is parallel to $\vec{v}_J$. A further detailed analysis of the kinematics of this situation (requiring nothing other than the principles outlined in this chapter) reveals that the final spacecraft velocity can be almost twice its initial velocity, as viewed in the inertial frame. This type of assistance from gravity, sometimes called the "slingshot effect", should be of tremendous value in future exploration of the solar system.

4.3 THE EQUIVALENT ONE-DIMENSIONAL MODEL AND THE VIRIAL THEOREM

There are many interesting and important properties of central fields which can be determined without the necessity of solving the equations of motion. Often, a general qualitative description of the orbit is all that is desired, and this can be obtained quite simply through the equivalent one-dimensional model which we shall now discuss.

The radial equation of motion, eq. (4.8), can be written

$$m\ddot{r} = -\frac{\partial V(r)}{\partial r} + \frac{\mathscr{L}^2}{mr^3}.$$

In view of the fact that $\mathscr{L}$ is a known constant, the θ dependence is entirely absent from this equation; it is, in effect, a one-dimensional relation for a particle whose position is specified by the single coordinate r. Indirectly, the angular motion makes its presence felt through the term

$$\frac{\mathscr{L}^2}{mr^3} = \frac{(mr^2\dot\theta)^2}{mr^3} = mr\dot\theta^2 = \frac{mv_\theta^2}{r}, \qquad v_\theta \equiv r\dot\theta,$$

which is the usual centrifugal force expression and corresponds to a potential $\mathscr{L}^2/2mr^2$. If we define an equivalent one-dimensional potential $V_e(r)$ by superposing regular and centrifugal potentials,

$$V_e(r) \equiv V(r) + \frac{\mathscr{L}^2}{2mr^2}, \qquad (4.16)$$

then the equation of motion becomes

$$m\ddot{r} = -\frac{\partial V_e}{\partial r}.$$

Furthermore, from eq. (4.9) we can write

$$\tfrac{1}{2}m\dot{r}^2 = E - V(r) - \frac{\mathscr{L}^2}{2mr^2} = E - V_e(r). \tag{4.17}$$

That is, the difference between the constant E and the effective potential V_e (at a given value of r) directly measures the radial part of the kinetic energy.

These results are most easily grasped by a graphical approach; we shall examine the inverse square law once again, our understanding of the present method being facilitated by comparison with the known solutions in that case (Figure 4.5). First, plot the effective potential

$$V_e = -\frac{k}{r} + \frac{\mathscr{L}^2}{2mr^2}$$

as a function of r. Then, select some fixed value of energy, represented by a horizontal dashed line (several typical values are depicted in the figure). Motion is only possible for those values of r where the energy line lies *above* (or on) the V_e curve, which ensures by eq. (4.17) that the radial kinetic energy is positive (or zero). A typical positive energy is denoted by E_+ in Figure 4.5; $E_+ \geqslant V_e$ for all $r \geqslant r_+$. The particle can move out to infinity with this great an energy, but can move in no closer to the force center than r_+; as we know from the previous section, the orbit is hyperbolic. The angular part of the kinetic energy is $\mathscr{L}^2/2mr^2$, which is just the difference $V_e - V$; this quantity rises to its maximum at the turning point r_+. The description of motion in the parabolic orbit with $E = 0$ is similar.

A typical negative energy is represented by E_-. Here, there are *two* turning points, at r_1 and r_2, where E_- and V_e are equal. Thus, the orbit is bounded and

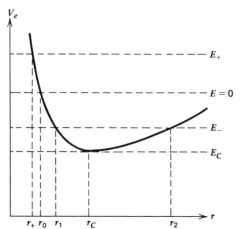

Figure 4.5 Equivalent one-dimensional potential for inverse square law.

(from the previous section) elliptical. From graphical analysis alone, we cannot say that the orbit is *closed*, so as to repeat itself indefinitely, but only that r is bounded. An ellipse slowly precessing, for example, is a case of boundedness with no closure.

Finally, E_C represents the case of the lowest possible energy for which an orbit is possible. Since $E_C = V_e$ at $r = r_C$, $\dot{r} = 0$ by eq. (4.17) and the radius is fixed, so we have the case of circular motion.

The qualitative analysis of the orbit proceeds similarly for other laws of force. With a little practice, the student can quickly get a feeling for the possible orbits from a rough plot of the equivalent potential.

The virial theorem is another aid to understanding; it is used to relate the time-averaged kinetic and potential energies of a physical system. Consider that particular function M of the position vectors of a system of N particles which is defined by

$$M \equiv \tfrac{1}{2} \sum_j m_j \vec{r}_j \cdot \vec{r}_j.$$

Upon differentiating twice with respect to time, we obtain

$$\ddot{M} = \sum_j m_j \dot{\vec{r}}_j \cdot \dot{\vec{r}}_j + \sum_j m_j \vec{r}_j \cdot \ddot{\vec{r}}_j.$$

To find the time average of a quantity such as $\ddot{M}$ over a long period τ (which will be denoted by a bar placed over the quantity), we form the expression

$$\overline{\ddot{M}} \equiv \frac{1}{\tau} \int_0^\tau \ddot{M}\, dt = \frac{1}{\tau}\left[\dot{M}(\tau) - \dot{M}(0)\right].$$

But as τ becomes large, it is necessary for $\overline{\ddot{M}}$ to vanish, because

$$\dot{M} = \sum_j m_j \vec{r}_j \cdot \dot{\vec{r}}_j$$

and both $\vec{r}_j$ and $\dot{\vec{r}}_j$ are bounded, therefore so is $\dot{M}$. (At least $\vec{r}_j$ is bounded if we assume some finite volume for the system such as containment within a vessel, and $\dot{\vec{r}}_j$ is bounded because particles in a system with finite energy cannot attain infinite velocities.) In fact, $\ddot{M}$ will vanish even for "short" times τ if the motion of the system is periodic and τ is some integral multiple of the period, since $\dot{M}(\tau)$ will then return to the value $\dot{M}(0)$. But the general expression for $\ddot{M}$, with $\overline{\ddot{M}} = 0$, implies that

$$\overline{\sum_j m_j \dot{\vec{r}}_j \cdot \dot{\vec{r}}_j} = -\overline{\sum_j m_j \vec{r}_j \cdot \ddot{\vec{r}}_j}$$

Multiplying by $\frac{1}{2}$ on each side, we get the time average of the kinetic energy on the left, while on the right $m_j \ddot{\vec{r}}_j$ can be replaced by the force $\vec{F}_j$ on the jth particle to yield

$$\overline{T} = -\tfrac{1}{2}\overline{\sum_j \vec{r}_j \cdot \vec{F}_j}. \tag{4.18}$$

This result is known as the **virial theorem**, the quantity on the right having been called the **virial** by Clausius.

A common application of the theorem relates the average kinetic and potential energies for a single particle in a central force field. Taking the force in power law form as

$$\vec{F} = kr^n\left(\frac{\vec{r}}{r}\right),$$

the potential will be

$$V(r) = -\left(\frac{k}{n+1}\right)r^{n+1},$$

giving from eq. (4.18)

$$\overline{T} = -\tfrac{1}{2}\overline{(kr^{n+1})} = \left(\frac{n+1}{2}\right)\overline{V},$$

which for the inverse square law becomes the famous relation

$$\overline{T} = -\tfrac{1}{2}\overline{V} \qquad (n = -2).$$

Classically, the virial theorem is used mainly in the kinetic theory of gases and in the study of central fields. In atomic physics, a quantum virial theorem can be established also, and it is useful for investigating the properties of wave functions. At the other end of the size spectrum, the theorem has even had interesting applications of late in astronomy. There, the amount of mass known to be present in the galaxies is too small to satisfy the average potential energy requirements of the theorem, leading astronomers to suspect that there is much more matter in existence than is actually being observed—the celebrated problem of the "missing mass."

One might also consider a satellite in a circular orbit, with energy

$$E = T + V = \overline{T} + \overline{V} = \tfrac{1}{2}\overline{V} = -\overline{T},$$

using

$$\overline{T} = -\tfrac{1}{2}\overline{V}$$

(since $T = \bar{T}$, $V = \bar{V}$; these quantities are fixed for a stable circular orbit). But if the frictional contact with the atmosphere slowly converts mechanical energy into heat, E must decrease; the fact that $E = \frac{1}{2}\bar{V} = -\frac{1}{2}(k/r)$ means that r decreases also, while $E = -\bar{T}$ makes it necessary for T to *increase*. Therefore, the satellite speeds up as a result of the drag on it, a rather unusual situation!

Incidentally, the power law form of the potential, $V(r) \sim r^{n+1}$, has been treated quite thoroughly in the literature. The inverse square case has already been examined in detail. Later in the chapter or in the problems, orbits for linear ($n = +1$) and inverse cube ($n = -3$) laws will be examined, along with some other special cases which can be worked out in terms of simple trigonometric functions. By means of elliptic integrals, a few other power laws with small integral or fractional exponents can also be handled, but these are of less interest because of lack of familiarity with the elliptic functions and lack of applications. It has not been possible to obtain a general orbit solution for arbitrary n.

Note also that the *same* force law often leads to orbits which are at first sight quite different in nature. We have seen how the inverse square law can give hyperbolic, parabolic, and elliptical orbits, superficially distinct from one another yet in reality all of the same polar form (differing only in eccentricity). Similar situations can arise with other force laws. Consider, for example, the spiral orbit $r = 1/\theta$; then

$$\mu \equiv \frac{1}{r} = \theta, \qquad \frac{d^2\mu}{d\theta^2} = 0,$$

and eq. (4.10) gives

$$F\left(\frac{1}{\mu}\right) = -\frac{\mathscr{L}^2\mu^2}{m}(0 + \mu) = -\frac{(\mathscr{L}^2/m)}{r^3}.$$

The spiral takes place in an attractive inverse cube field of force. That is, it is a *possible* orbit in such a field. To find the most general type of orbit for that law, apply eq. (4.10) once more with $F(r) = -k/r^3$ given:

$$-\frac{k}{r^3} = -k\mu^3 = -\frac{\mathscr{L}^2\mu^2}{m}\left(\frac{d^2\mu}{d\theta^2} + \mu\right),$$

or

$$\frac{d^2\mu}{d\theta^2} + \left(1 - \frac{mk}{\mathscr{L}^2}\right)\mu = 0,$$

which has the general solution

$$\mu = \frac{1}{r} = A \sin\left(\sqrt{1 - \frac{mk}{\mathscr{L}^2}}\,\theta\right) + B \cos\left(\sqrt{1 - \frac{mk}{\mathscr{L}^2}}\,\theta\right), \qquad (4.19)$$

where A and B are integration constants. Clearly, the spiral is a special solution of this differential equation obtained by setting $mk = \mathscr{L}^2$, but it bears little relationship to the circular function solutions though it results from the same basic force law.

There is one other very interesting point to be made in connection with power laws of force in a central field. Given arbitrary values of angular momentum and energy (the latter negative for an attractive case), Bertrand[1] was able to prove in 1873 that the only possible force laws which can give *closed orbits* are the inverse square law and the direct first power law (as for a radial spring), a result now known as **Bertrand's theorem**. Other force laws may give closed orbits under certain special conditions, but not for any arbitrary values of energy and angular momentum that might be selected. The proof of the theorem is straightforward but algebraically tedious, so we shall avoid it; we do want to deduce a corollary, however. As motion proceeds in a central field orbit, the particle moves in and out radially as θ increases; the fact that the orbit is closed means that the periods of the two motions (in r and θ) are equal or perhaps rational multiples of one another. Most bound celestial bodies move in essentially closed orbits as just described, and this fact in conjunction with Bertrand's theorem suggests that the law of gravity must be inverse square, regardless of the detailed shape of the orbit.

4.4 SCATTERING AND CROSS SECTIONS

One very important part of central force theory is the scattering of a beam of particles from a center of force. The general nature of the problem can be investigated by classical methods, which in many cases will lead to conclusions that are roughly accurate, even though a quantum mechanical treatment is really required.

Consider a beam of many identical particles moving along a common direction toward a fixed force center; all such particles will have the same energy initially and will differ only in the perpendicular distance between the initial direction of a particle's velocity and the line parallel to this velocity through the origin, or force center (see Figure 4.6). This distance is called the **impact parameter b**, and it is clearly proportional to the angular momentum. Assuming that the force decreases

---

[1] J. Bertrand, *Comptes Rendus* **77**, 849 (1873).

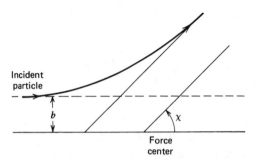

Figure 4.6 Impact parameter and scattering angle.

to zero at infinity, a particle initially moving in a particular direction will deviate more and more from this direction as it approaches closer to the force center and "feels" the effect of the force. Later, as it moves far out once more, it again will approach a straight-line trajectory, but in the process of scattering it will have been deflected through some angle χ.

It is clear that the *smaller* the impact parameter, the *greater* will be the effect of the force and hence the scattering angle. If we call the initial magnitude of velocity v_0 for the beam, then

$$E = \tfrac{1}{2}mv_0^2, \quad \text{or} \quad v_0 = \sqrt{\frac{2E}{m}},$$

and

$$\mathscr{L} = mv_0 b = \sqrt{2mE}\, b$$

in terms of the energy E. Those particles falling in the range of impact parameters between b and $(b + db)$ will be scattered through angles between χ and $(\chi + d\chi)$, or into the solid angle $d\Omega$ about the direction of χ. The **differential scattering cross section** $\sigma(\Omega)$ is therefore defined by

$$\sigma(\Omega)\, d\Omega \equiv \frac{\text{no. of particles scattered into } d\Omega \text{ per unit time}}{\text{incident beam intensity}}, \quad (4.20)$$

where the beam intensity $I \equiv$ no. of particle crossing a unit area perpendicular to the beam velocity in unit time. (Because of the area factor in the definition of I, $\sigma(\Omega)$ does have the dimensions of area.) The number of particles per unit time crossing a ring of area $2\pi b\, db$ (about the central axis and with plane perpendicular to the beam) is $(2\pi b\, db)I$; these particles will all be scattered into $d\Omega$, the number by eq. (4.20) being

$$I\sigma(\Omega)\, d\Omega = I\sigma(\Omega) 2\pi \sin\chi\, d\chi.$$

Equating the numbers,
$$I\sigma(\Omega)2\pi \sin \chi \, d\chi = -2\pi b \, db \, I,$$
or
$$\sigma(\Omega) = -\frac{b}{\sin \chi}\frac{db}{d\chi}, \tag{4.21}$$

where the minus sign is required because χ increases as b decreases. To use eq. (4.21), however, we must still relate b to the angle χ, and it is this relation which varies from one force law to another. Once the differential cross section has been calculated from (4.21), where it is expressed as a function of χ, the total cross section can be found by integration over all angles, for its definition is

$$\sigma_{\text{total}} \equiv \int_{\text{(all }\Omega)}\sigma(\Omega) \, d\Omega = 2\pi\int_0^\pi \sigma(\chi)\sin\chi \, d\chi. \tag{4.22}$$

The total cross section, however, is not always a useful concept, for sometimes it diverges. The Coulomb repulsion of like charges is an example discussed in many textbooks; it is included in the problem set and leads to the famous Rutherford scattering formula. The Coulomb $1/r$ potential is said to be a "long-range" potential, in that it can scatter particles to some extent even at very large distances, thereby causing σ_{total} to be infinite. Since the total cross section exists in classical mechanics only when the force field cuts off sharply with distance, it is usually the differential cross section that is of interest.

One further point deserves some attention. The scattering angle that is measured gives the deflection of the incident beam particles, but the angle χ discussed here is the angle between the final and initial directions of the relative vector between the beam and target particles. Because of the recoil of the target particle, these angles are not the same, so χ is not what is actually observed. But in the CM system, as seen in the collision analysis of Chapter 1, the particles maintain equal but opposite momenta so that *either* particle scatters through the same angle (the angle χ for the deflection of the relative vector). The transformations of Chapter 1 then allow us to calculate the laboratory scattering angle from χ in the CM system; nothing further need be said here (see Problem 1.6).

Example 1 Scattering in an Inverse Cube Repulsive Force

In the investigation of an inverse cube repulsive force, we can take over the result (4.19) by merely changing the sign of the force constant k, giving

$$\frac{1}{r} = A \sin\left(\sqrt{1 + \frac{mk}{\mathcal{L}^2}}\,\theta\right) + B \cos\left(\sqrt{1 + \frac{mk}{\mathcal{L}^2}}\,\theta\right).$$

It will be more convenient, however, if we write this equation in the alternative form

$$\frac{1}{r} = C \sin\left[\sqrt{1 + \frac{mk}{\mathscr{L}^2}}\,(\theta - \theta')\right],$$

where the constants C and θ' could easily be related to A and B if desired; in this form we need only work with a single trigonometric function. Letting

$$\alpha \equiv \sqrt{1 + \frac{mk}{\mathscr{L}^2}}\,(\theta - \theta'),$$

we have $1/r = C \sin \alpha$. Clearly, $r_{\min}$ occurs for $\alpha = \pi/2$, while $r = \infty$ arises for $\alpha = 0$ or π. Consulting Figure 4.7, we see that

$$\chi = \pi - 2[\theta(\alpha = \pi/2) - \theta(\alpha = 0)] = \pi - 2\left(\frac{\pi}{2\sqrt{1 + \frac{mk}{\mathscr{L}^2}}} + \theta' - \theta'\right)$$

$$= \pi\left(1 - \frac{\mathscr{L}}{\sqrt{\mathscr{L}^2 + mk}}\right) = \pi\left(1 - \frac{\sqrt{2mE}\,b}{\sqrt{2mEb^2 + mk}}\right)$$

$$= \pi\left(1 - \sqrt{\frac{2Eb^2}{2Eb^2 + k}}\right),$$

where, by using $\mathscr{L} = \sqrt{2mE}\,b$, we have obtained an expression for χ in terms of the impact parameter b and some appropriate constants. Upon differentiating,

$$\frac{d\chi}{db} = -\frac{\sqrt{2E}\,\pi k}{(2Eb^2 + k)^{3/2}},$$

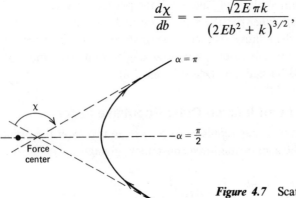

Figure 4.7 Scattering by inverse cube repulsion.

whereupon by eq. (4.21) one obtains

$$\sigma(\Omega) = -\frac{b}{\sin \chi}\frac{1}{(d\chi/db)} = +\frac{b}{\sin \chi}\frac{(2Eb^2 + k)^{3/2}}{\pi k\sqrt{2E}}.$$

Finally, the χ relation above can be solved for b in terms of χ so that the cross section can be expressed in terms of χ alone. Performing this relatively simple algebra yields

$$b = \sqrt{\frac{k}{2E\chi(2\pi - \chi)}}(\pi - \chi)$$

and, therefore,

$$\sigma(\Omega) = \frac{\pi^2 k}{2E}\frac{(\pi - \chi)}{\chi^2 \sin \chi(2\pi - \chi)^2}$$

upon substitution of the expression for b and algebraic simplification. This is the desired expression for the differential cross section. Note that it has a dependence on the energy.

4.5 THE MANY-BODY PROBLEM

In the previous section, we have considered the interaction of two bodies through a central force, effectively reducing the problem to that of a single particle by means of the reduced mass. Now the two-body problems have closed mathematical solutions, as has been shown; that is, given their initial values, the kinematical quantities can be predicted with accuracy at any later time. (It is fortunate that many common problems can be treated adequately with two-body analysis or with perturbation theory. The sun and planets, for example, constitute a many-body system in principle; but in practice, the sun dominates this system with its large mass, so that the motions of a particular planet are not greatly affected by the other planets, and two-body analysis often suffices.) With N bodies, where $N > 2$, the situation is not so simple; in general, such closed solutions cannot be found. We wish now to examine some of the main characteristics of the many-body problem.

Consider a system of N particles, the ith one as usual having mass m_i and position vector $\vec{r}_i$ relative to a Newtonian frame. The only forces acting are the mutual gravitational attractions. Let $\vec{r}_{ij}$ be the vector from the ith to the jth particle; the equations of motion become

$$m_i \ddot{\vec{r}}_i = G\sum_j \frac{m_i m_j}{r_{ij}^3}\vec{r}_{ij}.$$

The subscripts i and j run over all N particles, with the natural restriction that $i \neq j$. In view of the fact that $\vec{r}_{ij} = -\vec{r}_{ji}$, $\sum_i m_i \ddot{\vec{r}}_i = 0$ because of the cancellation of pairs of terms on the right side. Integration twice then produces

$$\sum_i m_i \vec{r}_i = \vec{C}_1 t + \vec{C}_2,$$

where $\vec{C}_1$ and $\vec{C}_2$ are vector constants of integration. (Alternatively, one could integrate $\sum_i m_i \ddot{x}_i = 0$ and get two scalar constants, repeating the process for the y- and z-components. The vector integration yields the six constants simultaneously.) Equation (1.8) specifies the position vector $\vec{R}$ of the CM as

$$\vec{R} = \frac{1}{M} \sum_i m_i \vec{r}_i = \frac{(\vec{C}_1 t + \vec{C}_2)}{M} \qquad \left(M = \sum_i m_i \right). \qquad (4.23)$$

As expected from the theorems of Chapter 1, the CM must either be at rest or moving uniformly, since $M\dot{\vec{R}} = \vec{C}_1$.

We also know from Chapter 1, eq. (1.11), that the angular momentum $\vec{L} = L_x \hat{i} + L_y \hat{j} + L_z \hat{k}$ is a constant of the motion. This means that each component, such as

$$\sum_i m_i (y_i \dot{z}_i - z_i \dot{y}_i) = L_x,$$

is a constant. With two bodies moving in a plane orbit, the radius vector sweeps out area at a constant rate $dA/dt = \frac{1}{2} r^2 \dot{\theta}$. Assigning a direction (normal to the plane) to the area, we can write the law of areas in vector form for the equivalent one-body problem as

$$\frac{d\vec{A}}{dt} = \tfrac{1}{2} \vec{r} \times \vec{v},$$

the x-component then being $dA_x/dt = \frac{1}{2}(y\dot{z} - z\dot{y})$. Upon substitution and generalization to N bodies, $L_x = 2\sum_i m_i \dot{A}_{x_i}$, so integration gives

$$\sum_i m_i A_{x_i} = \frac{L_x}{2} t + C_x \qquad (C_x = \text{constant}), \qquad (4.24)$$

similarly for the y- and z-components. The interpretation of this result is straightforward: A_{x_i} is the projection on the yz-plane of the area swept out by the

The Many-Body Problem

radius vector $\vec{r}_i$. The sum of the products of mass times projected area will vary linearly with the time; this is the generalization of the law of areas.

Next, take the dot product of the velocity $\dot{\vec{r}}_i$ with the force vector for the ith particle:

$$\dot{\vec{r}}_i \cdot (m_i \ddot{\vec{r}}_i) = G \sum_j \frac{m_i m_j}{r_{ij}^3} (\dot{\vec{r}}_i \cdot \vec{r}_{ij}).$$

But we can write

$$\vec{r}_{ij} = \vec{r}_j - \vec{r}_i = r_{ij} \hat{\epsilon}_{ij}$$

and

$$\dot{\vec{r}}_{ij} = \dot{r}_{ij} \hat{\epsilon}_{ij} + r_{ij} \dot{\hat{\epsilon}}_{ij} = \dot{\vec{r}}_j - \dot{\vec{r}}_i,$$

where $\hat{\epsilon}_{ij}$ is a unit vector along the direction of $\vec{r}_{ij}$. For any unit vector $\hat{\epsilon}$,

$$\frac{d}{dt}(\hat{\epsilon} \cdot \hat{\epsilon}) = 2\hat{\epsilon} \cdot \dot{\hat{\epsilon}} = \frac{d}{dt}(1) = 0,$$

hence $\hat{\epsilon}$ is perpendicular to $\dot{\hat{\epsilon}}$ unless the latter is zero. Summing the dot product on i and making use of these vector relations in the lines to follow, one finds

$$P \equiv \sum_i m_i \dot{\vec{r}}_i \cdot \ddot{\vec{r}}_i = G \sum_{i,j} \frac{m_i m_j}{r_{ij}^2} \dot{\vec{r}}_i \cdot \hat{\epsilon}_{ij}.$$

(Note that for each pair of numbers i and j, there is a pair of terms of the form $\dot{\vec{r}}_i \cdot \hat{\epsilon}_{ij}$ and $\dot{\vec{r}}_j \cdot \hat{\epsilon}_{ji}$.) We obtain

$$P = G \sum_{i<j} \frac{m_i m_j}{r_{ij}^2} (\dot{\vec{r}}_i - \dot{\vec{r}}_j) \cdot \hat{\epsilon}_{ij} = -G \sum_{i<j} \frac{m_i m_j}{r_{ij}^2} (\dot{r}_{ij} \hat{\epsilon}_{ij} + r_{ij} \dot{\hat{\epsilon}}_{ij}) \cdot \hat{\epsilon}_{ij}$$

$$= -G \sum_{i<j} \frac{m_i m_j}{r_{ij}^2} \dot{r}_{ij},$$

which may be written

$$\frac{d}{dt}\left[\tfrac{1}{2} \sum_i m_i \dot{\vec{r}}_i^{\,2} - G \sum_{i<j} \frac{m_i m_j}{r_{ij}}\right] = \frac{d}{dt}(T + V) = 0,$$

or

$$T + V = E = \text{constant}. \qquad (4.25)$$

In other words, conservation of energy holds for the system of particles.

The relations (4.23), (4.24), and (4.25) produce a total of 10 constants of integration (one from energy, three from angular momentum, and six from the CM). Early in this chapter, we found that after separation of the CM motion, any two-body central field problem was equivalent to a one-body problem in a plane. The energy and angular momentum alone were then sufficient for the determination of the orbit equation, in which two further constants denoting initial radial and angular positions appeared as parameters; a total of four constants was then required for a single body with two degrees of freedom.

In general, however, a total of $6N$ quantities are involved in the specification of the motion of N particles in three dimensions (position and momentum components for each particle), and we have found only 10 constants of the motion by integration. Almost a century ago, both Bruns and Poincaré were separately able to show that no integrals of the motion could be obtained in algebraic form other than these 10, or some combination of them. It is still possible that another type of solution, such as one in transcendental form, may yet be found, but at the present time no neat analytical solution to the N-body equations of motion is available for $N > 2$. The number of known constants of the motion is simply not sufficient for the elimination of variables needed to put the equations into a form which can be managed.

A computerized solution remains as a distinct possibility. In addition, certain special cases have been examined in some degree of approximation. An example is the **restricted three-body problem**[2], a situation (such as a spacecraft in Earth–moon orbit) wherein one mass is much smaller that the other two; we shall discuss it later in this chapter. Next, we investigate some exact special solutions first obtained by Lagrange.

In the three-body problem, Lagrange in 1772 found two solutions which are called **stationary**. By this term it is meant that the geometrical configuration of the three masses does not change in time. The masses might be located in a plane, for example, and as a whole group undergo rotation in the plane about the CM, the relative distances of the masses from one another being unchanged. Suppose all three bodies are rotating in circular orbits about the CM at a constant angular velocity ω. We want to find the conditions that must be satisfied in order to have a solution of this kind for the equations of motion.

---

[2] See A. Wintner, *The Analytical Foundations of Celestial Mechanics*, Princeton Univ. Press, Princeton, N.J., 1947; or J. Bartlett, *Classical and Modern Mechanics*, Univ. of Alabama Press, University, Alabama, 1975.

The Many-Body Problem

The position vectors relative to the CM are $\vec{r}_1 = r_1\hat{\epsilon}_1$, $\vec{r}_2 = r_2\hat{\epsilon}_2$, and $\vec{r}_3 = r_3\hat{\epsilon}_3$, where the $\vec{r}_i$'s are constant for the circular orbits and the $\hat{\epsilon}_i$'s are unit vectors. The centripetal acceleration for the ith mass is $\ddot{\vec{r}}_i = -\omega^2 r_i \hat{\epsilon}_i$, so Newton's law becomes

$$m_i \ddot{\vec{r}}_i = -\omega^2 m_i r_i \hat{\epsilon}_i = G\sum_j \frac{m_i m_j}{r_{ij}^3} \vec{r}_{ij}$$

$$= G\sum_j \frac{m_i m_j}{r_{ij}^3}(\vec{r}_j - \vec{r}_i) = G\sum_j \frac{m_i m_j}{r_{ij}^3}(r_j \hat{\epsilon}_j - r_i \hat{\epsilon}_i).$$

Writing out these relations explicitly (recall that $i \neq j$), we have

$$-\omega^2 r_1 \hat{\epsilon}_1 = G\left[\frac{m_2}{r_{12}^3}(r_2 \hat{\epsilon}_2 - r_1 \hat{\epsilon}_1) + \frac{m_3}{r_{13}^3}(r_3 \hat{\epsilon}_3 - r_1 \hat{\epsilon}_1)\right],$$

$$-\omega^2 r_2 \hat{\epsilon}_2 = G\left[\frac{m_1}{r_{12}^3}(r_1 \hat{\epsilon}_1 - r_2 \hat{\epsilon}_2) + \frac{m_3}{r_{23}^3}(r_3 \hat{\epsilon}_3 - r_2 \hat{\epsilon}_2)\right],$$

$$-\omega^2 r_3 \hat{\epsilon}_3 = G\left[\frac{m_1}{r_{13}^3}(r_1 \hat{\epsilon}_1 - r_3 \hat{\epsilon}_3) + \frac{m_2}{r_{23}^3}(r_2 \hat{\epsilon}_2 - r_3 \hat{\epsilon}_3)\right].$$

Since the vectors $\vec{r}_i$ are defined relative to the CM, we also have

$$m_1 r_1 \hat{\epsilon}_1 + m_2 r_2 \hat{\epsilon}_2 + m_3 r_3 \hat{\epsilon}_3 = 0.$$

This is not a fourth independent relationship but can be obtained by adding the three equations of motion after multiplying each in turn by one of the masses. Its advantage is that it may be used instead of, say, the third equation of motion to simplify the algebra a bit.

Now, the unit vectors $\hat{\epsilon}_i$ are rotating, so their components on fixed axes will be changing from one moment to the next. But in a frame rotating at the rate ω, these unit vectors are constant, so we may write the equations of motion in terms of fixed x- and y-components in such a frame (i.e., $x_1 \equiv \vec{r}_1 \cdot \hat{i}$, etc., where $\hat{i}$ and $\hat{j}$ are rectangular unit vectors of the rotating frame). This gives

$$-\omega^2 x_1 + \frac{Gm_2}{r_{12}^3}(x_1 - x_2) + \frac{Gm_3}{r_{13}^3}(x_1 - x_3) = 0,$$

$$-\omega^2 x_2 + \frac{Gm_1}{r_{12}^3}(x_2 - x_1) + \frac{Gm_3}{r_{23}^3}(x_2 - x_3) = 0, \qquad (4.26)$$

$$m_1 x_1 + m_2 x_2 + m_3 x_3 = 0.$$

Similar expressions hold for the y-components. These relations are still somewhat complicated in general, so let us simplify them a bit more by assuming that the masses are located at the vertices of an equilateral triangle, giving $r_{ij} = r =$ constant for all i and j values. The six equations then become

$$(-\omega^2 r^3 + Gm_2 + Gm_3)x_1 - Gm_2 x_2 - Gm_3 x_3 = 0,$$

$$-Gm_1 x_1 + (-\omega^2 r^3 + Gm_1 + Gm_3)x_2 - Gm_3 x_3 = 0,$$

$$m_1 x_1 + m_2 x_2 + m_3 x_3 = 0,$$

$$(-\omega^2 r^3 + Gm_2 + Gm_3)y_1 - Gm_2 y_2 - Gm_3 y_3 = 0,$$

$$-Gm_1 y_1 + (-\omega^2 r^3 + Gm_1 + Gm_3)y_2 - Gm_3 y_3 = 0,$$

$$m_1 y_1 + m_2 y_2 + m_3 y_3 = 0.$$

This set of six homogeneous linear equations in the six unknowns x_i and y_i will only have a nontrivial solution (see Appendix C) provided that the determinant of the coefficients vanishes: (We have multiplied the third and sixth equations, the CM relations, by G so that *all* masses in the determinant are multiplied by G. In doing the algebra, the G factors can be ignored if desired and reintroduced at the end as a factor multiplying each mass.)

$$\begin{vmatrix} -\omega^2 r^3 + Gm_2 + Gm_3 & -Gm_2 & -Gm_3 & 0 & 0 & 0 \\ -Gm_1 & -\omega^2 r^3 + Gm_1 + Gm_3 & -Gm_3 & 0 & 0 & 0 \\ Gm_1 & Gm_2 & Gm_3 & 0 & 0 & 0 \\ 0 & 0 & 0 & -\omega^2 r^3 + Gm_2 + Gm_3 & -Gm_2 & -Gm_3 \\ 0 & 0 & 0 & -Gm_1 & -\omega^2 r^3 + Gm_1 + Gm_3 & -Gm_3 \\ 0 & 0 & 0 & Gm_1 & Gm_2 & Gm_3 \end{vmatrix} = 0$$

If the first and third, then second and third, then fourth and sixth, and finally fifth and sixth rows are added (the sums replacing the first, second, fourth, and fifth rows, respectively), which will not change the value of the determinant according to the properties developed in Appendix C, we have

$$\begin{vmatrix} -\omega^2 r^3 + GM & 0 & 0 & 0 & 0 & 0 \\ 0 & -\omega^2 r^3 + GM & 0 & 0 & 0 & 0 \\ Gm_1 & Gm_2 & Gm_3 & 0 & 0 & 0 \\ 0 & 0 & 0 & -\omega^2 r^3 + GM & 0 & 0 \\ 0 & 0 & 0 & 0 & -\omega^2 r^3 + GM & 0 \\ 0 & 0 & 0 & Gm_1 & Gm_2 & Gm_3 \end{vmatrix} = 0,$$

where $M = m_1 + m_2 + m_3$. This form leads at once by inspection to the equation

$$G^2 m_3^2 (-\omega^2 r^3 + GM)^4 = 0,$$

or

$$\omega = \pm \sqrt{GM/r^3}. \tag{4.27}$$

In words, the equations of motion *do* have a meaningful physical solution for the triangle, the rotational angular velocity being determined solely by the total mass and the separation of the vertices.

In addition to the triangle solution, Lagrange also obtained a straight-line solution to eq. (4.26). The choice $y_1 = y_2 = y_3 = 0$ satisfies the y-equations. With $x_3 > x_2 > x_1$, let us set $x_3 - x_2 \equiv r$ and $x_2 - x_1 \equiv 1$, since the length scale can be adjusted to make some length equal to unity for algebraic convenience. Then $x_3 - x_1 = 1 + r$, and eqs. (4.26) become

$$-\omega^2 x_1 + Gm_2(-1) + \frac{Gm_3}{(1+r)^3}(1+r)(-1) = 0$$

$$= -\omega^2 x_1 - Gm_2 - \frac{Gm_3}{(1+r)^2},$$

$$-\omega^2(1 + x_1) + Gm_1 + \frac{Gm_3}{r^3}(-r) = 0 = -\omega^2(1 + x_1) + Gm_1 - \frac{Gm_3}{r^2},$$

$$m_1 x_1 + m_2(1 + x_1) + m_3(1 + r + x_1) = 0.$$

From the first of these relations,

$$\omega^2 = -\frac{G}{x_1}\left[m_2 + \frac{m_3}{(1+r)^2}\right]$$

is obtained. Substituting this result into the second equation gives

$$\frac{G}{x_1}(1 + x_1)\left[m_2 + \frac{m_3}{(1+r)^2}\right] + Gm_1 - \frac{Gm_3}{r^2} = 0,$$

or

$$x_1 = \frac{\left[m_2(1+r)^2 + m_3\right]r^2}{m_3[1 + 2r] - (m_1 + m_2)r^2(1+r)^2}.$$

Finally, inserting this x_1 expression into the third equation yields, after some simplification,

$$(m_1 + m_2)r^5 + (3m_1 + 2m_2)r^4 + (3m_1 + m_2)r^3 -$$

$$(m_2 + 3m_3)r^2 - (2m_2 + 3m_3)r - (m_2 + m_3) = 0.$$

This fifth-degree equation can only be solved by numerical approximation methods. It is easily seen from the theory of equations to have only one real and positive root for r. Thus, m_3 can be located relative to m_1 and m_2, which can be placed at whatever points desired, so long as the scale is adjusted to make unit length between them.

The solutions obtained by Lagrange in the form of a triangle or a straight line are, it should be emphasized, in the *rotating* frame of reference, where gravitational and "fictitious" centrifugal forces of rotation are balanced so as to maintain a stationary configuration. The points at the triangle vertices or on the line solution are called **Lagrangian points**; masses located there have a certain stability and, if initially at rest, will tend to remain at those points.

More general matters of stability for a mechanical system are discussed in Chapter 7. At this point, we shall only note that the Lagrangian points on the line are usually designated L_1, L_2, and L_3 (see Figure 4.8), and that they are actually unstable under even a small displacement from their normal positions. By contrast, the triangle vertex points, designated L_4 and L_5, exhibit stability in some cases. For example, in the restricted three-body problem, stability under *small* displacements results when one mass is approximately 25 times (or more)

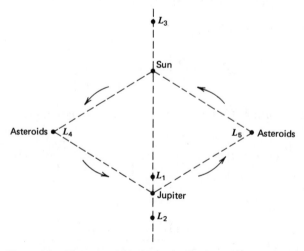

Figure 4.8 The asteroid belts in the Jupiter reference frame.

greater than the other of the two larger masses in the system. (The situation is difficult to analyze for sizable displacements.)

The classic example of the Lagrangian point theory is the Trojan asteroid belt. Imagine a frame of reference moving with Jupiter in revolution around the sun (or better, rotation about a common CM). There are two stable Lagrangian points, each of which forms an equilateral triangle with the sun and Jupiter at the other two vertices. A small group of asteroids is located at each point; given the proper initial conditions, each asteroid moves in a small elliptical orbit around the Lagrangian point, in accordance with the above ideas.

PROBLEMS

4.1 A spherical asteroid having a uniform density slowly accumulates rock fragments of the same density. After several billion years, its radius has thereby increased by a factor of α. Suppose on the average the matter is acquired radially, and that the asteroid rotated initially with an angular velocity ω_0. What is its final rate of rotation?

4.2 Let a perpendicular be drawn from the force center at the origin to the line of the velocity vector (tangent to the curve at any point of a central field orbit). Work out an expression for the angle which this perpendicular makes with the x-axis (in terms of r, θ, $\dot{r}$, and $\dot{\theta}$), and from this expression show that the magnitude of the perpendicular distance is inversely proportional to the magnitude of the velocity.

4.3 A particle of mass m and angular momentum $\mathscr{L}$ is in a central field with orbit equation $r = K \sin(n\theta)$, where K and n are constants. Find the radial dependence of the central force acting and show that the total energy is zero in this field.

4.4 A particle moves in an elliptical orbit of major axis $2a$ and minor axis $2b$, with the origin at the center of the ellipse. If the radius vector to the particle sweeps out area at a constant rate as usual, find the law of force in terms of the mass m and period P of the motion. If the particle were restricted to the x-axis under this same force law, what kind of motion would result?

4.5 A comet follows a parabolic trajectory, which brings it to a distance d from the sun at its closest point. The orbit of the comet is in the plane of the Earth's orbit, taken as a circle of radius R. Find the fraction of a year which the comet spends inside the orbit of the Earth as a function of the ratio d/R, and show that this fraction can never exceed $2/3\pi$.

4.6 A particle of mass m moves in a central field under a law of force given by $f(r) = -k/r^4$, k = constant. First analyze this motion by the method of the effective one-dimensional potential. Then show that the orbit equation can be satisfied by $r = A(1 + \cos\theta)$, $A = $ constant. Finally, find the total energy of the particle.

4.7 The gravitational potential for a Newtonian force is $-k/r$. Suppose a small perturbation δ/r^2 is added to the potential. Find the general orbit equation. Show that, if $\delta \ll \mathscr{L}^2/2m$, the orbit is given by an ellipse with major axis precessing slowly, having angular velocity of precession given by $\delta/(\mathscr{L}a^2\sqrt{1-\varepsilon^2})$ in the notation of the text; δ is a constant. (This problem indicates how even a small deviation from a pure inverse square field could be detected easily from orbit characteristics. Later, in the discussion of general relativity, we shall refer to this situation.)

4.8 (a) Prove that the speed of a planet (or satellite) in an elliptical orbit, at that point when the planet is at its maximum distance from the major axis, is equal to the geometrical mean of the maximum and minimum orbital speeds.
(b) Show that the ratio of extreme orbital speeds (at perihelion and aphelion) is $(1 + e)/(1 - e)$, where e is the eccentricity.
(c) Take the Earth's eccentricity as 0.0167 and that of Halley's comet as 0.967; calculate the ratio in part (b) for each. (It will then be clear why comets spend little time in the vicinity of the sun.)

4.9 The relationship

$$d\theta = \frac{\mathscr{L}\, dr}{mr^2 \sqrt{\frac{2}{m}\left[E - V(r) - \frac{\mathscr{L}^2}{2mr^2}\right]}}$$

can be integrated in many cases to yield the orbit equation. Assuming a power law potential of the type $V = Cr^{n+1}$, show that for $n = -5$ the expression can be integrated in terms of the Legendre elliptic integral of the first kind, defined in indefinite form as

$$J_0(k) \equiv \int \frac{dx}{\sqrt{(1-x^2)(1-k^2x^2)}}$$

(Take the parameters as being of whatever sign you need to make radicals real.)

4.10 Consider the scattering of charged particles by a repulsive inverse square force $F(r) = k/r^2$, as in the classical Rutherford scattering of alpha particles. Find the relation between impact parameter b and scattering angle χ and use it to obtain the differential cross section. (Quantum mechanics in the Schrödinger formulation provides the same result for $\sigma(\chi)$.)

4.11 In modern physics, one often makes use of a spherical "well" potential, of the form
$$V = -V_0 \quad (r \leqslant a),$$
$$V = 0 \quad (r > a)$$
Analyze the scattering from such a well, obtaining the differential cross section as a function of the scattering angle χ. Then show that the total cross section has the value expected on the basis of geometrical intuition.

Chapter Five

MATRICES, TENSORS, AND ROTATION OPERATORS

5.1 LINEAR TRANSFORMATIONS; MATRICES

In this chapter, we develop more sophisticated mathematical tools, the applications of which are many and varied. The main emphasis, however, will be on those calculational devices which are useful in the analysis of the rotation of rigid bodies.

Let us begin by considering a simple rotation in the xy-plane. The rotation takes place through an angle θ about the z-axis, carrying the original xy-axes into new $x'y'$-axes. The location of a point P in the plane can be described equally well in terms of either the original coordinates x and y or the new primed coordinates x' and y', and there is a simple relationship between these two sets of coordinates. The nature of this relation is easily obtained from the geometry of Figure 5.1, by concentrating attention on the triangles CAP and COB. The result

$$x' = AP + OB = (\cos\theta)x + (\sin\theta)y,$$
$$y' = CB - CA = -(\sin\theta)x + (\cos\theta)y, \qquad (5.1)$$

shows that the primed coordinates are each linear functions of the unprimed coordinates, the coefficients being determined solely by the angle of rotation. The equations can be inverted to give x and y in terms of x' and y' by solving the two relations simultaneously or by further geometrical analysis, leading in either case to

$$x = (\cos\theta)x' - (\sin\theta)y',$$
$$y = (\sin\theta)x' + (\cos\theta)y'. \qquad (5.2)$$

Equations (5.1) and (5.2) furnish examples of **linear transformations** from one set of quantities to another. While a rotation always produces a linear transformation of coordinates, it is not always possible to interpret such a transformation as

Linear Transformations; Matrices

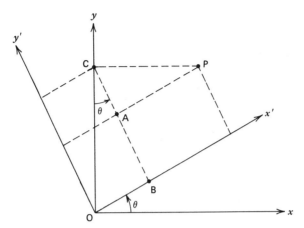

Figure 5.1 Geometry for rotation in a plane.

corresponding to a physical rotation (for example, some of the coefficients might be greater than unity in magnitude, or even imaginary, and so could not be associated with real angles θ).

It turns out that eqs. (5.1) and (5.2) can be obtained even more simply by a vectorial method which is readily generalized to three dimensions. Consider the position vector $\vec{r}$ from the origin to point P. Now, $\vec{r}$ can be expressed in either set of coordinates:

$$\vec{r} = x\hat{i} + y\hat{j} = x'\hat{i}' + y'\hat{j}',$$

making use of the familiar unit vectors in Cartesian coordinates. To find the expression for x, take the dot product of $\vec{r}$ with $\hat{i}$ giving

$$\vec{r} \cdot \hat{i} = (x\hat{i} + y\hat{j}) \cdot \hat{i} = x = x'(\hat{i}' \cdot \hat{i}) + y'(\hat{j}' \cdot \hat{i}).$$

But $\hat{i}' \cdot \hat{i}$ is just the cosine of the angle between the x'- and x-axes, or $\cos\theta$, and $\hat{j}' \cdot \hat{i} = \cos(90° + \theta) = -\sin\theta$, so once again we arrive at eq. (5.2). The point, however, is that each coefficient in the rotational transformation is merely the cosine of the angle between a pair of axes (one primed and one unprimed). Generalizing this approach to three dimensions, we have

$$x' = \cos(\hat{i}', \hat{i})x + \cos(\hat{i}', \hat{j})y + \cos(\hat{i}', \hat{k})z,$$

$$y' = \cos(\hat{j}', \hat{i})x + \cos(\hat{j}', \hat{j})y + \cos(\hat{j}', \hat{k})z,$$

$$z' = \cos(\hat{k}', \hat{i})x + \cos(\hat{k}', \hat{j})y + \cos(\hat{k}', \hat{k})z, \qquad (5.3)$$

in an obvious notation. The inverse equations can be written similarly. If the direction cosines of the x'-axis with respect to the three unprimed axes are denoted by α_1, α_2, and α_3, respectively, then $\hat{\imath}' = \alpha_1 \hat{\imath} + \alpha_2 \hat{\jmath} + \alpha_3 \hat{k}$ and $\hat{\imath}' \cdot \hat{\imath} = \cos(\hat{\imath}', \hat{\imath}) = \alpha_1$, so that the coefficients in eqs. (5.3) are direction cosines for primed axes relative to unprimed axes.

The nine coefficients, or direction cosines, are not all independent; it will be shown later that only three independent quantities are needed to specify the orientation of a rigid body. To see the kind of conditions the cosines must satisfy, it is convenient to change the notation somewhat. Let us set

$$x_1 \equiv x, \qquad x_2 \equiv y, \qquad x_3 \equiv z,$$

$$a_{11} \equiv \cos(\hat{\imath}', \hat{\imath}), \qquad a_{12} \equiv \cos(\hat{\imath}', \hat{\jmath}), \qquad \text{etc.}$$

In this new notation, eqs. (5.3) become

$$x_1' = a_{11}x_1 + a_{12}x_2 + a_{13}x_3,$$
$$x_2' = a_{21}x_1 + a_{22}x_2 + a_{23}x_3,$$
$$x_3' = a_{31}x_1 + a_{32}x_2 + a_{33}x_3.$$

Each a-coefficient is the cosine of an angle between two axes, the first subscript denoting the primed and the second the unprimed axis. In general, we may now write rather succinctly

$$x_i' = \sum_{j=1}^{3} a_{ij} x_j, \qquad i = 1, 2, 3. \tag{5.4}$$

In view of the fact that the squared magnitude of the position vector is the same in either set of coordinates, we can write

$$\sum_j x_j^2 = \sum_j x_j'^2.$$

Upon substitution of eq. (5.4) on the right, being careful to include two sums with independently running indices for the two factors of $x_j'^2$, we have

$$\sum_j x_j^2 = \sum_j \left(\sum_k a_{jk} x_k \right) \left(\sum_l a_{jl} x_l \right) = \sum_{k,l} x_k x_l \left(\sum_j a_{jk} a_{jl} \right).$$

But we are summing x_j^2, so there can be no terms of the form $x_k x_l$ arising unless $k = l$. For the two sides to agree, it is necessary to have

$$\sum_j a_{jk} a_{jl} = \delta_{k,l}, \tag{5.5}$$

Linear Transformations; Matrices

where the **Kronecker delta function** $\delta_{k,l}$ is defined by

$$\delta_{k,l} \equiv \begin{Bmatrix} 1 & k = l. \\ 0 & k \neq l. \end{Bmatrix} \tag{5.6}$$

Equation (5.5) is known as the **orthogonality condition** on the direction cosines, and the transformation (5.4) is consequently called an **orthogonal transformation**. The term orthogonality arises from the fact that the relation (5.5) has the form of the dot product of two unit vectors, in which a sum is carried out over products of corresponding rectangular components, the sum giving unity for two identical vectors and zero for different vectors that are perpendicular to one another. Noting the symmetry between the indices k and l, eq. (5.5) furnishes a set of six conditions among the nine quantities a_{ij} so that only three are independent, as desired.

In the two-dimensional example with which we began this section, we can choose $k = l = 1$ in eq. (5.5), to get

$$a_{11}a_{11} + a_{21}a_{21} = 1 = (\cos\theta)^2 + (-\sin\theta)^2,$$

using eq. (5.1). In addition, if we choose $k = 1$ and $l = 2$, then

$$a_{11}a_{12} + a_{21}a_{22} = 0 = \cos\theta \sin\theta + (-\sin\theta)\cos\theta,$$

and so on. Clearly the trigonometric relationships satisfy eq. (5.5) in two dimensions. In that case, there are only four a_{ij} values, and the orthogonality conditions furnish three equations, so just one independent parameter (the angle θ) is needed.

The chief advantage of formulating linear transformations through the use of the a_{ij} values in (5.4) and (5.5) is that we are now able to cast the theory into matrix form, which simplifies the form of the equations and expedites calculation. Thus, we define the **transformation matrix A** as an array of the nine a_{ij}'s

$$\mathbf{A} \equiv \begin{pmatrix} a_{11} & a_{12} & a_{13} \\ a_{21} & a_{22} & a_{23} \\ a_{31} & a_{32} & a_{33} \end{pmatrix}. \tag{5.7}$$

We also define the column vector, or column matrix, **X** as

$$\mathbf{X} \equiv \begin{pmatrix} x_1 \\ x_2 \\ x_3 \end{pmatrix}, \quad \text{with } \mathbf{X}' \equiv \begin{pmatrix} x_1' \\ x_2' \\ x_3' \end{pmatrix}. \tag{5.8}$$

Inasmuch as the modern student is usually familiar with the basic properties of matrices, we shall only refer the reader who feels deficient in this area to the details of Appendix C. Here, the elements of a product **C** of two matrices **A** and **B** are defined by

$$c_{ij} \equiv \sum_k a_{ik} b_{kj}, \qquad (5.9)$$

where **C** = **AB**. Unless certain symmetries are present in the matrices concerned, the order of the matrices in the product is vitally important; in general, the multiplication is not commutative (**AB** ≠ **BA**).

We shall also need the unit matrix,

$$\mathbf{1} \equiv \begin{pmatrix} 1 & 0 & 0 \\ 0 & 1 & 0 \\ 0 & 0 & 1 \end{pmatrix}, \qquad (5.10)$$

with elements $(\mathbf{1})_{ij} = \delta_{i,j}$, which has the properties (for any **A**)

$$\mathbf{A1} = \mathbf{A} = \mathbf{1A}, \quad \text{and} \quad \mathbf{AA}^{-1} = \mathbf{1} = \mathbf{A}^{-1}\mathbf{A},$$

provided that the inverse of **A**, written $\mathbf{A}^{-1}$, exists.

With these ideas in mind, it is easy to see that one obtains a new column vector through multiplication of **A** in eq. (5.7) with **X** of (5.8) according to the rule of (5.9). The ith element of the product is

$$(\mathbf{AX})_i = \sum_k a_{ik} x_k.$$

(The column vectors have another index implicitly present but suppressed in the notation.) Inasmuch as k is merely a dummy index, (5.4) tells us that

$$x'_i = (\mathbf{AX})_i,$$

or in matrix terms,

$$\mathbf{X'} = \mathbf{AX}. \qquad (5.11)$$

The new column vector obtained by the product has elements which are the primed components. In effect, the matrix **A** is a rotation operator which performs a linear transformation of coordinates.

Next, let the matrix $\mathbf{A}^{-1}$ be called **D**, with elements d_{ij}. From the fact that $\mathbf{AA}^{-1} = \mathbf{1}$, eq. (5.9) produces

$$\sum_k a_{ik} d_{kj} = (\mathbf{1})_{ij} = \delta_{i,j}.$$

Multiplication by a_{il} and summation on i gives

$$\sum_{i,k} a_{il}(a_{ik}d_{kj}) = \sum_i a_{il}\delta_{i,j} = a_{jl},$$

whereas with use of eq. (5.5),

$$\sum_{i,k}(a_{il}a_{ik})d_{kj} = \sum_k \delta_{l,k}d_{kj} = d_{lj}.$$

Therefore,

$$a_{jl} = (\mathbf{A})_{jl} = d_{lj} = (\mathbf{A}^{-1})_{lj}.$$

Thus, the inverse of a rotation matrix is found simply by an interchange of its rows and columns, a procedure which defines the **transpose** of $\mathbf{A}$, designated as $\tilde{\mathbf{A}}$ or $\mathbf{A}^T$:

$$\mathbf{A}^{-1} = \tilde{\mathbf{A}} = \mathbf{A}^T. \tag{5.12}$$

A matrix $\mathbf{A}$ which satisfies eq. (5.12) is called an **orthogonal** matrix, for its elements must also satisfy the orthogonality conditions stated above. In general, the inverse of a matrix cannot be obtained so easily.

Note, however, for later usage that if the product $\mathbf{C} = \mathbf{AB}$ of *any* two matrices is transposed, then

$$(\mathbf{AB})^T_{ij} = c_{ji} = \sum_k a_{jk}b_{ki} = \sum_k b_{ki}a_{jk} = \sum_k (\mathbf{B}^T)_{ik}(\mathbf{A}^T)_{kj} = (\mathbf{B}^T\mathbf{A}^T)_{ij},$$

or

$$(\mathbf{AB})^T = \mathbf{B}^T\mathbf{A}^T. \tag{5.13}$$

That is, the order of the factors is reversed when a product is transposed.

Proceeding now by means of eqs. (5.11) and (5.12), one can write

$$\mathbf{X} = \mathbf{A}^{-1}\mathbf{X}' = \mathbf{A}^T\mathbf{X}', \tag{5.14}$$

which is an inverse transformation law:

$$x_i = \sum_k (\mathbf{A}^T)_{ik}x'_k = \sum_k a_{ki}x'_k. \tag{5.15}$$

In addition, the knowledge that

$$\mathbf{AA}^{-1} = \mathbf{1} = \mathbf{AA}^T$$

produces

$$\sum_k (\mathbf{A})_{ik}(\mathbf{A}^T)_{kj} = \delta_{i,j} = \sum_k a_{ik} a_{jk}, \qquad (5.16)$$

which is another orthogonality condition analogous to (5.5). The sum is carried out here over identical *second* indices.

As an example of these principles, consider once more the two-dimensional rotation of (5.1) about the z-axis. The rotation matrix is

$$\mathbf{A} = \begin{pmatrix} \cos\theta & \sin\theta & 0 \\ -\sin\theta & \cos\theta & 0 \\ 0 & 0 & 1 \end{pmatrix}, \qquad (5.17)$$

with

$$\mathbf{A}^T = \mathbf{A}^{-1} = \begin{pmatrix} \cos\theta & -\sin\theta & 0 \\ \sin\theta & \cos\theta & 0 \\ 0 & 0 & 1 \end{pmatrix}. \qquad (5.17')$$

The correctness of the inverse is readily demonstrated by direct multiplication.

One final point should be made. The discussion thus far has left the physical vector **X** (or any other vector **V** for that matter; the components of **V** will transform in the same way as **X** under rotation) "untouched" by the transformation, the *coordinate system* being rotated so that primed components are given in terms of unprimed through the elements of **A**. Such a procedure is known as the **passive view** of the rotation. But it is equally reasonable (and often more desirable physically) to consider the matrix **A** as acting on the original (unprimed) components of **X** and rotating the *vector* such that these components take on new (primed) values, with both sets of components expressed in the "one and only" original coordinate system. This latter interpretation is called the **active view** of the rotation. As seen in Figure 5.2, a counterclockwise rotation (system) viewed

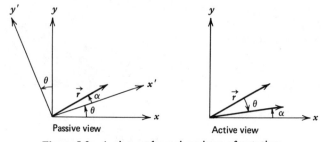

Figure 5.2 Active and passive views of rotation.

passively must correspond to a clockwise rotation (vector) viewed actively, and vice versa. With simple cases such as those of (5.17), a *counterclockwise* vector rotation in the active view is easily obtained by transposing **A**, or just changing the sign in some of its elements. No great confusion of mathematics should result; it is a matter of keeping sign conventions straight. The distinction between the two viewpoints is important, however, for both are frequently seen in the literature.

5.2 THE EULER ANGLES AND THE FULL ROTATION MATRIX

At this point, we should try to understand why only six independent coordinates (and not $3N$) are required in order to specify the positions of all N particles making up a rigid body. For each pair of particles i and j, there is a constraint relation which fixes the distance r_{ij} between the pair at some constant value. Inasmuch as each of the N particles is "connected" to $N - 1$ others by such relations but each pair is to be counted only once, the constraints must be $\frac{1}{2}N(N - 1)$ in number. Clearly, all the equations are not independent of one another.

A simple procedure for establishing the number of independent coordinates needed can be seen as follows (Figure 5.3): Consider particle 1 fixed at some point in the rigid body, assigning to it the usual three coordinates (this point might be chosen as the CM, for example). Some other particle (number 2) must be located at a constant distance C_{12} from the first, so it will lie on the surface of a sphere of radius C_{12} about point 1; its position can therefore be determined by the two angular coordinates θ and ϕ for a spherical system. Now introduce a third particle (not in a line with the other two) at distance C_{13} from the first and C_{23} from the second. For convenience in the analysis, let us put particle 1 at the origin and particle 2 at the point $(C_{12}, 0, 0)$. Particle 3 at (x_3, y_3, z_3) must simulta-

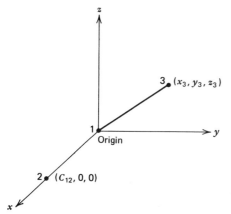

Figure 5.3 Geometry for the determination of independent coordinates.

neously satisfy the equations

$$x_3^2 + y_3^2 + z_3^2 = C_{13}^2,$$

$$(x_3 - C_{12})^2 + y_3^2 + z_3^2 = C_{23}^2.$$

Substituting the first relation into the second, one obtains

$$C_{13}^2 + C_{12}^2 - 2C_{12}x_3 = C_{23}^2,$$

or

$$x_3 = \frac{C_{12}^2 + C_{13}^2 - C_{23}^2}{2C_{12}},$$

whereupon

$$y_3^2 + z_3^2 = C_{13}^2 - \left(\frac{C_{12}^2 + C_{13}^2 - C_{23}^2}{2C_{12}}\right)^2 = \text{constant}.$$

This is the equation of a circle centered on the line between particles 1 and 2 and in a plane with fixed x_3 value; it is the intersection of the two spheres of radius C_{13} and C_{23}. Particle 3 must lie somewhere on this intersection, and its position can be specified by a single angular coordinate around the circle. Thus, a total of six coordinates have been required for the pinpointing of locations for these three particles. The remaining $N - 3$ particles are determined in position by their distances from the first three, by extension of these considerations; only the CM and three angular coordinates are required for the positions of all N particles to be specified.

The traditional way of introducing the three angular coordinates in applications however, is not as described above but rather by means of what are called the **Euler angles**, defined as follows:

1. Rotate the xyz-axes counterclockwise by angle ϕ about z, forming a new set of axes $x'y'z'$ (z' will be the same as z, of course).
2. Rotate the $x'y'z'$-axes counterclockwise by angle θ about x', forming $x''y''z''$-axes. The intersection of the xy- and $x''y''$-planes is called the **line of nodes**.
3. Rotate the $x''y''z''$-axes counterclockwise by angle ψ about z'', forming $x'''\,y'''\,z'''$-axes.

Angles ϕ, θ, and ψ are often chosen as the Euler angles, though the definitions vary a little from one text to another. They are the independent angular coordinates needed, and are illustrated in Figure 5.4. (The Euler angles have been defined so as to be convenient in the analysis of the motion of a top.)

Expressing the orientation of a rigid body in terms of the three Euler angles is very difficult from a strictly geometrical viewpoint, but it becomes easy if we proceed step by step with orthogonal matrices. Consider two distinct rigid body rotations, the first given by a matrix $\mathbf{R}_1$ and the second by $\mathbf{R}_2$. ($\mathbf{R}_1$ rotates from unprimed to primed coordinates, $\mathbf{R}_2$ from primed to double primed.) With the aid of (5.11), we have

$$\mathbf{X}' = \mathbf{R}_1 \mathbf{X}$$

and

$$\mathbf{X}'' = \mathbf{R}_2 \mathbf{X}' = \mathbf{R}_2(\mathbf{R}_1 \mathbf{X}) = (\mathbf{R}_2 \mathbf{R}_1)\mathbf{X}.$$

The last step follows from matrix associativity. This result shows that two successive linear transformations are equivalent to a third which goes directly from unprimed to double primed sets through a matrix which is the product $\mathbf{R}_2 \mathbf{R}_1$. Obviously, this scheme can be continued if necessary.

Carrying these ideas further, we follow step 1 of the Euler angle definitions and observe that $\mathbf{R}_1$ must have the form

$$\mathbf{R}_1 \equiv \begin{pmatrix} \cos\phi & \sin\phi & 0 \\ -\sin\phi & \cos\phi & 0 \\ 0 & 0 & 1 \end{pmatrix}$$

by analogy with (5.17). For the second step, the θ-rotation is about the x'-axis (the transformation involves a linear combination of y' and z'), so that

$$\mathbf{R}_2 \equiv \begin{pmatrix} 1 & 0 & 0 \\ 0 & \cos\theta & \sin\theta \\ 0 & -\sin\theta & \cos\theta \end{pmatrix}.$$

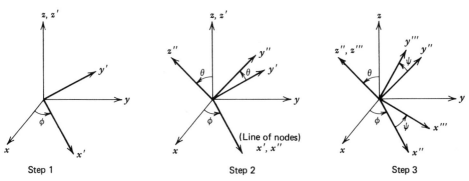

Figure 5.4 Steps in the determination of the Euler angles.

Finally, the third step requires

$$\mathbf{R}_3 \equiv \begin{pmatrix} \cos\psi & \sin\psi & 0 \\ -\sin\psi & \cos\psi & 0 \\ 0 & 0 & 1 \end{pmatrix}.$$

Putting all of this together,

$$\mathbf{X}''' = \mathbf{R}_3 \mathbf{X}'' = (\mathbf{R}_3 \mathbf{R}_2 \mathbf{R}_1)\mathbf{X},$$

the final transformation from $\mathbf{X}$ to $\mathbf{X}'''$ being achieved by a single matrix $\mathbf{R}$ which is the product of those in the three steps:

$$\mathbf{R} \equiv \mathbf{R}_3 \mathbf{R}_2 \mathbf{R}_1.$$

We shall refer to $\mathbf{R}$ as the **full rotation matrix**; some of its properties will be explored later.

Carrying out the multiplication, the full rotation matrix is found to be

$$\mathbf{R} = \begin{pmatrix} \cos\phi\cos\psi - \sin\phi\cos\theta\sin\psi & \sin\phi\cos\psi + \cos\phi\cos\theta\sin\psi & \sin\theta\sin\psi \\ -\cos\phi\sin\psi - \sin\phi\cos\theta\cos\psi & -\sin\phi\sin\psi + \cos\phi\cos\theta\cos\psi & \sin\theta\cos\psi \\ \sin\phi\sin\theta & -\cos\phi\sin\theta & \cos\theta \end{pmatrix}.$$

(5.18)

Note that this matrix reduces to the identity when all the Euler angles are zero.

5.3 CARTESIAN TENSORS AND DYADICS

Before proceeding further with the analysis of rotations, we must learn a little about some mathematical entities which are closely related to orthogonal matrices. Many physical applications of these concepts can be found in areas other than classical dynamics, as will be evident later.

Given a rotation matrix $\mathbf{A}$ with elements a_{ij}, a Cartesian tensor T in three-dimensional space is defined as a quantity which transforms according to the rule

$$T'_{lmn\ldots} \equiv \sum_{\substack{i,j,k,\ldots \\ =1}}^{3} a_{li} a_{mj} a_{nk} \ldots T_{ijk\ldots} \tag{5.19}$$

in a rotation from unprimed to primed coordinates. The $T_{ijk\ldots}$ are called the **components** of the tensor, and they are some functions of the unprimed coordinates, the $T'_{lmn\ldots}$ being the corresponding components in the primed system.

Cartesian Tensors and Dyadics

There is an element of **A** (direction cosine) as a coefficient in the sum for every tensor component index, the total number of indices N being called the **rank** of the tensor. As each index in three-dimensional space can take on the values 1, 2, or 3, the tensor must have 3^N components. A zero-rank tensor, then, has only one component with no summation indices; that component is a scalar invariant under orthogonal transformations. It is, however, still generally a function of the coordinates; the square of the position vector is a good example of such an invariant, as demonstrated below eq. (5.4). A first-rank tensor satisfies

$$T'_l = \sum_i a_{li} T_i,$$

which has the same form as eq. (5.4). Thus, a first-rank tensor, having three components, is simply a vector.

Suppose that we have two vectors, $\vec{A}$ and $\vec{B}$, which transform as

$$A'_l = \sum_i a_{li} A_i \quad \text{and} \quad B'_l = \sum_j a_{lj} B_j.$$

Writing the dot product of these two vectors in the primed system, one obtains

$$\vec{A}' \cdot \vec{B}' = \sum_l A'_l B'_l = \sum_{l,i,j} (a_{li} A_i)(a_{lj} B_j)$$

$$= \sum_{i,j} \left(\sum_l a_{li} a_{lj} \right) A_i B_j = \sum_{i,j} (\delta_{i,j}) A_i B_j = \sum_i A_i B_i = \vec{A} \cdot \vec{B},$$

using the orthogonality condition (5.5). In other words, the dot product of any two vectors is an invariant under the rotation. Put another way, the product is of zero rank even though the factors $\vec{A}$ and $\vec{B}$ are of first rank. When tensor indices are set equal and summed over (as l or i above), the process is known as **contraction**. Invariants are often obtained through such a procedure, and they are naturally of importance in the analysis of physical systems.

Incidentally, many authors frequently employ the **Einstein summation convention**, which requires an automatic summation over a repeated index in a tensorial expression (such as $A_i B_i$) and thus eliminates the need for a summation sign. While this convention is a very useful one in practice, we feel it can make the basic algebraic steps obscure to a student unfamiliar with tensors, and we therefore shall not adopt it.

The rank of the tensor that is of greatest interest to us is the second, which has nine components, transforming as

$$T'_{lm} = \sum_{i,j} a_{li} a_{mj} T_{ij}. \tag{5.20}$$

These nine elements can *always* be written in the form of a 3 × 3 matrix, but in general the elements of such a matrix do *not* form a tensor because they will fail to satisfy eq. (5.20) for a given pair of coordinate systems. In this sense, the matrix concept is broader than that of the tensor.

Consider as an example $T_{ij} \equiv x_i x_j$. In matrix form, this is

$$\begin{pmatrix} x^2 & xy & xz \\ yx & y^2 & yz \\ zx & zy & z^2 \end{pmatrix}.$$

But does this quantity transform as a second-rank tensor? (The fact that it has two indices and is called T_{ij} does not mean anything!) To find the answer, we note from (5.4) that

$$\sum_{i,j} a_{li} a_{mj} T_{ij} = \left[\sum_i a_{li} x_i \right] \left[\sum_j a_{mj} x_j \right]$$

$$= x'_l x'_m \equiv T'_{lm}.$$

Thus, by (5.20), T_{ij} must indeed be a legitimate tensor. It should be clear that other functions of the coordinates could have been chosen for T_{ij}, but many would not have satisfied the transformation (5.20).

The tensor properties just examined have a direct relevance for physics. The motion of a physical system can be described equally well in some particular coordinate system or in another system rotated with respect to the first, for the orientation of the axes cannot be of significance. In order that the equations of motion will have the same form as functions of whatever coordinate axes are selected, it is necessary that the terms of the equations transform according to the tensor rules. This requirement puts a severe limitation on the type of function that is permissible in a basic equation of physics.

In the same way that a vector $\vec{A}$ can be specified by means of its components (A_1, A_2, A_3) or by the unit vector expression

$$\vec{A} = A_1 \hat{i} + A_2 \hat{j} + A_3 \hat{k},$$

a second-rank tensor can be given by listing its nine components, or by an expression involving *pairs* of unit vectors. A pair of vectors in a definite order, such as $\hat{i}\hat{j}$, is called a **dyad**, and a linear combination of dyads is known as a **dyadic**. A tensor can be cast into dyadic form by writing

$$\vec{\vec{T}} = T_{11} \hat{i}\hat{i} + T_{12} \hat{i}\hat{j} + T_{13} \hat{i}\hat{k} + T_{21} \hat{j}\hat{i} + \cdots,$$

where the subscripts in each term correspond to the unit vector pairs.

In a similar fashion, two arbitrary vectors $\vec{A}$ and $\vec{B}$ placed next to one another give rise to a dyadic

$$\vec{A}\vec{B} \equiv A_x B_x \hat{i}\hat{i} + A_x B_y \hat{i}\hat{j} + A_x B_z \hat{i}\hat{k}$$
$$+ A_y B_x \hat{j}\hat{i} + A_y B_y \hat{j}\hat{j} + A_y B_z \hat{j}\hat{k} \qquad (5.21)$$
$$+ A_z B_x \hat{k}\hat{i} + A_z B_y \hat{k}\hat{j} + A_z B_z \hat{k}\hat{k}.$$

This dyadic relation can be operated on from *either* side by taking the dot or cross product with a third vector, $\vec{C}$. In multiplying from the left, one uses the unit vector on the left in each term for the vector operation, while in operating from the right, one uses the right unit vector instead. Thus, we have

$$\vec{C} \cdot (\vec{A}\vec{B}) = C_x \left(A_x B_x \hat{i} + A_x B_y \hat{j} + A_x B_z \hat{k} \right)$$
$$+ C_y \left(A_y B_x \hat{i} + A_y B_y \hat{j} + A_y B_z \hat{k} \right)$$
$$+ C_z \left(A_z B_x \hat{i} + A_z B_y \hat{j} + A_z B_z \hat{k} \right)$$
$$= (\vec{C} \cdot \vec{A})\vec{B},$$

while $(\vec{A}\vec{B}) \cdot \vec{C} = \vec{A}(\vec{B} \cdot \vec{C})$ in a similar way. Taking the dot product of a dyadic with a vector gives a new vector in both direction and magnitude, the result depending on the side from which the multiplication is done. Only if the dyadic is symmetric (the tensor from which it is formed having $T_{ij} = T_{ji}$) does the dot product give the same result from either side. By analogy with the unit matrix, there is a unit dyadic

$$\vec{1} \equiv \hat{i}\hat{i} + \hat{j}\hat{j} + \hat{k}\hat{k} \qquad (5.22)$$

such that the dot product taken with any vector from either side leaves the vector unchanged.

Now suppose that two physically different vectors **U** and **V** are related by a transformation through a matrix **T**:

$$\mathbf{U} = \mathbf{TV}.$$

(In dyadic terms, $\vec{U} = \vec{T} \cdot \vec{V}$.) This relation merely states that the components of $\vec{U}$ are linear functions of the $\vec{V}$ components. Suppose next that a rotation is performed with an orthogonal matrix **A**, carrying the coordinates of $\vec{U}$ and $\vec{V}$ to a new primed set of axes. Noting that $\mathbf{A}^{-1}\mathbf{A} = \mathbf{1}$, we have

$$\mathbf{U}' = \mathbf{AU} = \mathbf{ATV} = \mathbf{AT}(\mathbf{A}^{-1}\mathbf{A})\mathbf{V} = (\mathbf{ATA}^{-1})\mathbf{V}',$$

where $\mathbf{V}' = \mathbf{AV}$. The relation will have the same form in the primed axes as in the original ones,

$$\mathbf{U}' = \mathbf{T}'\mathbf{V}',$$

provided one identifies

$$\mathbf{T}' \equiv \mathbf{ATA}^{-1}. \tag{5.23}$$

A column vector is transformed by $\mathbf{A}$ alone, but a matrix such as $\mathbf{T}$ must be acted on by $\mathbf{A}$ and $\mathbf{A}^{-1}$, one from each side; the action on $\mathbf{T}$ specified by (5.23) is called a **similarity transformation**.

Let us examine the elements of the matrices in (5.23). From eq. (5.9), one can write

$$(\mathbf{T}')_{lm} = \sum_{i,j} (\mathbf{A})_{li}(\mathbf{T})_{ij}(\mathbf{A}^{-1})_{jm} = \sum_{i,j} a_{li} a_{mj} (\mathbf{T})_{ij},$$

which agrees precisely with (5.20). In other words, preservation under rotation of a linear vector relationship means that $\mathbf{T}$ must undergo a similarity transformation, which is the same as saying that the elements of $\mathbf{T}$ form a second-rank tensor.

Example 1

Consider a mass m at the origin attached to three long perpendicular springs placed along the negative rectangular axes (fastened rigidly at their distant ends). A force applied to the mass along the positive x-direction will displace it along that axis, with

$$F_x = -k_x x$$

as the force expression (k_x is the force constant of the x-spring). If the y- and z-springs are long enough or the displacement x is small enough, the other two springs will not appreciably change in length and thereby produce other forces in addition.

In similar ways, forces applied along either y- or z-axes alone would be given by

$$F_y = -k_y y \quad \text{or} \quad F_z = -k_z z,$$

respectively. If the mass is displaced arbitrarily so that all components act simultaneously, then

$$\vec{F} = -\left(k_x x \hat{i} + k_y y \hat{j} + k_z z \hat{k}\right) = -\left[k_x \hat{i}(\hat{i} \cdot \vec{r}) + k_y \hat{j}(\hat{j} \cdot \vec{r}) + k_z \hat{k}(\hat{k} \cdot \vec{r})\right]$$

$$= -\vec{\vec{k}} \cdot \vec{r},$$

where

$$\vec{\vec{k}} \equiv k_x \hat{i}\hat{i} + k_y \hat{j}\hat{j} + k_z \hat{k}\hat{k}$$

is a typical dyadic. Observe that for identical springs with $k_x = k_y = k_z \equiv k$, one has $\vec{\vec{k}} = k\vec{\vec{1}}$, so $\vec{F}$ is oppositely directed to $\vec{r}$.

The simple diagonal form of $\vec{\vec{k}}$ in general is due to the fact that the springs were oriented along the axes. Suppose that the springs are left untouched while the axes are rotated to a new primed set. The linear transformation between force and displacement represented by the dyadic or tensor relation $\vec{F} = -\vec{\vec{k}} \cdot \vec{r}$ will still hold in the new axes, but, in accordance with (5.23), the transformed dyadic $\vec{\vec{k}}$ will now have off-diagonal terms (terms in $\hat{i}'\hat{j}'$, etc.). Alternatively, one can observe from eq. (5.15) that

$$\hat{i} = a_{11}\hat{i}' + a_{21}\hat{j}' + a_{31}k', \text{ etc.}$$

Substitution for the unit vector pairs in the $\vec{\vec{k}}$ expression will lead at once to the more complicated transformed components of the dyadic.

5.4 THE TENSOR OF INERTIA

Let us again direct our attention to the rotation of rigid bodies. A set of rectangular **body axes** is normally chosen as fixed within the body and moving with it relative to **space axes** which are fixed instead in some inertial system. Three of the six coordinates needed for locating all particles of the rigid body will be specified by the position of the origin, which is taken to coincide in the two sets of axes (and might be the CM, for example). In the following chapter it will be shown that the most general possible motion of a rigid body is a rotation about the instantaneous axis of the angular velocity $\vec{\omega}$, provided the origin (on the axis) is held fixed so no translation can occur. Three orientational coordinates such as the Euler angles will then completely specify the position of all body particles. Of course, the rectangular coordinates of a particular particle are always fixed quantities in the body axes; it is in the space set that they vary in time as the object moves.

Taking $\vec{r}_i$ as the position vector to the ith particle of the rigid body, its velocity $\vec{v}_i$ in the space axes is given by the familiar elementary relation (see Figure 5.5)

$$\vec{v}_i = \vec{\omega} \times \vec{r}_i. \tag{5.24}$$

(A vector equation such as this one can have instantaneous components taken along *either* set of axes; the velocity is zero in the body set, for the coordinates

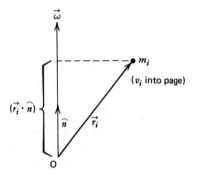

Figure 5.5 Space velocity in rigid body rotation.

there are not changing in time, yet we can still project the space velocity $\vec{v}_i$ along the body axes at each instant. A similar situation often occurs for other vectors.) The angular momentum about the fixed point now becomes

$$\vec{L} = \sum_i m_i \vec{r}_i \times \vec{v}_i = \sum_i m_i \vec{r}_i \times (\vec{\omega} \times \vec{r}_i)$$

$$= \sum_i m_i \left[r_i^2 \vec{\omega} - \vec{r}_i (\vec{r}_i \cdot \vec{\omega}) \right] \quad (5.25)$$

upon expanding the triple product. We can now introduce the **tensor of inertia** $\vec{\vec{I}}$ in dyadic form as

$$\vec{\vec{I}} \equiv \sum_i m_i \left(r_i^2 \vec{\vec{1}} - \vec{r}_i \vec{r}_i \right) \quad (5.26)$$

so that

$$\vec{L} = \vec{\vec{I}} \cdot \vec{\omega}. \quad (5.27)$$

The inertia tensor gives the elements of the linear transformation which relates angular velocity and momentum. It is clear, therefore, that in general $\vec{L}$ and $\vec{\omega}$ differ in direction as well as magnitude; elementary applications often do not consider such a complication. Writing out (5.26) and (5.27) more fully,

$$L_x = I_{xx}\omega_x + I_{xy}\omega_y + I_{xz}\omega_z,$$

$$L_y = I_{yx}\omega_x + I_{yy}\omega_y + I_{yz}\omega_z,$$

$$L_z = I_{zx}\omega_x + I_{zy}\omega_y + I_{zz}\omega_z,$$

The Tensor of Inertia

with

$$I_{xx} \equiv \sum_i m_i(r_i^2 - x_i^2),$$

$$I_{xy} \equiv -\sum_i m_i x_i y_i,$$

and so on. (The generalization to a continuous body with a density function is trivial.)

The diagonal elements of $\overset{\leftrightarrow}{I}$ are called the **moment of inertia coefficients**, and the off-diagonal elements are known as **products of inertia**. It is convenient to evaluate them (and therefore the $\vec{L}$ components) in the set of body axes where the coordinates such as x_i are all constants, giving another set of constants for the elements of $\overset{\leftrightarrow}{I}$. More significantly, the mass and the inertia tensor are the vital quantities that determine the dynamical properties of a rigid body.

Letting the unit vector $\hat{n}$ have the instantaneous direction of the angular velocity (see Fig. 5.5), $\vec{\omega} = \omega\hat{n}$, we define the scalar I, the **moment of inertia** about the direction of $\hat{n}$, as

$$I \equiv \hat{n} \cdot \overset{\leftrightarrow}{I} \cdot \hat{n} = \sum_i m_i\left[r_i^2 - (\vec{r}_i \cdot \hat{n})^2\right]. \tag{5.28}$$

This is the familiar quantity of elementary physics, the expression in the bracket giving the squared distance of the ith particle from the rotation axis $\hat{n}$, the fixed origin being on that axis.

The kinetic energy is easily written in terms of the inertia tensor:

$$\begin{aligned} T &= \tfrac{1}{2}\sum_i m_i \vec{v}_i \cdot \vec{v}_i = \tfrac{1}{2}\sum_i m_i \vec{v}_i \cdot (\vec{\omega} \times \vec{r}_i) \\ &= \tfrac{1}{2}\vec{\omega} \cdot \left[\sum_i m_i(\vec{r}_i \times \vec{v}_i)\right] = \tfrac{1}{2}\vec{\omega} \cdot \vec{L} \\ &= \tfrac{1}{2}\vec{\omega} \cdot \overset{\leftrightarrow}{I} \cdot \vec{\omega} = \tfrac{1}{2}\omega^2 \hat{n} \cdot \overset{\leftrightarrow}{I} \cdot \hat{n} \\ &= \tfrac{1}{2}I\omega^2, \end{aligned} \tag{5.29}$$

where we have used eqs. (5.24), (5.27), and (5.28). The last equality should come as no surprise; the elementary form results even from our more sophisticated analysis.

A generalized form of the celebrated parallel axis theorem is also readily established. As in Chapter 1, we denote the position of the ith particle relative to the origin O by the vector $\vec{r}_i$ and the location of the CM by $\vec{R}$, while the vector from the CM to the ith particle is $\vec{r}_i''$, giving

$$\vec{r}_i = \vec{r}_i'' + \vec{R}.$$

From (5.26), the inertia tensor about O is then

$$\vec{\vec{I}}_O = \sum_i m_i [(\vec{r}_i'' + \vec{R}) \cdot (\vec{r}_i'' + \vec{R}) - (\vec{r}_i'' + \vec{R})(\vec{r}_i'' + \vec{R})]$$

$$= \sum_i m_i [r_i'^2 \vec{\vec{1}} - \vec{r}_i'' \vec{r}_i''] + \left(\sum_i m_i\right)[R^2 \vec{\vec{1}} - \vec{R}\vec{R}]$$

$$+ 2\left[\vec{R} \cdot \left(\sum_i m_i \vec{r}_i''\right)\right]\vec{\vec{1}} - \vec{R}\left(\sum_i m_i \vec{r}_i''\right) - \left(\sum_i m_i \vec{r}_i''\right)\vec{R}.$$

The last three summation terms on the right will vanish because $\sum_i m_i \vec{r}_i'' = 0$ by the CM definition; the first term on the right is the inertia tensor $\vec{\vec{I}}_c$ relative to the CM position, so that we have (letting $\sum_i m_i \equiv M$)

$$\vec{\vec{I}}_O = \vec{\vec{I}}_c + M[R^2 \vec{\vec{1}} - \vec{R}\vec{R}]. \tag{5.30}$$

The inertia tensor relative to an arbitrary origin is that relative to the CM, plus a tensor corresponding to the entire mass as if concentrated at the CM but evaluated relative to O.

5.5 EIGENVALUE EQUATIONS; PRINCIPAL MOMENTS AND AXES OF THE INERTIA TENSOR

Let us briefly digress from the analysis of the inertia tensor in order to examine the properties of equations such as

$$\mathbf{AX} = \lambda \mathbf{X}, \tag{5.31}$$

where $\mathbf{A}$ is an arbitrary square matrix, $\mathbf{X}$ is a column vector called the *eigenvector*, and λ is a constant known as the **eigenvalue** of $\mathbf{A}$. Only certain special values of λ and $\mathbf{X}$ will satisfy this eigenvalue equation. We need to know how it is solved, for the equation has many interesting physical applications in both classical and modern physics. (Presumably, the reader has already encountered eigenvalue equations; if not, further details can be found in Appendix C.)

Writing (5.31) as

$$(\mathbf{A} - \lambda \mathbf{1})\mathbf{X} = \mathbf{0},$$

or explicitly as

$$(a_{11} - \lambda)x_1 + a_{12}x_2 + a_{13}x_3 = 0,$$
$$a_{21}x_1 + (a_{22} - \lambda)x_2 + a_{23}x_3 = 0, \quad (5.32)$$
$$a_{31}x_1 + a_{32}x_2 + (a_{33} - \lambda)x_3 = 0,$$

there results a set of three homogeneous linear equations in the x's (in three dimensions). A nontrivial solution is possible only if the determinant of the coefficients of the x-components vanishes:

$$|\mathbf{A} - \lambda \mathbf{1}| = 0. \quad (5.33)$$

This relation, called the **characteristic** or **secular** equation, yields the permissible values of λ. In the case of the set (5.32), it is a cubic with three roots for λ; substitution of each one of these eigenvalues into (5.32) will produce the *ratio* of components (hence the direction) of the corresponding eigenvector. (The magnitude of the vector $\mathbf{X}$ is not specified or needed, for if a given $\mathbf{X}$ is multiplied by any constant, (5.31) will still be satisfied.)

In this regard, a **Hermitian** matrix, defined as having elements which satisfy

$$A_{ij} = A_{ji}^*, \quad (5.34)$$

is of particular interest to physicists. There is a mathematical theorem[1] (often quoted in quantum mechanics) which states that the eigenvalues of such a matrix are real and its eigenvectors are orthogonal, provided that all the roots of (5.33) are distinct.

The inertia tensor in matrix form clearly satisfies a simpler version of (5.34), for its elements are always entirely real, and with

$$I_{ij} = I_{ji}, \quad (5.34')$$

there are only six independent components. Such a matrix is called **symmetric**. The symmetric property is preserved under rotation, as is readily shown by similarity transforming the inertia matrix $\mathbf{I}$ by any orthogonal matrix. We shall now investigate the properties of the eigenvectors of symmetric matrices.

It is well known that any such matrix with distinct characteristic roots can be similarity transformed into a diagonal form, where the eigenvalues ($\lambda_1, \lambda_2, \lambda_3$ in three dimensions) are the diagonal elements. That is, for any matrix $\mathbf{I}$ that appears in (5.31), we have

$$\mathbf{IX} = \lambda \mathbf{X}.$$

---
[1] G. Arfken, *Mathematical Methods for Physicists*, Academic, New York, 1966, p. 337.

Upon transforming by a particular rotation matrix $\mathbf{R}$, this becomes

$$\mathbf{RIX} = (\mathbf{RIR}^{-1})(\mathbf{RX}) = \mathbf{R}(\lambda \mathbf{X}) = \lambda(\mathbf{RX}). \tag{5.35}$$

The rotated eigenvectors $\mathbf{RX}$ will turn out to be orthogonal to one another when $\mathbf{I}$ is symmetric, or can easily be selected so as to satisfy that condition. The eigenvalues of the original $\mathbf{I}$ are the same λ values as for the transformed matrix $\mathbf{RIR}^{-1} \equiv \mathbf{I}'$ (as is apparent from the last equation), but the latter will be in an especially simple diagonal form for the proper choice of $\mathbf{R}$.

To reach these conclusions, consider a unit vector $\boldsymbol{\varepsilon}$ which in column form is given by

$$\boldsymbol{\varepsilon} = \begin{pmatrix} \varepsilon_1 \\ \varepsilon_2 \\ \varepsilon_3 \end{pmatrix}, \qquad \tilde{\boldsymbol{\varepsilon}} = (\varepsilon_1 \; \varepsilon_2 \; \varepsilon_3).$$

Next, compute the scalar function

$$f \equiv \tilde{\boldsymbol{\varepsilon}} \mathbf{I} \boldsymbol{\varepsilon},$$

formed with any symmetric $\mathbf{I}$. (This is analogous to taking the dot product of a vector with $\hat{\varepsilon}$ in order to obtain its projection along the $\hat{\varepsilon}$-direction.) As $\boldsymbol{\varepsilon}$ is varied in direction, there will be some orientation for which f will take on its maximum value; the new (rotated) x'-axis or $\hat{i}'$ will be chosen along this direction by definition. Now, in the plane perpendicular to $\hat{i}'$, again vary $\boldsymbol{\varepsilon}$ in direction until f reaches a maximum, calling this direction that of $\hat{j}'$. Of course, $\hat{k}'$ is determined by the condition $\hat{i}' \times \hat{j}' = \hat{k}'$, so that the orientation of the primed axes has been fully established.

Upon differentiation of f, one obtains

$$df = d\tilde{\boldsymbol{\varepsilon}} \mathbf{I} \boldsymbol{\varepsilon} + \tilde{\boldsymbol{\varepsilon}} \mathbf{I} \, d\boldsymbol{\varepsilon},$$

since $\mathbf{I}$ does not change when $\boldsymbol{\varepsilon}$ is varied. Using the fact that f is maximized when $\boldsymbol{\varepsilon}$ is along $\hat{i}'$, and recognizing that $d\boldsymbol{\varepsilon}$ must be perpendicular to $\boldsymbol{\varepsilon}$ as the latter is a unit vector, let us take $d\boldsymbol{\varepsilon}$ along $\hat{j}'$ to get

$$df = 0 = (0 \; d\varepsilon_2 \; 0)\mathbf{I}\begin{pmatrix} 1 \\ 0 \\ 0 \end{pmatrix} + (1 \; 0 \; 0)\mathbf{I}\begin{pmatrix} 0 \\ d\varepsilon_2 \\ 0 \end{pmatrix} = (I_{21} + I_{12}) \, d\varepsilon_2 = 2I_{12} \, d\varepsilon_2,$$

the last step coming from the symmetry of $\mathbf{I}$. Since $d\varepsilon_2 \neq 0$, we conclude that $I_{12} = I_{21}$ must be zero.

In a similar way, taking $d\varepsilon$ along $\hat{k}'$ leads to the result that I_{13} and I_{31} are zero, and the secondary maximizing condition in the $y'z'$-plane (ε along $\hat{j}'$, $d\varepsilon$ along $\hat{k}'$) produces $I_{23} = I_{32} = 0$. Therefore, *in the specified axes*, only the diagonal elements of **I** are nonzero; we shall designate the matrix as **I′** in this coordinate system.

Calling the diagonal elements I_1, I_2, and I_3 for simplicity, we note that

$$\mathbf{I'} = \begin{pmatrix} I_1 & 0 & 0 \\ 0 & I_2 & 0 \\ 0 & 0 & I_3 \end{pmatrix} \quad \text{and} \quad \mathbf{I'}\begin{pmatrix} 1 \\ 0 \\ 0 \end{pmatrix} = I_1 \begin{pmatrix} 1 \\ 0 \\ 0 \end{pmatrix}.$$

Similar results hold for the unit vectors along the y'- or z'-axes.

The eigenvalues of an inertia matrix are known as the **principal moments of inertia**, and the primed axes for which **I** is diagonal are called **principal axes**. The vectors along these directions are the eigenvectors of the diagonal matrix **I′**, in accordance with eq. (5.35) (the diagonal elements of **I′** are the eigenvalues). The eigenvectors are then clearly orthogonal.

For this conclusion to hold, the eigenvalues must all be distinct. If two of them, say, I_1 and I_2, were equal, then any linear combination ($A\hat{i}' + B\hat{j}'$) in the $x'y'$-plane would be an eigenvector with the common eigenvalue $I_1 = I_2$. But in that case, two orthogonal vectors in the plane could be *chosen* as the eigenvectors.

An arbitrary inertia matrix is usually not given in diagonal form. The proof above merely indicates that it can always be diagonalized. In order to do so, one must find the eigenvalues and eigenvectors with the aid of eqs. (5.33) and (5.32). The principal axes are then given by the directions of the three eigenvectors. Furthermore, if a 3 × 3 matrix **V** is constructed with these eigenvectors (in the original coordinates) making up its columns, then clearly

$$\mathbf{IV} = \mathbf{VI'},$$

or

$$\mathbf{I'} = \mathbf{V}^{-1}\mathbf{IV}. \tag{5.36}$$

This means that the transformation matrix **R** of eq. (5.35) is simply $\mathbf{V}^{-1}$. (We assume here that the vectors in **V** have been normalized to unit magnitude, so that **V** will be an orthogonal matrix.)

This formalism is really not as difficult in practice as it might seem; the following example should illustrate the salient points. Suppose there are three point masses of 1, 2, and 3 mass units placed respectively at the points $(1, 1, 0)$, $(1, -1, 0)$, and $(-1, 1, 0)$ (Figure 5.6). Let us work out the elements of the inertia

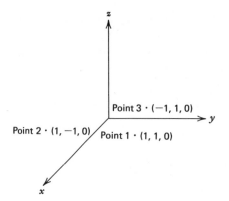

Figure 5.6 Distribution of masses in space.

tensor about the origin, using eq. (5.26):

$$I_{xx} = 1(1) + 2(1) + 3(1) = 6,$$

$$I_{yy} = 1(1) + 2(1) + 3(1) = 6,$$

$$I_{zz} = 1(1 + 1) + 2(1 + 1) + 3(1 + 1) = 12,$$

$$I_{xy} = -1(1) - 2(-1) - 3(-1) = 4 = I_{yx},$$

$$I_{xz} = I_{zx} = I_{yz} = I_{zy} = 0,$$

so that

$$\mathbf{I} = \begin{pmatrix} 6 & 4 & 0 \\ 4 & 6 & 0 \\ 0 & 0 & 12 \end{pmatrix}.$$

The characteristic equation is

$$(12 - \lambda)\left[(6 - \lambda)^2 - 16\right] = 0,$$

with roots $\lambda = 2$, 10, and 12. For the first root, the relations (5.32) are

$$(6 - 2)x_1 + 4x_2 = 0, \quad \text{or} \quad x_1 + x_2 = 0,$$

$$4x_1 + (6 - 2)x_2 = 0, \quad \text{or} \quad x_1 + x_2 = 0,$$

$$(12 - 2)x_3 = 0, \quad \text{or} \quad x_3 = 0.$$

Therefore, the eigenvector for this root is along the line $y = -x$ in the xy-plane.

Similarly, for the second root 10, the vector is along $y = +x$ in that plane; and for the last root 12, the eigenvector requires $x = y = 0$, and so is along the z-axis.

These results seem natural in view of the geometrical configuration of the masses. (Notice that the eigenvectors form an orthogonal system.) Normalizing the eigenvectors to unity, the matrix $\mathbf{V}$ is found to be

$$\mathbf{V} = \begin{pmatrix} 1/\sqrt{2} & 1/\sqrt{2} & 0 \\ -1/\sqrt{2} & 1/\sqrt{2} & 0 \\ 0 & 0 & 1 \end{pmatrix},$$

$$\mathbf{V}^{-1} = \mathbf{R} = \begin{pmatrix} 1/\sqrt{2} & -1/\sqrt{2} & 0 \\ 1/\sqrt{2} & 1/\sqrt{2} & 0 \\ 0 & 0 & 1 \end{pmatrix},$$

and also

$$\mathbf{I}' = \mathbf{V}^{-1}\mathbf{I}\mathbf{V} = \begin{pmatrix} 2 & 0 & 0 \\ 0 & 10 & 0 \\ 0 & 0 & 12 \end{pmatrix}$$

is the expected diagonal matrix. In addition, applying $\mathbf{R}$ to the first eigenvector (root = 2) yields

$$\mathbf{R} \begin{pmatrix} 1/\sqrt{2} \\ -1/\sqrt{2} \\ 0 \end{pmatrix} = \begin{pmatrix} 1 \\ 0 \\ 0 \end{pmatrix}.$$

The $\mathbf{R}$ matrix rotates by $-45°$ about the z-axis, orienting the first vector along the x'-axis.

From a physical point of view, when principal axes are employed, one always has from eq. (5.17) that

$$L_x = I_1\omega_x, \qquad L_y = I_2\omega_y, \qquad L_z = I_3\omega_z,$$

which is a great simplification in the angular momentum. Furthermore, if $\vec{\omega}$ is taken along one of the principal axes, it follows that $\vec{L}$ is also along that same axis. Such a situation is often realized, as with the spinning wheels encountered in elementary physics.

Principal axes can often be detected by inspection without a need for detailed calculation. Thus, if the xy-plane is a plane of symmetry for some rigid body, for every mass element at (x, y, z), there will be a corresponding element at

$(x, y, -z)$, so $I_{xz} = 0 = I_{yz}$. This will be true for any line parallel to the z-axis where it intersects the xy-plane.

Theorem *If a body possesses a symmetry plane, any axis perpendicular to this plane is a principal axis where it intersects the plane.*

By similar reasoning, we also have:

Theorem *If a body is one of revolution about a given axis, then that axis is a principal one at all points along it.*

There is a third theorem of this type, the proof of which is left for the problem set:

Theorem *If a straight line is a principal axis at the center of mass, it is a principal axis at all points along it.*

Example 2

The practical importance of the principal axis concept can be seen from a simple physical application: the static and dynamic balancing of automobile tires. Consider a body rotating about a fixed axis, with no applied forces or torques. Each mass element m_i must be supplied an inward centripetal force; the vector sum of all the centrifugal reactions on the axis is

$$\vec{F}_C = -\sum_i m_i \vec{\omega} \times (\vec{\omega} \times \vec{r}_i).$$

(The origin of the body axes is on the axis, and we have omitted primes on the coordinates for simplicity.) Let us take the z-axis along the angular velocity vector, that is, $\vec{\omega} = \omega_z \hat{k}$, giving

$$\vec{F}_C = -\omega_z^2 \sum_i m_i \hat{k} \times (\hat{k} \times \vec{r}_i) = -\omega_z^2 \sum_i m_i \hat{k} \times (x_i \hat{j} - y_i \hat{i})$$

$$= \omega_z^2 \sum_i m_i (x_i \hat{i} + y_i \hat{j}) = M\omega_z^2 (R_x \hat{i} + R_y \hat{j}),$$

where $M = \sum_i m_i$ is the total mass and $\vec{R}$ is the vector to the CM. The meaning of this relation is that there will be a net reaction force on the axis unless $R_x = 0 = R_y$; that is, unless the CM lies on the axis of rotation. If it does not, vibrations will be imparted to a car and excessive wear on the tires will result. To correct the situation, lead weights are placed on the rim of the tire to shift the CM toward the axis, a procedure known as **static balancing**.

The net torque due to the centrifugal reactions is

$$\vec{N}_C = -\sum_i m_i \vec{r}_i \times [\vec{\omega} \times (\vec{\omega} \times \vec{r}_i)]$$

$$= \omega_z^2 \sum_i m_i \vec{r}_i \times (x_i \hat{i} + y_i \hat{j})$$

$$= \omega_z^2 \left[-\left(\sum_i m_i y_i z_i\right)\hat{i} + \left(\sum_i m_i x_i z_i\right)\hat{j} \right]$$

$$= \omega_z^2 (I_{yz}\hat{i} - I_{xz}\hat{j}).$$

Clearly, $\vec{N}_C = 0$ only if both products of inertia vanish; that is, if the z-axis of rotation is a principal axis. When that is the case, $\vec{L}$ will lie along $\vec{\omega}$ by (5.27); otherwise, $\vec{L}$ will vary in direction as time passes, causing the wheel to wobble. Further weights can be added to correct the wobble, a procedure called **dynamic balancing**.

Suppose that a tire has a heavy "lump" of rubber on the outer rim of the *exterior* side (where $y > 0$, $z > 0$, with origin at the tire center). A weight can be placed on the opposite and *interior* side of the tire ($y < 0$, $z < 0$), and a static "bubble" balancing machine will show that the CM condition is satisfied. Unfortunately, however, I_{yz} is certainly not zero, and the torque along the x-axis will produce a dynamic imbalance. The lead weight should really be added to the opposite side of the *exterior* ($y < 0$, $z > 0$) to correct the imbalance, aesthetic considerations notwithstanding. In practice, a dynamic balance is achieved by spinning the tire at high speeds until the proper location for the weights is determined.

5.6 ROTATION OPERATORS

Earlier in this chapter, we saw that a complicated rotational sequence expressed in terms of Euler angles could actually be treated rather straightforwardly by means of the full rotation matrix. For many problems, this formulation is convenient. But it is often preferable to take the active point of view with regard to the rotation of rigid bodies, letting a physical vector rotate from one orientation to another in a fixed coordinate system. More importantly, it will be shown in the next chapter that any sequence of rotations in a particular order can always be replaced by a single equivalent rotation through an angle β about some axis specified by a unit vector $\hat{n}$. It is desirable to express the rotational formalism solely in terms of β and $\hat{n}$. The dyadic notation will be most useful, though matrices could also be employed.

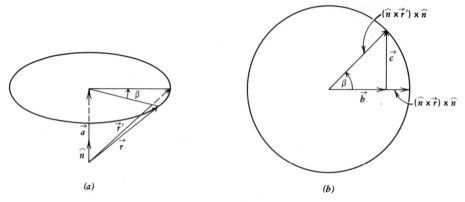

Figure 5.7 Vectors for rotation by angle β: (a) in three dimensions; (b) in circle plane.

Consider, then, the position vector $\vec{r}$ rotated by angle β into $\vec{r}'$; the rotation takes place along the arc of a circle in the plane perpendicular to $\hat{n}$. Let $\vec{a}$ be a vector form of the component of $\vec{r}$ (or $\vec{r}'$) parallel to $\hat{n}$, so that

$$\vec{a} \equiv \hat{n}(\hat{n} \cdot \vec{r}),$$

with $\vec{b}$ and $\vec{c}$ as vectors in the circle plane designed in such a way (see Figure 5.7) that

$$\vec{r}' = \vec{a} + \vec{b} + \vec{c}.$$

The radius of the circle is given very simply by

$$|\hat{n} \times \vec{r}| = |(\hat{n} \times \vec{r}) \times \hat{n}| = |(\hat{n} \times \vec{r}') \times \hat{n}|,$$

whereupon

$$\vec{b} = [(\hat{n} \times \vec{r}) \times \hat{n}] \cos \beta \quad \text{and} \quad \vec{c} = (\hat{n} \times \vec{r}) \sin \beta.$$

Finally, we obtain

$$\vec{r}' = \hat{n}(\hat{n} \cdot \vec{r}) + \cos \beta (\hat{n} \times \vec{r}) \times \hat{n} + \sin \beta (\hat{n} \times \vec{r})$$

$$= (1 - \cos \beta)\hat{n}(\hat{n} \cdot \vec{r}) + (\cos \beta)\vec{r} + \sin \beta (\hat{n} \times \vec{r}), \qquad (5.37)$$

upon expanding the triple product. Defining a **dyadic rotation operator** $\overset{\leftrightarrow}{R}(\hat{n}, \beta)$ such that

$$\vec{r}' = \overset{\leftrightarrow}{R}(\hat{n}, \beta) \cdot \vec{r}, \qquad (5.38)$$

one has

$$\overset{\leftrightarrow}{R}(\hat{n}, \beta) = (1 - \cos \beta)\hat{n}\hat{n} + \cos \beta \overset{\leftrightarrow}{1} + \sin \beta (\hat{n} \times \overset{\leftrightarrow}{1}). \qquad (5.39)$$

The entity $\overset{\leftrightarrow}{R}$ is *coordinate free*, depending only on the angle β and direction of $\hat{n}$; upon specification of a basis set, the elements of this dyadic can be calculated in detail. Operators such as $\overset{\leftrightarrow}{R}$ were first introduced by J. W. Gibbs[2] about the turn of the century and have recently been popularized by the work of C. Leubner and others (see Reference section).

Certain special cases of eq. (5.39) are worthy of mention. Note that as $\beta \to 0$, $\overset{\leftrightarrow}{R} \to \overset{\leftrightarrow}{1}$ for all $\hat{n}$. For all β values, $\hat{n}$ taken along $\hat{k}$ produces

$$\overset{\leftrightarrow}{R}(\hat{k}, \beta) = \hat{k}\hat{k} + \cos\beta(\hat{i}\hat{i} + \hat{j}\hat{j}) + \sin\beta(\hat{j}\hat{i} - \hat{i}\hat{j}),$$

which agrees with the matrix result (5.17) for **A**, provided that β (or θ) is replaced there by $-\beta$, for here in the active view we are effectively rotating the coordinate system clockwise.

The rotation dyadics satisfy a number of desirable properties which can readily be verified from (5.39). These include

$$\overset{\leftrightarrow}{R}(\hat{n}, 0) = \overset{\leftrightarrow}{1}, \tag{5.40}$$

$$\overset{\leftrightarrow}{R}(\hat{n}, \beta) \cdot \hat{n} = \hat{n}, \tag{5.41}$$

$$\overset{\leftrightarrow}{R}(\hat{n}, \alpha) \cdot \overset{\leftrightarrow}{R}(\hat{n}, \beta) = \overset{\leftrightarrow}{R}(\hat{n}, \alpha + \beta), \tag{5.42}$$

$$\overset{\leftrightarrow}{R}(\hat{n}, \beta) \cdot \overset{\leftrightarrow}{R}^T(\hat{n}, \beta) = \overset{\leftrightarrow}{1}, \tag{5.43}$$

where $\overset{\leftrightarrow}{R}^T(\hat{n}, \beta) \equiv \overset{\leftrightarrow}{R}(\hat{n}, -\beta)$ is the transpose or inverse of $\overset{\leftrightarrow}{R}$ in the usual sense.

Suppose we want to rotate a vector such as $\vec{r}$ into a new orientation $\vec{r}'$ by means of some rotation dyadic $\overset{\leftrightarrow}{R}_1$. Then,

$$\vec{r}' = \overset{\leftrightarrow}{R}_1 \cdot \vec{r}.$$

If $\vec{r}'$ is next rotated to $\vec{r}''$ by means of a dyadic $\overset{\leftrightarrow}{R}_2$, then

$$\vec{r}'' = \overset{\leftrightarrow}{R}_2 \cdot \vec{r}' = \overset{\leftrightarrow}{R}_2 \cdot \overset{\leftrightarrow}{R}_1 \cdot \vec{r}. \tag{5.44}$$

Since the process can obviously be continued in this vein, the dyadic analogue of the earlier Euler rotation matrix result becomes

$$\vec{r}''' = \overset{\leftrightarrow}{R}_3(\hat{k}'', \psi) \cdot \overset{\leftrightarrow}{R}_2(\hat{i}', \theta) \cdot \overset{\leftrightarrow}{R}_1(\hat{k}, \phi) \cdot \vec{r} \equiv \overset{\leftrightarrow}{R} \cdot \vec{r}. \tag{5.45}$$

The interpretation of equations (5.18) and (5.45) is somewhat different, however. In eq. (5.18), the relation gives the coordinates of the original vector $\vec{r}$ in terms of the final (triple primed) system. But here in the active view, eq. (5.45) gives the components of the rotated vector $\vec{r}'''$ in the original (unprimed) system.

---

[2] J. W. Gibbs, *Vector Analysis*, Scribner, New York, 1901, Chap. VI.

As a further application, let us find out how dyadics of the same angle are related when their direction axes are different. If a unit vector $\hat{n}$ is rotated into $\hat{n}'$ by $\vec{\vec{R}}(\hat{m}, \alpha)$, then

$$\hat{n}' = \vec{\vec{R}}(\hat{m}, \alpha) \cdot \hat{n}.$$

We wish to determine how $\vec{\vec{R}}(\hat{n}', \beta)$ is related to $\vec{\vec{R}}(\hat{n}, \beta)$. Now, from (5.39),

$$\vec{\vec{R}}(\hat{n}', \beta) = \vec{\vec{R}}(\vec{\vec{R}}(\hat{m}, \alpha) \cdot \hat{n}, \beta)$$

$$= (1 - \cos\beta)[\vec{\vec{R}}(\hat{m}, \alpha) \cdot \hat{n}][\vec{\vec{R}}(\hat{m}, \alpha) \cdot \hat{n}]$$

$$+ \cos\beta \vec{\vec{1}} + \sin\beta [\vec{\vec{R}}(\hat{m}, \alpha) \cdot \hat{n}] \times \vec{\vec{1}}.$$

Noting that in the last factor of the first term we can insert

$$\vec{\vec{R}}(\hat{m}, \alpha) \cdot \hat{n} = \hat{n} \cdot \vec{\vec{R}}^T(m, \alpha), \tag{5.46}$$

as is clear from the dyadic definition, and that

$$(\vec{\vec{R}} \cdot \vec{V}) \times \vec{\vec{1}} = \vec{\vec{R}} \cdot (\vec{V} \times \vec{\vec{1}}) \cdot \vec{\vec{R}}^T \tag{5.47}$$

(the proof of which is discussed in the problem set) for any $\vec{V}$ and $\vec{\vec{R}}$, we see with the aid of eq. (5.43) that

$$\vec{\vec{R}}(\hat{n}', \beta) = \vec{\vec{R}}(\hat{m}, \alpha) \cdot \vec{\vec{R}}(\hat{n}, \beta) \cdot \vec{\vec{R}}^T(\hat{m}, \alpha). \tag{5.48}$$

This is really just a similarity transformation in dyadic form.

By means of the relation (5.48), it becomes possible to shift the axis of a rotation. Returning to the notation employed in the earlier discussion of Euler angles, one can write in dyadic terms

$$\vec{\vec{R}}_2(\hat{i}', \theta) = \vec{\vec{R}}_1(\hat{k}, \phi) \cdot \vec{\vec{R}}_2(\hat{i}, \theta) \cdot \vec{\vec{R}}_1^T(\hat{k}, \phi)$$

as an example. But then taking the dot product with $\vec{\vec{R}}_1(\hat{k}, \phi)$ from the right gives

$$\vec{\vec{R}}_2(\hat{i}', \theta) \cdot \vec{\vec{R}}_1(\hat{k}, \phi) = \vec{\vec{R}}_1(\hat{k}, \phi) \cdot \vec{\vec{R}}_2(\hat{i}, \theta).$$

In a similar way,

$$\vec{\vec{R}}_3(\hat{k}'', \psi) = [\vec{\vec{R}}_2(\hat{i}', \theta) \cdot \vec{\vec{R}}_1(\hat{k}, \phi)] \cdot \vec{\vec{R}}_3(\hat{k}, \psi) \cdot [\vec{\vec{R}}_1^T(\hat{k}, \phi) \cdot \vec{\vec{R}}_2^T(\hat{i}', \theta)],$$

Rotation Operators

and so on; multiplying from the right by $\vec{R}_2 \cdot \vec{R}_1$,

$$\vec{R}_3(\hat{k}'', \psi) \cdot \vec{R}_2(\hat{i}', \theta) \cdot \vec{R}_1(\hat{k}, \phi) = \vec{R}_2(\hat{i}', \theta) \cdot \vec{R}_1(\hat{k}, \phi) \cdot \vec{R}_3(\hat{k}, \psi)$$
$$= \vec{R}_1(\hat{k}, \phi) \cdot \vec{R}_2(\hat{i}, \theta) \cdot \vec{R}_3(\hat{k}, \psi), \quad (5.49)$$

the last step following from the $\vec{R}_2 \cdot \vec{R}_1$ relation obtained previously. This astonishing result shows that for *any* Euler angles ϕ, θ, or ψ the full rotation dyadic in (5.45) can be obtained purely by rotations about the *unprimed* axes, so long as these are taken in the *opposite* of the usual order!

We shall pursue the implications of this "rotation reversal" theorem in the chapter to follow, but for the moment we wish to add only one further thought. In rigid body dynamics, normally the unprimed axes are those of the space (inertial) set, the triple primed axes are those of the body. The position vector can be written

$$\sum_j x_j \hat{\varepsilon}_j = \vec{r} = \sum_k x_k''' \hat{\varepsilon}_k'''$$

in terms of components and unit vectors in the two frames. Multiplication by $\hat{\varepsilon}_i'''$ yields

$$\sum_k x_k''' (\hat{\varepsilon}_k''' \cdot \hat{\varepsilon}_i''') = x_i''' = \sum_j x_j (\hat{\varepsilon}_j \cdot \hat{\varepsilon}_i''') = \sum_j x_j (\hat{\varepsilon}_j \cdot \vec{R} \cdot \hat{\varepsilon}_i)$$

from (5.45). Returning to the linear transformation (5.4), we see that here the full rotation matrix has elements which can be identified as

$$a_{ij} \equiv \hat{\varepsilon}_j \cdot \vec{R} \cdot \hat{\varepsilon}_i.$$

Using (5.49), this can be written (inserting $\vec{1} = \sum_k \hat{\varepsilon}_k \hat{\varepsilon}_k$)

$$a_{ij} = \sum_{k,l} [\hat{\varepsilon}_j \cdot \vec{R}_1(\hat{k}, \phi) \cdot \hat{\varepsilon}_k][\hat{\varepsilon}_k \cdot \vec{R}_2(\hat{i}, \theta) \cdot \hat{\varepsilon}_l][\hat{\varepsilon}_l \cdot \vec{R}_3(\hat{k}, \psi) \cdot \hat{\varepsilon}_i],$$

an expression that is easily evaluated and found to agree with (5.18). The relation to the product matrices of Section 5.2 should be apparent.

Finally, we should like to put the rotation dyadic into an exponential form. With an infinitesimally small rotation angle $\Delta\beta$, eq. (5.39) becomes

$$\vec{R}(\hat{n}, \Delta\beta) \simeq \vec{1} + \Delta\beta(\hat{n} \times \vec{1})$$

to first order in $\Delta\beta$. Repeating this rotation m times about the axis $\hat{n}$, the overall operator is

$$\vec{R}(\hat{n}, \beta) \simeq [\vec{1} + \Delta\beta(\hat{n} \times \vec{1})]^m = \left[\vec{1} + \frac{\beta(\hat{n} \times \vec{1})}{m}\right]^m,$$

where m is a large integer such that $m \cdot \Delta\beta = \beta$. Taking the limit as $m \to \infty$ and $\Delta\beta \to 0$ in the last expression (β remaining finite), the calculus definition of the base of the natural logarithms suggests that

$$\vec{\vec{R}}(\hat{n}, \beta) \to e^{\beta(\hat{n} \times \vec{\vec{1}})} \equiv e^{-i\beta(\hat{n} \times i\vec{\vec{1}})}. \tag{5.50}$$

Alternatively, one can expand this exponential in series form, getting (5.39) after some simplification of algebra.

Next, define three new operators $\vec{\vec{J}}_k$ with the aid of the unit vectors $\hat{\varepsilon}_k$ along the axes:

$$\vec{\vec{J}}_k \equiv \hat{\varepsilon}_k \times i\vec{\vec{1}}.$$

Inasmuch as $\hat{n}$ can be written in terms of its components n_j,

$$\hat{n} = \sum_j n_j \hat{\varepsilon}_j,$$

we have

$$\vec{\vec{R}}(\hat{n}, \beta) = e^{-i\beta \sum_j n_j \vec{\vec{J}}_j} = e^{-i\beta \hat{n} \cdot \vec{\vec{J}}}$$

in general. Should $\hat{n}$ be directed along the kth axis, the expression reduces to

$$\vec{\vec{R}}(\hat{\varepsilon}_k, \beta) = e^{-i\beta \vec{\vec{J}}_k}.$$

The latter expression is precisely the result obtained from the angular momentum formalism of quantum mechanics.[3] In that theory, the $\vec{\vec{J}}_k$ are angular momentum operators which satisfy the familiar commutation rule

$$\vec{\vec{J}}_j \cdot \vec{\vec{J}}_k - \vec{\vec{J}}_k \cdot \vec{\vec{J}}_j = i \sum_l \delta_{jkl} \vec{\vec{J}}_l.$$

[The δ_{jkl} take on the values ± 1 and zero; they are defined in eq. (6.13) of the following chapter.] The dyadic operators introduced here obey the same relations (see the problem set).

5.7 THE STRESS–STRAIN TENSORS

As a final application of the dyadic–tensor formalism, we should like to examine some interesting results from the theory of elasticity. Suppose that the vector $\vec{r}$ designates the position of a point A in a homogeneous elastic medium (Figure

---

[3] M. E. Rose, *Elementary Theory of Angular Momentum*, John Wiley, New York, 1957, p. 51.

The Stress–Strain Tensors

5.8). If forces are applied so as to strain the medium, this point will be displaced a small amount to A', as represented by the vector

$$\vec{\rho} \equiv \rho_1 \hat{i} + \rho_2 \hat{j} + \rho_3 \hat{k},$$

which in general is dependent upon the position $\vec{r}$. A neighboring point B at a position $\vec{r} + d\vec{r}$ is displaced a corresponding amount $\vec{\rho} + d\vec{\rho}$ to B'. The "shift," or difference of the displacements at A and B, is given by

$$d\vec{\rho} = \frac{\partial \vec{\rho}}{\partial x} dx + \frac{\partial \vec{\rho}}{\partial y} dy + \frac{\partial \vec{\rho}}{\partial z} dz = d\vec{r} \cdot \nabla\vec{\rho}, \qquad (|d\vec{\rho}| \ll |d\vec{r}|)$$

where

$$\nabla\vec{\rho} \equiv \frac{\partial \rho_1}{\partial x} \hat{i}\hat{i} + \frac{\partial \rho_2}{\partial x} \hat{i}\hat{j} + \frac{\partial \rho_3}{\partial x} \hat{i}\hat{k} + \frac{\partial \rho_1}{\partial y} \hat{j}\hat{i} + \frac{\partial \rho_2}{\partial y} \hat{j}\hat{j} + \frac{\partial \rho_3}{\partial y} \hat{j}\hat{k}$$

$$+ \frac{\partial \rho_1}{\partial z} \hat{k}\hat{i} + \frac{\partial \rho_2}{\partial z} \hat{k}\hat{j} + \frac{\partial \rho_3}{\partial z} \hat{k}\hat{k} \qquad (5.51)$$

is the **strain dyadic**.

Let us define some additional strain parameters:

$$a_1 \equiv \frac{\partial \rho_1}{\partial x}, \qquad a_2 \equiv \frac{\partial \rho_2}{\partial y}, \qquad a_3 \equiv \frac{\partial \rho_3}{\partial z},$$

$$g_1 \equiv \frac{1}{2}\left(\frac{\partial \rho_2}{\partial z} + \frac{\partial \rho_3}{\partial y}\right), \qquad g_2 \equiv \frac{1}{2}\left(\frac{\partial \rho_3}{\partial x} + \frac{\partial \rho_1}{\partial z}\right), \qquad g_3 \equiv \frac{1}{2}\left(\frac{\partial \rho_1}{\partial y} + \frac{\partial \rho_2}{\partial x}\right),$$

$$h_1 \equiv \frac{1}{2}\left(\frac{\partial \rho_3}{\partial y} - \frac{\partial \rho_2}{\partial z}\right), \qquad h_2 \equiv \frac{1}{2}\left(\frac{\partial \rho_1}{\partial z} - \frac{\partial \rho_3}{\partial x}\right), \qquad h_3 \equiv \frac{1}{2}\left(\frac{\partial \rho_2}{\partial x} - \frac{\partial \rho_1}{\partial y}\right).$$

$$(5.52)$$

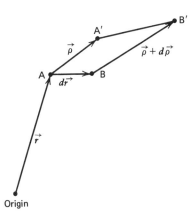

Figure 5.8 Strain in an elastic medium.

The strain dyadic is given as a sum of symmetric and antisymmetric dyadics in terms of these parameters:

$$\nabla \vec{\rho} = \overset{\leftrightarrow}{\Phi} + \overset{\leftrightarrow}{\Omega}, \tag{5.53}$$

with

$$\overset{\leftrightarrow}{\Phi} \equiv a_1 \hat{i}\hat{i} + g_3 \hat{i}\hat{j} + g_2 \hat{i}\hat{k} \qquad \overset{\leftrightarrow}{\Omega} \equiv + h_3 \hat{i}\hat{j} - h_2 \hat{i}\hat{k}$$

$$+ g_3 \hat{j}\hat{i} + a_2 \hat{j}\hat{j} + g_1 \hat{j}\hat{k} \qquad - h_3 \hat{j}\hat{i} + h_1 \hat{j}\hat{k}$$

$$+ g_2 \hat{k}\hat{i} + g_1 \hat{k}\hat{j} + a_3 \hat{k}\hat{k}, \qquad + h_2 \hat{k}\hat{i} - h_1 \hat{k}\hat{j},$$

where $\overset{\leftrightarrow}{\Phi}$ is called the **pure-strain dyadic**, and $\overset{\leftrightarrow}{\Omega}$ is the **rotation dyadic**.

Let us now investigate the physical meaning of these relations. First, orient the x-axis so that the distance $d\vec{r}$ from A to B lies along it. In that event, clearly

$$d\vec{\rho} = (dx\,\hat{i}) \cdot \nabla\vec{\rho} = \left(\frac{\partial \rho_1}{\partial x}\hat{i} + \frac{\partial \rho_2}{\partial x}\hat{j} + \frac{\partial \rho_3}{\partial x}\hat{k}\right) dx.$$

If $\vec{\rho}$ corresponds to a simple **elongation** per unit length, or a stretching along the x-direction, it will only have a ρ_1-component, and the other two terms will vanish. But suppose there is instead a **shear** or sidewise motion along the y-direction, so that $\vec{\rho}$ has the sole component ρ_2. If it is further assumed that ρ_2 is independent of y or z, only the component $\partial \rho_2/\partial x$ survives in the dyadic, and it is therefore defined as the **shearing strain**. Note that a shear involved displacement in one direction per unit of length in another direction. These two examples demonstrate the salient points: the diagonal elements of $\nabla \vec{\rho}$ correspond to elongations along the three axes, while the off-diagonal elements represent shears (Figure 5.9). (An elongation produces a change of volume but not of shape; a shear does just the reverse.)

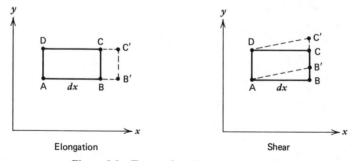

Figure 5.9 Types of strains on a rectangle.

The Stress–Strain Tensors 159

We note further that both $\partial\rho_2/\partial x$ and $\partial\rho_1/\partial y$ refer to shears in the *xy*-plane, so to speak. The *sum* of these two derivatives gives the total shearing strain about the *z*-axis; it also appears in the symmetric g_3-parameter of $\vec{\Phi}$. On the other hand, the *h*-parameters involve *differences* of these same kind of derivatives; the *h*-parameters, in fact, are just the components of $\frac{1}{2}$ curl $\vec{\rho}$. Defining the vector

$$\vec{h} \equiv \tfrac{1}{2} \text{ curl } \vec{\rho},$$

we have

$$d\vec{r}\cdot\vec{\vec{\Omega}} = -\vec{\vec{\Omega}}\cdot d\vec{r} = -\left[h_1(\hat{j}\hat{k} - \hat{k}\hat{j}) + h_2(\hat{k}\hat{i} - \hat{i}\hat{k}) + h_3(\hat{i}\hat{j} - \hat{j}\hat{i})\right]\cdot d\vec{r}$$

$$= \left[h_1(\hat{i}\times\vec{1}) + h_2(\hat{j}\times\vec{1}) + h_3(\hat{k}\times\vec{1})\right]\cdot d\vec{r} = \vec{h}\times d\vec{r}.$$

The conclusion reached in this way is that the antisymmetric contribution to the displacement $d\vec{\rho}$ is simply $\vec{h}\times d\vec{r}$, which strongly resembles eq. (5.24). Dividing by a unit time factor, $\vec{h}$ has the nature of an angular velocity, while the antisymmetric contribution to $d\vec{\rho}$ becomes a linear velocity. (The full connection of antisymmetric matrices and rotations is explored in the next chapter.) Some motions of the medium can then be considered simply as pure rotations, while the symmetric $\vec{\vec{\phi}}$ contribution to $d\vec{\rho}$ is one of "pure strain" without rotation.

To be more explicit, under strain, the particle originally at B will be displaced by the vector $d\vec{r} + d\vec{\rho}$ from that particle initially at A. This is

$$d\vec{r} + d\vec{\rho} = d\vec{r}\cdot(\vec{\vec{1}} + \vec{\vec{\Phi}} + \vec{\vec{\Omega}}),$$

of which the symmetric part is

$$d\vec{r}\cdot(\vec{\vec{1}} + \vec{\vec{\Phi}}) = dx\left[(1 + a_1)\hat{i} + g_3\hat{j} + g_2\hat{k}\right] + dy, dz \text{ terms}.$$

Now, if $d\vec{r}$ is fixed in magnitude but varied in direction, the locus of points B is originally a sphere around A; after the strain is present, however, the terms in $\vec{\vec{\Phi}}$ are added so that in general the locus becomes a rotated ellipsoid. The physical manifestation of the strain is then always given by a squeezing of points on a sphere into an ellipsoidal shape plus an added rotation.

Let us now consider the forces which are responsible for the strain. Imagine a cube of material within a uniform medium such that its faces are oriented perpendicular to the rectangular axes (Figure 5.10a). By the term **stress** one means a force per unit area exerted by the medium across the plane of a face of the cube, a pull being taken as positive and a push as negative (opposite to the outward normal from the cube). The tangential components on each plane are known as the **shearing stresses**, while the normal components are called **tension** stresses when positive and **pressures** when negative. There are three possible orientations of the faces, and for each the stress has three components, so with nine elements a second-rank tensor or dyadic is immediately suggested.

Let us find the stress $\vec{f}$ on a tetrahedron face ABC of area A, which is *not* perpendicular to a coordinate axis (Figure 5.10b). Introducing

$$\hat{n} \equiv n_1 \hat{i} + n_2 \hat{j} + n_3 \hat{k}$$

as a unit normal to the face ABC, then the areas of the faces OBC, OAC, and OAB are given by the projections of ABC, namely, $n_1 A, n_2 A, n_3 A$. Assuming that the tetrahedron as a whole is in equilibrium, the force $A\vec{f}$ on ABC must balance the forces on the other faces; hence,

$$A\vec{f} = \left(F_{xx}\hat{i} + F_{xy}\hat{j} + F_{xz}\hat{k}\right)n_1 A + \left(F_{yx}\hat{i} + F_{yy}\hat{j} + F_{yz}\hat{k}\right)n_2 A$$
$$+ \left(F_{zx}\hat{i} + F_{zy}\hat{j} + F_{zz}\hat{k}\right)n_3 A,$$

where F_{xy} is the force per unit area in the y-direction on a face perpendicular to the x-axis, and so on. On simplifying this expression,

$$\vec{f} = \hat{n} \cdot \vec{\vec{F}} \qquad (5.54)$$

where

$$\vec{\vec{F}} \equiv F_{xx}\hat{i}\hat{i} + F_{xy}\hat{i}\hat{j} + F_{xz}\hat{i}\hat{k} + F_{yx}\hat{j}\hat{i} + \cdots \qquad (5.55)$$

is the **stress dyadic**.

Now consider the forces acting on the cube faces that are perpendicular to the x- and y-axes. The forces F_{yx} on opposite sides will produce a torque (couple)

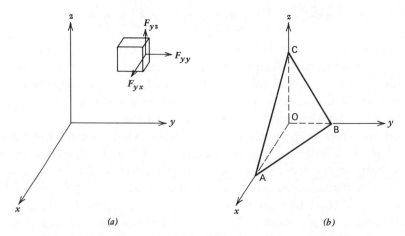

Figure 5.10 Stress analysis for (*a*) cube and (*b*) tetrahedron.

The Stress–Strain Tensors

tending to rotate the cube about the z-axis; in an opposite sense, forces F_{xy} on other faces will do the same. The net torque on a small cube of volume $dx\,dy\,dz$ is just

$$(F_{xy}\,dy\,dz)\,dx - (F_{yx}\,dx\,dz)\,dy = (F_{xy} - F_{yx})\,dx\,dy\,dz.$$

This net torque must be equated to the moment of inertia of the cube times its angular acceleration about the z-direction. But the inertia factor involves the medium density times the volume element times the square of an infinitesimal length factor (the radius of gyration). In the limit at a point in the medium, canceling the cube volume and reducing the radius of gyration to zero, we must conclude that $F_{xy} = F_{yx}$; similar considerations lead to the same results for other axes. The stress tensor is therefore symmetric. The importance of this conclusion lies in the fact that it can then always be diagonalized using the appropriate set of principal axes.

There is one other important consequence of this symmetry. Since both the stress and the pure-strain tensors share this property, each can have only six independent components. According to Hooke's law, for small strains (such that the elastic limit of the material is not exceeded) each element of the stress tensor can be written as a linear function of the nine elements of the pure-strain tensor; the coefficients in the relation are the **moduli of elasticity**. As a result of the symmetrical property, there are only 6×6, or 36, of these coefficients to be considered, rather than 9×9, or 81. In fact, energy relations give further elastic identities which limit the number of independent moduli to only 21 in general.[4] If additional symmetries are present as well (as in the cubic structure of a crystal), there may be only a very few independent coefficients.

Suppose the axes have been selected so as to render the stress dyadic diagonal, eliminating all shears. In an isotropic medium, the pure-strain dyadic must then also be diagonal. Recognizing that a tension stress along x, as an example, produces an elongation in that direction and a compression along y and z (the same by symmetry along either of these axes), the xx-element of the stress tensor can be written in terms of the diagonal elements of the pure-strain dyadics as

$$F_{xx} = c_1 a_1 + c_2 a_2 + c_2 a_3 = (c_1 - c_2)a_1 + c_2(a_1 + a_2 + a_3).$$

Now

$$a_1 + a_2 + a_3 = \frac{\partial \rho_1}{\partial x} + \frac{\partial \rho_2}{\partial y} + \frac{\partial \rho_3}{\partial z} = \nabla \cdot \vec{\rho}$$

is the change in volume (produced by the stress) per unit original volume $dx\,dy\,dz$,

---

[4] W. Band, *Introduction to Mathematical Physics*, Van Nostrand, Princeton, N.J., 1959, p. 67.

seen easily by noting that $[(\partial \rho_1/\partial x)\,dx]\,dy\,dz$ is the added volume from the x-elongation, and so on. Writing similar expressions for the other elements of $\ddot{F}$ and setting $c_2 \equiv k - \tfrac{2}{3}\mu$ and $c_1 \equiv c_2 + 2\mu$ to agree with common notation in the literature, we have

$$\ddot{F} = (k - \tfrac{2}{3}\mu)(\nabla \cdot \vec{\rho})\ddot{1} + 2\mu\ddot{\Phi}. \tag{5.56}$$

This is a tensor equation which is valid in any set of axes.

As simple examples to aid in understanding the meaning of the constants, let us consider the following. First, suppose that $F_{xx} = F_{yy} = F_{zz}$. By eq. (5.56), we must have $a_1 = a_2 = a_3 = (\nabla \cdot \vec{\rho}/3)$, so that

$$\ddot{F} = (k - \tfrac{2}{3}\mu)(\nabla \cdot \vec{\rho})\ddot{1} + 2\mu(\tfrac{1}{3}\nabla \cdot \vec{\rho})\ddot{1} = k(\nabla \cdot \vec{\rho})\ddot{1} = F_{xx}\ddot{1}.$$

Then k is the ratio of stress to change of volume per unit volume or, in other words, the **bulk modulus** of the material. μ can be shown to be the **shear modulus** in a more detailed analysis.

Secondly, let us consider a piece of wire under tension along the x-axis; then,

$$F_{xx} = (k - \tfrac{2}{3}\mu)(\nabla \cdot \vec{\rho}) + 2\mu a_1.$$

Inasmuch as $F_{yy} = F_{zz} = 0$, eq. (5.56) tells us that

$$a_2 = a_3 = -\frac{1}{2\mu}\left(k - \frac{2}{3}\mu\right)(\nabla \cdot \vec{\rho}) = -\frac{1}{2\mu}\left(k - \frac{2}{3}\mu\right)(a_1 + a_2 + a_3),$$

or

$$a_2 = a_3 = -\frac{1}{2}\frac{k - \tfrac{2}{3}\mu}{k + \tfrac{1}{3}\mu}a_1,$$

and

$$\nabla \cdot \vec{\rho} = \frac{\mu}{k + \tfrac{1}{3}\mu}a_1,$$

hence,

$$F_{xx} = \frac{\mu(k - \tfrac{2}{3}\mu)}{k + \tfrac{1}{3}\mu}a_1 + 2\mu a_1 = \frac{3\mu k}{k + \tfrac{1}{3}\mu}a_1.$$

Young's modulus for the wire is then defined as

$$Y \equiv \frac{F_{xx}}{a_1} = \frac{3\mu k}{k + \tfrac{1}{3}\mu}$$

in terms of the bulk and shear moduli.

The Stress–Strain Tensors

The stress tensor has numerous applications in addition to the study of elastic moduli. One that can be pursued without much difficulty is the analysis of wave motion in an isotropic medium. Considering again the cube of Figure (5.10), the net force along the x-direction on the two faces perpendicular to that direction is certainly given by

$$\left(\frac{\partial F_{xx}}{\partial x} dx\right) dy\, dz$$

with x-contributions of

$$\left(\frac{\partial F_{yx}}{\partial y} dy\right) dx\, dz \quad \text{and} \quad \left(\frac{\partial F_{zx}}{\partial z} dz\right) dx\, dy$$

from the other pairs of faces. The total force along the x-axis is then

$$\left(\frac{\partial F_{xx}}{\partial x} + \frac{\partial F_{yx}}{\partial y} + \frac{\partial F_{zx}}{\partial z}\right) dx\, dy\, dz\, \hat{i} \equiv (\nabla \cdot \overset{\leftrightarrow}{F})_x\, dx\, dy\, dz.$$

Inasmuch as the other components behave similarly, the net vector force per unit volume on the cube is just the dyadic divergence $\nabla \cdot \overset{\leftrightarrow}{F}$.

In the limit as the volume is reduced to zero, we can equate this volume force to the product of the medium density D at the cube and its acceleration:

$$D \frac{\partial^2 \vec{\rho}}{\partial t^2} = \nabla \cdot \overset{\leftrightarrow}{F}. \tag{5.57}$$

In the isotropic case, the application of eq. (5.56) now, using $\nabla \cdot \overset{\leftrightarrow}{1} = \nabla$, yields

$$D \frac{\partial^2 \vec{\rho}}{\partial t^2} = (k - \tfrac{2}{3}\mu)\nabla(\nabla \cdot \vec{\rho}) + 2\mu \nabla \cdot \overset{\leftrightarrow}{\Phi}.$$

But

$$\nabla \cdot \overset{\leftrightarrow}{\Phi} = \nabla \cdot \tfrac{1}{2}\left[\nabla\vec{\rho} + (\nabla\vec{\rho})^T\right] = \tfrac{1}{2}\nabla \cdot (\nabla\vec{\rho}) + \tfrac{1}{2}\nabla(\nabla \cdot \vec{\rho})$$

is a dyadic identity which can be established by writing out the components in detail, and it gives

$$D \frac{\partial^2 \vec{\rho}}{\partial t^2} = (k + \tfrac{1}{3}\mu)\nabla(\nabla \cdot \vec{\rho}) + \mu \nabla^2 \vec{\rho} \tag{5.58}$$

as a consequence.

Newton's second law has led directly to eq. (5.58); it is a wave equation for the particle displacement in the medium and has two important kinds of solutions. To understand them, note that any vector such as $\vec{\rho}$ can be decomposed into an irrotational and a solenoidal part, according to a well-known theorem of Helmholtz.[5] Thus,

$$\vec{\rho} \equiv \vec{\rho}_{irr} + \vec{\rho}_{sol}, \qquad (5.59)$$

where curl $\vec{\rho}_{irr} = 0$, div $\vec{\rho}_{sol} = 0$. Both parts of $\vec{\rho}$ must separately satisfy the wave equation. For the solenoidal part, we have simply

$$D\frac{\partial^2 \vec{\rho}_{sol}}{\partial t^2} = \mu \nabla^2 \vec{\rho}_{sol}, \qquad (5.60)$$

an equation describing a transverse plane wave of speed $\sqrt{\mu/D}$. On the other hand, using the identity

$$\text{curl curl } \vec{\rho}_{irr} = 0 = \nabla(\nabla \cdot \vec{\rho}_{irr}) - \nabla^2 \vec{\rho}_{irr},$$

one finds for the irrotational part that

$$D\frac{\partial^2 \vec{\rho}_{irr}}{\partial t^2} = \left(k + \tfrac{1}{3}\mu\right)\nabla^2 \vec{\rho}_{irr} + \mu \nabla^2 \vec{\rho}_{irr} = \left(k + \tfrac{4}{3}\mu\right)\nabla^2 \vec{\rho}_{irr},$$

the equation for a longitudinal plane wave of speed $\sqrt{(k + \tfrac{4}{3}\mu)/D}$.

Let us verify these conclusions for the solenoidal part, which is a little easier to treat. Assuming a plane wave traveling along x with the speed $\sqrt{\mu/D}$, the wave equation must be

$$D\frac{\partial^2 \vec{\rho}_{sol}}{\partial t^2} = \mu \frac{\partial^2 \vec{\rho}_{sol}}{\partial x^2},$$

with each component of $\vec{\rho}_{sol}$ being of the form $f(x - \sqrt{\mu/D}\, t)$ but not dependent on either y or z. In view of the solenoidal condition,

$$\nabla \cdot \vec{\rho}_{sol} = 0 = \frac{\partial \rho_1}{\partial x} + \frac{\partial \rho_2}{\partial y} + \frac{\partial \rho_3}{\partial z} = \frac{\partial \rho_1}{\partial x}.$$

Then, ρ_1 is a constant in x which can be taken as zero for convenience, leaving only transverse components ρ_2 and ρ_3 for the displacement. (Of course, if the wave is more complicated than a plane wave, it may not be strictly transverse either.)

---

[5] E. Harris, *Introduction to Modern Theoretical Physics*, Vol. I, John Wiley, New York, 1975, p. 20.

In general, we see that the wave speed depends only on the two elastic moduli and on the density for an isotropic medium. Inasmuch as a fluid is not capable of sustaining a shear, it has $\mu = 0$, and an elastic wave in a fluid can only be longitudinal. (In addition, for a gas, the bulk modulus must be calculated under adiabatic rather than isothermal conditions, owing to the fact that the rapidity of the oscillations does not allow time for temperature equilibrium to be established.) But in an isotropic solid, both kinds of waves are present, the longitudinal wave being somewhat the faster of the two.

In the case of seismic waves penetrating the Earth's interior regions, the longitudinal waves are called P-waves, while the transverse are known as S-waves; typical speeds are on the order of a few kilometers per second. As different types of rock strata affect the local density inside the Earth and hence the wave speeds, reflection takes place in different ways for the two kinds of waves. P-waves are capable of passing through the Earth's center and reaching a point on the surface 180° from the site of the earthquake which originated the disturbance, while S-waves are strongly absorbed in the core and always appear within a 105° region surrounding the originating point.[6] From the analysis of these waves as received at many seismic stations, it has been possible to construct reasonable hypotheses concerning the nature of the Earth's interior. The core is believed to consist of molten iron and nickel, with a region solidified at the center.

5.8 A MATRIX FORMULATION OF LINEAR COLLISION THEORY

The earlier parts of this chapter may have given the impression that matrices are only useful when treating problems of rotation. Such an idea is certainly not valid, as we now demonstrate.

Suppose we have two masses m_1 and m_2 ($m_1 \geq m_2$) which undergo elastic collisions with one another while both are constrained to one-dimensional motion. Two gliders on an air track will serve as an example; reflections at the ends of the track (without loss of energy) will reverse the direction of motion of the gliders so that they may collide repeatedly.

Following the notation of Chapter 1 (with $\hat{\varepsilon} = \vec{v}/v$), we see that the final primed velocities after a collision are given from eq. (1.20) as

$$v_1' = \frac{m_1 - m_2}{M}v_1 + \frac{2m_2}{M}v_2, \quad (M \equiv m_1 + m_2)$$

$$v_2' = \frac{2m_1}{M}v_1 - \frac{m_1 - m_2}{M}v_2,$$

in terms of the initial unprimed velocities along the track. These equations can be

---

[6]J. Jacobs, R. Russell, and J. Wilson, *Physics and Geology*, 2nd edition, McGraw-Hill, New York, 1974, p. 32.

written somewhat more compactly by the introduction of a new quantity k, defined as

$$k \equiv \frac{m_1 - m_2}{M} \quad (0 \leqslant k < 1). \tag{5.61}$$

The relations (1.20) now become

$$v_1' = kv_1 + (1 - k)v_2,$$
$$v_2' = (1 + k)v_1 - kv_2,$$

which have the form of a linear transformation from initial to final velocities. Writing these results in terms of matrices, we have

$$\mathbf{v}' = \mathbf{S}\mathbf{v}, \tag{5.62}$$

where

$$\mathbf{v}' \equiv \begin{pmatrix} v_1' \\ v_2' \end{pmatrix}, \quad \mathbf{v} \equiv \begin{pmatrix} v_1 \\ v_2 \end{pmatrix},$$

and

$$\mathbf{S} \equiv \begin{pmatrix} k & 1-k \\ 1+k & -k \end{pmatrix}$$

is the **collision matrix**.

Next, suppose that the glider on the right, of mass m_2, strikes the right end of the track. Such a collision reverses the velocity of m_2 but does not affect the velocity of m_1 and therefore can be represented by a matrix

$$\mathbf{R} \equiv \begin{pmatrix} 1 & 0 \\ 0 & -1 \end{pmatrix} \tag{5.63}$$

acting on $\mathbf{v}$. In a similar way, a collision of m_1 with the left end of the track is represented by a matrix

$$\mathbf{L} \equiv \begin{pmatrix} -1 & 0 \\ 0 & 1 \end{pmatrix}. \tag{5.63'}$$

A sequence of events, namely, any combination of collisions of the gliders with one another or with the ends of the track, corresponds to some product of the matrices $\mathbf{S}$, $\mathbf{R}$, and $\mathbf{L}$. Note that

$$\mathbf{S}^2 = \mathbf{L}^2 = \mathbf{R}^2 = \mathbf{1} \tag{5.64}$$

and that

$$\mathbf{LR} = \mathbf{RL} = -\mathbf{1}, \tag{5.64'}$$

while $\mathbf{L}$ and $\mathbf{R}$ do not commute with $\mathbf{S}$. These relations are useful in evaluation of matrix products. (Two successive events of the same type, as represented by a squared matrix, actually do not occur in practice, however.)

As an example, consider the product $\mathbf{P} = \mathbf{LSRS}$. It corresponds to a glider collision followed by a reflection of glider 2 at the right end, then another collision of gliders, and finally a reflection of m_1 at the left end. The sequence of events can be read from the product $\mathbf{P}$, beginning at the right side. The final velocities $\mathbf{v}'$ are given in terms of the initial velocities $\mathbf{v}$ from the matrix expression

$$\mathbf{v}' = \mathbf{Pv}.$$

Certain sequences are of particular interest. One of them is $\mathbf{S(RS)}^n$, where n is an integer specifying a number of collisions and reflections which take place prior to a final collision. Let us suppose further that glider 2 was initially at rest and will again be brought to rest after this sequence, while glider 1 moved with an initial velocity v_1 that will eventually be reversed by the sequence. In mathematical form, we mean that

$$\mathbf{v}' = \begin{pmatrix} -v_1 \\ 0 \end{pmatrix} = \mathbf{Pv} = \mathbf{S(RS)}^n \begin{pmatrix} v_1 \\ 0 \end{pmatrix}. \tag{5.65}$$

It is not difficult to obtain a general expression for $(\mathbf{RS})^n$. Observing that

$$\mathbf{RS} = 2k\mathbf{1} - \mathbf{SR}, \tag{5.66}$$

then

$$(\mathbf{RS})^2 = 4k^2\mathbf{1} - 4k\mathbf{SR} + \mathbf{SRSR}$$

$$= 4k^2\mathbf{1} - 4k\mathbf{SR} + \mathbf{S}(2k\mathbf{1} - \mathbf{SR})\mathbf{R} = (4k^2 - 1)\mathbf{1} - 2k\mathbf{SR}$$

with the aid of eq. (5.64). The result is a linear combination of $\mathbf{1}$ and $\mathbf{SR}$ which is similar to (5.66) but with different coefficients. Continuing to higher powers, we can obviously write

$$(\mathbf{RS})^n = a_n\mathbf{1} + b_n\mathbf{SR}, \tag{5.67}$$

where a_n and b_n are some functions of k. To determine these coefficients, note that

$$(\mathbf{RS})^{n+1} = \mathbf{RS}(\mathbf{RS})^n = \mathbf{RS}(a_n\mathbf{1} + b_n\mathbf{SR})$$

$$= b_n\mathbf{1} + a_n\mathbf{RS} = (b_n + 2ka_n)\mathbf{1} - a_n\mathbf{SR} \equiv a_{n+1}\mathbf{1} + b_{n+1}\mathbf{SR}$$

from eqs. (5.64) and (5.66). Equating coefficients produces

$$b_{n+1} = -a_n$$

and therefore

$$a_{n+1} = 2ka_n + b_n = 2ka_n - a_{n-1}, \qquad (5.68)$$

a recursion relation which in fact identifies the coefficients a_n as **Chebyshev polynomials of the second kind**, called $U_n(k)$ in the literature.[7]

The first few Chebyshev polynomials can be found by consulting tables of functions, or by direct evaluation of coefficients in the scheme outlined above. They are:

$$U_0(k) = 1,$$

$$U_1(k) = 2k,$$

$$U_2(k) = 4k^2 - 1,$$

$$U_3(k) = 8k^3 - 4k,$$

and so on. From (5.67), we can now write

$$\mathbf{S}(\mathbf{RS})^n = a_n \mathbf{S} + b_n \mathbf{R} \equiv U_n(k)\mathbf{S} - U_{n-1}(k)\mathbf{R}.$$

Using this expression, (5.65) yields the two equations

$$kU_n(k) - U_{n-1}(k) + 1 = 0 \quad \text{and} \quad (1 + k)U_n(k) = 0,$$

which can be solved simultaneously by setting

$$U_n(k) = 0. \qquad (5.69)$$

The solution (5.69) must be satisfied if a sequence of the type $\mathbf{S}(\mathbf{RS})^n$ is to occur for a given integer n. In other words, the roots of the polynomial specify certain values of k which are the only ones compatible with the sequence of events; this means that only particular mass ratios can generate such a sequence.

Thus, for $n = 1$, the relevant product is **SRS**, and we must set $U_1(k) = 0 = 2k$, or $k = 0$, implying that the masses are equal in such a sequence. The process is easy to visualize; glider 1 first strikes glider 2 (at rest), giving up all its kinetic

---

[7]*Handbook of Mathematical Functions*, National Bureau of Standards, Washington, D.C., 1964, pp. 773–802.

energy, whereupon glider 2 reflects off the right end and strikes glider 1 again, returning it to the left with its initial speed.

For $n = 2$, $U_2(k) = 4k^2 - 1 = 0$, so $k = \pm \frac{1}{2}$. The root $+ \frac{1}{2}$ corresponds to $m_1 = 3m_2$, while the $- \frac{1}{2}$ root is not possible with $m_1 \geqslant m_2$. The procedure can obviously be continued to higher n. The extraneous roots can be eliminated and the allowed k values expressed in a simple trigonometric form (see the problem set).

When n is a small integer, the straightforward algebraic techniques of Chapter 1 can provide the mass ratios needed for the sequences discussed. The matrix method makes the analysis simpler, however, thereby permitting its extension to higher n with little difficulty. It seems appropriate to consider a matrix treatment whenever linear transformations are involved in a physical problem (they often are).

PROBLEMS

5.1 Prove that the product of two orthogonal matrices must also be an orthogonal matrix, as was assumed in Section 5.2 of this chapter.

5.2 Show explicitly that the full rotation matrix (5.18) is an orthogonal matrix.

5.3 Find the coefficients in the characteristic equation for a 3 × 3 matrix, and show that they are invariant under a similarity transformation.

5.4 A matrix with elements satisfying $a_{ij} = a_{ji}$ is called **symmetric**, while one with $a_{ij} = -a_{ji}$ is known as **antisymmetric**. Show that the properties of symmetry and antisymmetry are preserved under any similarity transformation by an orthogonal matrix.

5.5 Negative integral powers of a matrix can also be found through application of the Cayley–Hamilton theorem (see Appendix C). Consider the matrix

$$\mathbf{A} \equiv \begin{pmatrix} 2 & -2 & 3 \\ 1 & 1 & 1 \\ 1 & 3 & -1 \end{pmatrix}.$$

Express $\mathbf{A}^3$ in lower powers of $\mathbf{A}$ by means of the theorem, then use your result to find $\mathbf{A}^{-1}$. Finally, find the eigenvalues and eigenvectors of $\mathbf{A}$.

5.6 Consider the dyadic $\overset{\leftrightarrow}{D} \equiv (\hat{i} + 3\hat{j} - 2\hat{k})(4\hat{i} - \hat{j} + 5\hat{k})$. Write this dyadic as a sum of two dyadics, one of which is symmetric and one which is antisymmetric.

5.7 Let $\vec{r}_0$ be a unit vector outward from the origin and $\vec{r}$ be the usual position vector; $\vec{r}$ is related to $\vec{r}_0$ through the equation

$$\vec{r} = \overset{\leftrightarrow}{S} \cdot \vec{r}_0,$$

where $\overset{\leftrightarrow}{S}$ is a symmetric dyadic. Show that as the tip of $\vec{r}_0$ moves to various points on the unit sphere, the tip of $\vec{r}$ traces out an ellipsoid.

5.8 A cube of mass m and edge length S is of uniform density and has one corner at the origin with the rectangular axes lying along its edges.
  (a) Calculate the elements of the tensor of inertia.
  (b) If the cube rotates at a constant rate about the x-axis, find the angular momentum.

5.9 A thin uniform rectangular sheet lies in the xy-plane with origin at the center and axes parallel to its edges. The mass of the rectangle is m, and the dimensions of the sides are $2a$ and $2b$ along the x- and y-directions, respectively.
  (a) Find the elements of the tensor of inertia.
  (b) Transform the above tensor to coordinates with the origin at one corner of the rectangle and axes parallel to the original ones. Do your results agree with the parallel axis theorem?

5.10 Consider again the rectangle of Problem 5.9 with the origin at the center.
  (a) Transform the tensor of inertia to coordinates where the x-axis is along a diagonal of the rectangle.
  (b) Calculate the angular momentum of the sheet when it rotates about this diagonal, assuming a constant angular velocity.

5.11 Prove that a straight line which is a principal axis at the center of mass is a principal axis at all points along it.

5.12 Verify the correctness of eqs. (5.40) through (5.43).

5.13 Prove that $\overset{\leftrightarrow}{A} \cdot (\vec{\omega} \times \overset{\leftrightarrow}{1}) \cdot \overset{\leftrightarrow}{A}^T = (\overset{\leftrightarrow}{A} \cdot \vec{\omega}) \times \overset{\leftrightarrow}{1}$ for any orthogonal dyadic $\overset{\leftrightarrow}{A}$ and any vector $\vec{\omega}$. (*Hint*: Much algebra can be avoided by a consideration of the volume formed with vectors $\vec{\omega}$, $\vec{\omega}_1$, and $\vec{\omega}_2$ as sides and the fact that such a volume is invariant under rotations. You need to multiply by other factors to get this form, however.)

5.14 Expand the rotation operator of eq. (5.50) in a series and show that (5.39) is obtained.

5.15 (a) Defining the dyadic operators $\overset{\leftrightarrow}{J}_k \equiv \hat{\varepsilon}_k \times i\overset{\leftrightarrow}{1}$ in terms of unit vectors $\hat{\varepsilon}_k$ along the coordinate axes, show that
$$\overset{\leftrightarrow}{J}_j \cdot \overset{\leftrightarrow}{J}_k - \overset{\leftrightarrow}{J}_k \cdot \overset{\leftrightarrow}{J}_j = i\sum_l \delta_{jkl} \overset{\leftrightarrow}{J}_l.$$
  (b) Defining a matrix representation for the operators by
$$(J_k)_{pq} \equiv \hat{\varepsilon}_p \cdot \overset{\leftrightarrow}{J}_k \cdot \hat{\varepsilon}_q,$$
evaluate the three matrices in detail. (Your results will be familiar from quantum mechanics.)

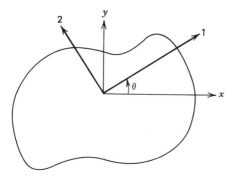

Figure 5.11 Axes for Problem 5.18.

5.16 Show by means of a simple physical situation (or by analysis of dyadic off-diagonal elements) that the coefficient μ of eq. (5.56) is the shear modulus of the medium, defined as the ratio of shearing stress to strain.

5.17 Show that the irrotational part of the solution to (5.58) can be used to describe a plane wave which must be longitudinal and of speed

$$\sqrt{\frac{k + \tfrac{4}{3}\mu}{D}}.$$

5.18 (a) The plane lamina shown in Figure 5.11 has principal moments I_1 and I_2 about its principal axes. Find (by a similarity transformation) the inertia coefficients and products of inertia about the xy-axes.

(b) Calculate

$$I_{av} \equiv \frac{I_{xx} + I_{yy}}{2} \quad \text{and} \quad R \equiv \sqrt{(I_{xx} - I_{av})^2 + I_{xy}^2}.$$

(c) Show that a plot of $(I_{xx} - I_{av})$ against I_{xy} for variable θ is a circle of radius R. This figure is known in engineering texts[8] as **Mohr's circle**. The associated two-dimensional analysis, as demonstrated in this problem, allows one to find moments and products of inertia in terms of the principal moments, the inverse of the usual eigenvalue calculation.

5.19 The Chebyshev polynomials have a trigonometric representation given by

$$U_n(k) = \frac{\sin\left[(n + 1)\arccos k\right]}{\sin(\arccos k)}.$$

(a) Show that this expression satisfies the recurrence relation (5.68).
(b) Show that eq. (5.69), which gives the allowed k values for the sequences discussed, is satisfied in general by $k = \cos[\pi/(n + 1)]$.

---

[8] J. Meriam, *Statics*, 2nd edition, John Wiley, New York, 1971.

5.20 In the linear collision analysis of the last section, let $r \equiv m_1/m_2$. Note that conservation of kinetic energy requires that $rv_1^2 + v_2^2$ be a collision invariant, while in terms of momenta, the invariant is instead $p_1^2 + rp_2^2$ ($p_1 = m_1v_1$, $p_2 = m_2v_2$). Neither $v_1^2 + v_2^2$ nor $p_1^2 + p_2^2$ is conserved in the collision. Define a modified velocity vector μ which is a compromise such that

$$\mu \equiv \begin{pmatrix} \mu_1 \\ \mu_2 \end{pmatrix} \equiv \begin{pmatrix} \sqrt{r}\, v_1 \\ v_2 \end{pmatrix},$$

the squared length of the vector $\mu_1^2 + \mu_2^2$ now itself being invariant under collisions between gliders. Find a matrix **T** which transforms **v** into μ, and show that **S** becomes an orthogonal matrix with determinant -1 under the transformation. What forms do **R** and **L** have in μ-space? (In the next chapter, it will become clear that the transformed **S** performs both a rotation and a reflection of an arbitrary vector μ in the modified velocity space.)

Chapter Six
THE PHYSICS OF ROTATION

6.1 EULER'S THEOREM

In the previous chapter, it was found that only six independent coordinates were required to fully specify the positions of all particles in a rigid body. Three of these six coordinates are often needed to give the location (usually within the body) of the origin for the body set of axes, a point which we shall take as coincident with the origin of the fixed (inertial) space set of axes. If the point of the body at the origin is held stationary, one can state **Euler's theorem** as follows:

The most general possible displacement of a rigid body with one point fixed is a rotation about some axis.

If the constraint of the fixed origin is removed, so that translation becomes possible as well, then one has an obvious consequence of Euler's theorem that is called **Chasles' theorem**:

The most general possible displacement of a rigid body is a translation combined with a rotation.

We shall now prove Euler's theorem. In order to do so, the rotation about the fixed origin will be described by means of the full rotation matrix **R** of eq. (5.18); the elements of this matrix are now functions of the time because the Euler angles vary continuously in time as the rotation proceeds. The matrix **R** gives the orientation of the body axes relative to the space set. If the two frames are coincident at the initial instant, then, at that moment, **R** is the identity matrix **1** (all Euler angles are zero), and its elements will evolve continuously in time thereafter.

It will be recalled from the previous chapter that any orthogonal matrix such as **R** must satisfy the condition $\mathbf{R}^{-1} = \tilde{\mathbf{R}}$, or,

$$\mathbf{R}\tilde{\mathbf{R}} = \mathbf{1}.$$

Forming the determinant of both sides of this relation and using the rule that the determinant of a matrix product is the product of the determinants, we have

$$|\mathbf{R}| \cdot |\tilde{\mathbf{R}}| = |\mathbf{1}| = 1,$$

and since $|\mathbf{R}| = |\tilde{\mathbf{R}}|$, then

$$|\mathbf{R}|^2 = 1, \quad \text{or} \quad |\mathbf{R}| = \pm 1. \tag{6.1}$$

The determinant of an orthogonal matrix must be of unit magnitude. For the rotations of interest in classical mechanics, it will always be positive as well; the case of the negative values will be interpreted shortly. But at this point, consider that $\mathbf{R}$ is initially equal to the unit matrix with determinant $+1$; as $\mathbf{R}$ evolves continuously in time, its determinant will remain at $+1$ and not suddenly undergo a change of sign.

The characteristic eq. (5.33) for any 3×3 matrix $\mathbf{R}$ can always be written

$$\lambda^3 - \lambda^2 Tr(\mathbf{R}) + \lambda Tr(\mathbf{C}) - |\mathbf{R}| = 0, \tag{6.2}$$

where

$$Tr(\mathbf{R}) \equiv \sum_i R_{ii} \tag{6.3}$$

is the trace of $\mathbf{R}$ and $\mathbf{C}$ is the matrix of the cofactors of $\mathbf{R}$.

Relation (6.2) is readily proved by writing out the terms of the eigenvalue equation for a 3×3 matrix, but for the reader unfamiliar with these concepts, a further discussion is given in Appendix C. Now in general the cofactor matrix $\mathbf{C}$ must satisfy

$$\mathbf{R}^{-1} = \frac{\tilde{\mathbf{C}}}{|\mathbf{R}|}. \tag{6.4}$$

For an orthogonal matrix with $|\mathbf{R}| = 1$, we have

$$\mathbf{R}^{-1} = \tilde{\mathbf{R}} = \frac{\tilde{\mathbf{C}}}{|\mathbf{R}|} = \tilde{\mathbf{C}}, \quad \text{or} \quad \mathbf{C} = \mathbf{R}.$$

The characteristic eq. (6.2) then becomes

$$\lambda^3 - (\lambda^2 - \lambda)Tr(\mathbf{R}) - 1 = 0. \tag{6.5}$$

We see by inspection that there is always a root $+1$; factoring out this root gives

$$(\lambda - 1)\{\lambda^2 - \lambda[Tr(\mathbf{R}) - 1] + 1\} = 0.$$

The quadratic factor in the braces has roots

$$\lambda = \frac{[Tr(\mathbf{R}) - 1] \pm \sqrt{[Tr(\mathbf{R}) - 1]^2 - 4}}{2}.$$

Should one of the remaining two roots have the value of unity, then $Tr(\mathbf{R}) = 3$ and *both* roots are unity. But in that event all three roots of (6.5) have unit value, so that the only way of satisfying the trace requirement and the orthogonality conditions (5.5) on the elements of $\mathbf{R}$ (really just direction cosines) is to put $R_{ij} \equiv \delta_{i,j}$, whereupon $\mathbf{R}$ becomes the identity matrix. In other words, the eigenvalue equation for $\mathbf{R}$ always has one and only one root of $+1$ except for the trivial case (as at the initial time) when $\mathbf{R}$ is the unit matrix.

In essence, these properties of $\mathbf{R}$ furnish the proof of Euler's theorem; all that is left is to interpret the results physically. Now, (6.5) is the characteristic equation for the matrix relation

$$\mathbf{X}' = \mathbf{R}\mathbf{X} = \lambda \mathbf{X},$$

which becomes

$$\mathbf{X}' = \mathbf{R}\mathbf{X} = \mathbf{X} \qquad (6.6)$$

for the unique (nontrivial) eigenvalue $+1$. But (6.6) simply states that there is a unique direction in space (that of the eigenvector $\mathbf{X}$ corresponding to the root $+1$) that is unaffected by the rotation, since $\mathbf{R}$ acting on $\mathbf{X}$ does not alter the components of $\mathbf{X}$. Of course, this direction would be that of the instantaneous axis of rotation. (A vector $\mathbf{V}$ along any other direction would have instantaneous components in the body axes given by $\mathbf{V}' = \mathbf{R}\mathbf{V}$, which would differ from the space components $\mathbf{V}$.) Once the eigenvector has been obtained, a single rotation about its direction will take the body from the initial to the final orientation. This last statement is just a rewording of Euler's theorem. (Note that the motion must be "frozen" at an instant of time and the direction of the axis or eigenvector calculated for that moment. A little later, the rotation will take the body to a new orientation and the axis may also have shifted in direction.)

As a final comment, let us again consider the possibility that $|\mathbf{R}| = -1$. A simple example of such a situation is the **inversion** operator, represented by the matrix

$$\mathbf{P} \equiv \begin{pmatrix} -1 & 0 & 0 \\ 0 & -1 & 0 \\ 0 & 0 & -1 \end{pmatrix} = -\mathbf{1}. \qquad (6.7)$$

This operator produces reflections through the origin, changing each of the

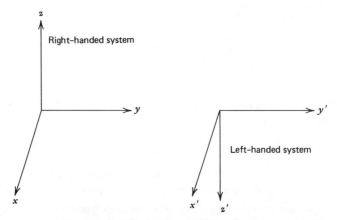

Figure 6.1 Handedness of Cartesian axes.

rectangular coordinates into its negative. A real physical rotation cannot accomplish such an inversion; for example, a rotation of 180° about the z-axis will change the signs of x and y but leave z as it was. The operation of **P** actually changes a right-handed coordinate system to a left-handed one, or vice versa, as illustrated (Figure 6.1). This alteration of the handedness can never be achieved by reorientation of a rigid body, and it is therefore not meaningful in classical mechanics. (The **P** matrix, of course, is the famous **parity** operator of quantum mechanics; in nuclear and particle physics, the handedness of the axes does have physical significance in connection with the momenta and the spins of elementary particles.)

The general matrix **R** of determinant -1 can be obtained from the product of **P** and a matrix of determinant $+1$. A matrix **R** with inversion included, called an "improper rotation" matrix, will be of little interest to us, inasmuch as a proper physical rotation matrix evolves continuously from the unit matrix and has a determinant $+1$.

6.2 VECTORIAL REPRESENTATION OF ROTATIONS; PSEUDOVECTORS

Consider a situation wherein a rigid body undergoes two successive rotations (possibly finite) about different axes (Figure 6.2). In accordance with the scheme of Euler angles, the first rotation is defined to be by the angle ϕ about the direction selected for the z-axis and is given by a dyadic operator of the form

$$\overset{\leftrightarrow}{R}_1 = (1 - \cos\phi)\hat{k}\hat{k} + \cos\phi \overset{\leftrightarrow}{1} + \sin\phi(\hat{k} \times \overset{\leftrightarrow}{1}),$$

from (5.39). The second rotation will be by ψ about a "new" z-axis (denoted by

Vectorial Representation of Rotations; Pseudovectors

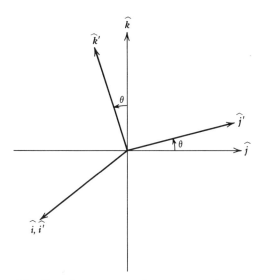

Figure 6.2 Coordinate systems for two successive rotations.

the unit vector $\hat{k}'$) which makes an angle θ with the original one. It will be specified by the dyadic

$$\vec{\vec{R}}_2 = (1 - \cos\psi)\hat{k}'\hat{k}' + \cos\psi \vec{\vec{1}} + \sin\psi(\hat{k}' \times \vec{\vec{1}}).$$

Note that the rigid body is definitely *not* being rotated through the angle θ, but only through ϕ and ψ. The z-axis is merely shifted by θ before the second (ψ) rotation takes place. If the directions of $\hat{k}$ and $\hat{k}'$, and hence the plane they determine, are known in advance, there will be no loss of generality in choosing the x-axis *after* the first (ϕ) rotation so as to be perpendicular to that plane. The shift by θ to the primed z-axis is therefore just a rotation in the yz-plane, so that

$$\hat{k}' = \vec{\vec{R}}(\hat{i}, \theta) \cdot \hat{k} = -(\sin\theta)\hat{j} + (\cos\theta)\hat{k}.$$

With the aid of this relationship, the terms of $\vec{\vec{R}}_2$ could be expressed if desired in the unprimed unit vectors, allowing a direct evaluation of the product $\vec{\vec{R}}_2 \cdot \vec{\vec{R}}_1$.

Now, according to Euler's theorem, there is a single equivalent rotation by angle β about an axis $\hat{n}$ which also will carry the rigid body to its final orientation. In this sense, (5.39) yields

$$(1 - \cos\beta)\hat{n}\hat{n} + \cos\beta \vec{\vec{1}} + \sin\beta(\hat{n} \times \vec{\vec{1}}) = \vec{\vec{R}}_2 \cdot \vec{\vec{R}}_1.$$

The importance of this expression lies in the fact that it enables us to obtain β and $\hat{n}$ for arbitrary values of ϕ, ψ, and θ, even though the product on the right is algebraically complicated. It turns out that the desired results can be derived with less labor if we examine only the antisymmetric parts of the two sides of the equation. Then, we have

$$\sin \beta (\hat{n} \times \vec{\vec{1}}) = \tfrac{1}{2}\left[\vec{\vec{R}}_2 \cdot \vec{\vec{R}}_1 - (\vec{\vec{R}}_2 \cdot \vec{\vec{R}}_1)^T \right]$$
$$= \tfrac{1}{2}\left[\vec{\vec{R}}_2 \cdot \vec{\vec{R}}_1 - \vec{\vec{R}}_1^T \cdot \vec{\vec{R}}_2^T \right]. \qquad (6.8)$$

When the terms of the right side are written out in detail, a number of dyad differences will result. These are readily simplified through the identities

$$\hat{k}\hat{j} - \hat{j}\hat{k} = \hat{i} \times \vec{\vec{1}}, \qquad \hat{i}\hat{k} - \hat{k}\hat{i} = \hat{j} \times \vec{\vec{1}}, \qquad \hat{j}\hat{i} - \hat{i}\hat{j} = \hat{k} \times \vec{\vec{1}}.$$

In addition, note that

$$(\hat{k} \times \vec{\vec{1}}) \cdot \hat{k} = \hat{k}' \times \hat{k} = -\sin\theta \hat{i},$$

and so on. Making use of these and other similar relationships (which we leave for the problem set) as well as trigonometric half-angle formulas, eq. (6.8) becomes

$$\sin \beta (\hat{n} \times \vec{\vec{1}}) = 2\left[\cos\tfrac{\psi}{2} \cos\tfrac{\phi}{2} - \sin\tfrac{\psi}{2} \sin\tfrac{\phi}{2} \cos\theta \right]$$
$$\cdot \left[\cos\tfrac{\psi}{2} \sin\tfrac{\phi}{2} (\hat{k} \times \vec{\vec{1}}) + \sin\tfrac{\psi}{2} \cos\tfrac{\phi}{2} (\hat{k}' \times \vec{\vec{1}}) \right.$$
$$\left. + \sin\tfrac{\psi}{2} \sin\tfrac{\phi}{2} (\hat{k}' \times \hat{k}) \times \vec{\vec{1}} \right]$$
$$= 2\sin\tfrac{\beta}{2} \cos\tfrac{\beta}{2} (\hat{n} \times \vec{\vec{1}}). \qquad (6.8')$$

Factored in this way, a consideration of special cases (such as $\theta = 0$) or a little further algebraic work will readily lead to the identifications

$$\cos\tfrac{\beta}{2} = \cos\tfrac{\psi}{2} \cos\tfrac{\phi}{2} - \sin\tfrac{\psi}{2} \sin\tfrac{\phi}{2} \cos\theta, \qquad (6.9)$$

$$\hat{n} = \frac{1}{\sin\beta/2}\left[\cos\tfrac{\psi}{2} \sin\tfrac{\phi}{2} \hat{k} + \sin\tfrac{\psi}{2} \cos\tfrac{\phi}{2} \hat{k}' + \sin\tfrac{\psi}{2} \sin\tfrac{\phi}{2} (\hat{k}' \times \hat{k}) \right]. \qquad (6.10)$$

Now, an interchange of ψ with ϕ and $\hat{k}'$ with $\hat{k}$ implies a reversal of the order of the two rotations. But the equivalent single rotation angle β is determined by eq. (6.9), which is symmetric in ψ and ϕ. Therefore, the order of rotation is irrelevant insofar as β is concerned.

Unfortunately, the matter is more complicated with regard to the direction of $\hat{n}$. Owing to the sign of the final $\hat{k}' \times \hat{k}$ term in eq. (6.10), the direction of the single equivalent rotation *does* depend on the order of the individual rotations.

It is easy to demonstrate this point. Hold a sheet of paper in a vertical plane, one of the shorter sides being turned toward you. Rotate first by 90° counterclockwise about this edge of the paper, then by a second 90° angle about an axis parallel to the (already rotated) long side; the sheet will end in a horizontal plane. Next, perform these two rotations in the reverse order about the *same* two directions in space; this time the paper will end up once more in a vertical plane.

The real meaning of this somewhat tedious discussion is that *finite* rotations cannot be represented by vectors because such entities would fail to commute on addition, corresponding to the fact that products of rotation matrices or dyadics fail to commute. But suppose that the ϕ and ψ rotations are infinitesimal, with $\sin \psi/2 \simeq \psi/2$, $\cos \psi/2 \simeq 1$, and so on. Neglecting the effect of higher order terms,

$$\hat{n} \simeq \frac{1}{\beta}[\phi\hat{k} + \psi\hat{k}'], \tag{6.10'}$$

an expression symmetric in the two rotations (as is β). It is then feasible to represent infinitesimal rotations as vector quantities, a fortunate situation in that the differential equations of physics often involve small rotations (as in time derivatives such as the angular velocity, for example).

The mathematical treatment of infinitesimal rotations is both simple and interesting. If we turn the vector $\vec{r}$ into $\vec{r}'$ by a small rotation $d\beta$ about the direction of $\hat{n}$, then, according to (5.38) and (5.39), we have

$$\vec{r}' \simeq \vec{r} + d\beta(\hat{n} \times \overset{\leftrightarrow}{1}) \cdot \vec{r}$$

to first order in $d\beta$. Setting

$$d\vec{r} \equiv \vec{r}' - \vec{r} = d\beta \hat{n} \times \vec{r} \tag{6.11}$$

and

$$\vec{\omega} \equiv \frac{d\beta}{dt}\hat{n},$$

then

$$\frac{d\vec{r}}{dt} = \frac{d\beta}{dt}\hat{n} \times \vec{r} = \vec{\omega} \times \vec{r} \tag{6.11'}$$

is the linear velocity of the tip of the rotating vector, in exact agreement with

(5.24). The relation for $d\vec{r}$ demonstrates all the essential features in the representation of infinitesimal rotations by vectors, but there is more to be learned from the viewpoint of matrices.

Solving (6.11) for $\vec{r}'$ and writing it in matrix form, we have

$$\begin{pmatrix} x' \\ y' \\ z' \end{pmatrix} = \begin{pmatrix} 1 & -d\beta\, n_3 & d\beta\, n_2 \\ d\beta\, n_3 & 1 & -d\beta\, n_1 \\ -d\beta\, n_2 & d\beta\, n_1 & 1 \end{pmatrix} \begin{pmatrix} x \\ y \\ z \end{pmatrix} \equiv (1 + \Delta)\mathbf{r},$$

where

$$\Delta \equiv d\beta \begin{pmatrix} 0 & -n_3 & n_2 \\ n_3 & 0 & -n_1 \\ -n_2 & n_1 & 0 \end{pmatrix}$$

is an antisymmetric matrix formed from the components of $\hat{n}$. It should come as no surprise that Δ is antisymmetric, as can be seen from a different kind of argument: The inverse of the rotation matrix $(1 + \Delta)$ is surely $(1 - \Delta)$ to first order, since

$$(1 + \Delta)(1 - \Delta) = 1 - \Delta^2 \simeq 1.$$

But the rotation matrix is orthogonal, with inverse equal to transpose, so

$$(1 + \Delta)^T = 1 + \Delta^T = 1 - \Delta, \quad \text{or} \quad \Delta^T = -\Delta \tag{6.12}$$

and Δ is therefore antisymmetric.

Many of the vectors that enter into rotational calculations, such as torque or angular momentum, are in the form of cross products. These vector products have some unusual transformation properties which can be explored in somewhat greater detail. Before doing so, we define the tensorial quantity δ_{ijk}:

$$\delta_{ijk} = \begin{cases} 1 \text{ if the indices } i, j, k \text{ form an even permutation of } 1, 2, 3 \\ -1 \text{ if the indices form an odd permutation} \\ 0 \text{ if any two of the indices are equal,} \end{cases} \tag{6.13}$$

which is called the **Levi–Civita density**. It is very convenient for simplification of many complicated expressions. For example, the determinant $|\mathbf{A}|$ of a 3×3 matrix $\mathbf{A}$ can be written as

$$|\mathbf{A}| = \tfrac{1}{2} \sum_{\substack{j,k, \\ l,m,n}} \delta_{ijk}\delta_{lmn} a_{il} a_{jm} a_{kn} \tag{6.14}$$

or found from

$$\delta_{ijk}|\mathbf{A}| = \sum_{l,m,n} \delta_{lmn} a_{il} a_{jm} a_{kn}, \tag{6.14'}$$

both of which can be established by writing out the various terms. (See also the definition of a determinant in Appendix C.) As a further illustration, the cross product of two vectors $\vec{A}$ and $\vec{B}$ can be expressed as

$$\vec{A} \times \vec{B} = \sum_{i,j,k} \delta_{ijk} A_j B_k \hat{\varepsilon}_i, \tag{6.15}$$

where $\hat{\varepsilon}_i$ is one of the rectangular unit vectors.

Now, an antisymmetric second-rank tensor is by definition one that has $T_{ij} = -T_{ji}$ and therefore has only three independent elements in three-dimensional space, since its diagonal elements vanish. The three components of *any* vector $\vec{v}$ can then be used to construct the matrix representation of such a tensor, with elements

$$T_{ij} \equiv \sum_k \delta_{ijk} v_k = -T_{ji} \qquad (T_{12} = v_3 = -T_{21}, \text{ etc.}) \tag{6.16}$$

or, alternatively,

$$v_k = \tfrac{1}{2} \sum_{i,j} \delta_{ijk} T_{ij}. \tag{6.16'}$$

The matrix formed with eq. (6.16) from an arbitrary vector may not obey the basic Cartesian transformation law (5.20), however.

But let us suppose we are given a genuine antisymmetric tensor T that *does* satisfy eq. (5.20):

$$T'_{ij} = \sum_{k,l} a_{ik} a_{jl} T_{kl}.$$

We can use it to construct a vector $\vec{v}$ according to eq. (6.16'), since it is the transformation properties of the vector in which we are really interested. Making use of the tensor transformation law under rotations, we can write in the primed coordinates (for some vector $\vec{v}$ formed from the tensor)

$$v'_m = \tfrac{1}{2} \sum_{i,j} \delta_{ijm} T'_{ij} = \tfrac{1}{2} \sum_{\substack{i,j \\ k,l}} \delta_{ijm} a_{ik} a_{jl} T_{kl}.$$

Substitution for T_{kl} from eq. (6.16) and use of (6.14') yields (see the problem set)

$$v'_m = \det \mathbf{A} \sum_n a_{mn} v_n. \tag{6.17}$$

The transformation law is exactly the same as in (5.4), except for the factor of det A in front.

The proof is perfectly general, except that the antisymmetric nature of the tensor was built in from the beginning through the use of eq. (6.16'). The meaning of the result is as follows. Under proper rotations, a vector constructed from such a tensor must transform in the usual way, for det $A = +1$. But if an inversion is included in A, then det $A = -1$ and the law has a minus sign in front which causes an unusual behavior to occur. Thus, for a pure inversion (6.7), $a_{mn} = -\delta_{m,n}$, and all coordinates would normally change sign; but because of the det A factor, they remain the same in sign.

Quantities that obey the transformation law (6.17) are known as **pseudovectors** or **axial vectors**; except for sign they transform exactly like the normal or **polar vector** embodied in (5.4). It is clear that the dot product of two pseudovectors is a normal scalar, whereas the dotting of a pseudovector with a polar vector gives a **pseudoscalar**, for a change of sign will occur under inversion.

Rotational calculations frequently involve cross products, which are axial vectors in nature. Thus, if $\vec{C} = \vec{A} \times \vec{B}$ is the cross product of two polar vectors $\vec{A}$ and $\vec{B}$, then $\vec{A}$ and $\vec{B}$ will both change sign under inversion by eq. (5.4), and therefore $\vec{C}$ cannot change sign, so it must be a pseudovector. A cross product has components which can be viewed as the elements of a (pseudo)vector or those of an antisymmetric tensor, a not too surprising conclusion in view of the form $C_1 = A_2 B_3 - A_3 B_2$, and so on. While the components of a polar vector could also be used in (6.16) to construct an antisymmetric matrix, the elements of such a matrix would not satisfy the tensor transformation law (5.20), but rather that of a pseudotensor. The latter will not be discussed here.

6.3 ROTATING REFERENCE SYSTEMS; ANGULAR VELOCITY

Consider now the motion of a particle as described in two rectangular coordinate systems. One is an inertial, or "space," set of axes x, y, z, and the other is a set x', y', z' moving and perhaps rotating relative to the space set. In the later analysis of rigid body motions, we take the second set as the "body" axes which are fixed within the body; for the moment, however, let us merely look at a particle of mass m instantaneously located at point P (Figure 6.3).

The position vector to P is $\vec{r}$ from the inertial origin O, and $\vec{r}'$ from the moving origin O'; $\vec{R}$ is the vector from O to O', so that

$$\vec{r} = \vec{R} + \vec{r}', \qquad (6.18)$$

where $\vec{r} = x\hat{i} + y\hat{j} + z\hat{k}$, and $\vec{r}' = x'\hat{i}' + y'\hat{j}' + z'\hat{k}'$. Upon differentiating with

Rotating Reference Systems; Angular Velocity

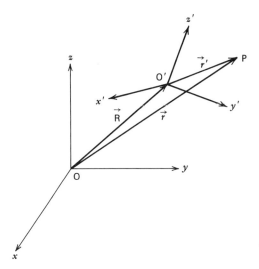

Figure 6.3 Motion of a particle in two different frames.

respect to the time, we have

$$\dot{\vec{r}} = \dot{\vec{R}} + \dot{\vec{r}}' = \dot{\vec{R}} + \left(\dot{x}'\hat{i}' + \dot{y}'\hat{j}' + \dot{z}'\hat{k}'\right) + \left(x'\dot{\hat{i}}' + y'\dot{\hat{j}}' + z'\dot{\hat{k}}'\right). \quad (6.19)$$

The velocity of P for an inertial frame observer then consists of three parts:

1. That due to the relative motion of the origin O', or $\dot{\vec{R}}$.
2. The triplet of terms in the first parentheses of (6.19), which is the **apparent velocity** of the particle relative to an observer in the "moving" (primed) axes.
3. The triplet of terms in the second parentheses, which arise from the rotation of the moving axes, without which $\hat{i}'$, $\hat{j}'$, $\hat{k}'$ would be fixed in direction and their time derivatives would vanish.

Let us focus attention for a moment on the last triplet of terms. For that purpose, let the origins coincide ($\vec{R} = 0$, $\vec{r} = \vec{r}'$) and let $\vec{r}'$ be a fixed vector in the primed frame, so that all but the final three terms of eq. (6.19) will vanish. While the notation being used here is somewhat different, the physical solution is clearly identical with that which gave rise to (6.11'). In the present analysis that equation could be written as

$$\frac{d\vec{r}}{dt} = \vec{\omega} \times \vec{r}',$$

where $\vec{\omega}$ is now interpreted as the angular velocity of the primed axes, the rotation of which is now solely responsible for the motion of the point P.

To understand this relationship better, suppose that $\vec{r}'$ is simply taken as the unit vector $\hat{i}'$ along the moving x'-axis. Then,

$$\frac{d\hat{i}'}{dt} = \vec{\omega} \times \hat{i}' = \omega_{z'}\hat{j}' - \omega_{y'}\hat{k}'$$

follows at once. Note that the time rate of change of $\hat{i}'$ is a linear combination of the *other* two unit vectors; the fact that $\hat{i}'$ is fixed in *magnitude* is responsible for this result. Second, observe that $d\hat{i}'/dt$ vanishes when $\vec{\omega}$ lies along the x'-axis, for $\hat{i}'$ is then a fixed vector along the axis of rotation.

Similar conclusions hold for the remaining unit vectors. Letting $\vec{v}'$ represent the apparent velocity relative to a primed observer, one could write in general that

$$x'\dot{\hat{i}}' + y'\dot{\hat{j}}' + z'\dot{\hat{k}}' = x'(\vec{\omega} \times \hat{i}') + y'(\omega \times \hat{j}') + z'(\omega \times \hat{k}')$$

$$= \vec{\omega} \times (x'\hat{i}' + y'\hat{j}' + z'\hat{k}') = \vec{\omega} \times \vec{r}'.$$

Hence

$$\vec{v} \equiv \dot{\vec{r}} = \dot{\vec{R}} + \vec{v}' + \vec{\omega} \times \vec{r}' \tag{6.20}$$

is the *true* velocity as seen in the inertial frame. It will be noted from this discussion that the time derivative of *any* vector (not simply $\vec{r}$ alone) in the space set of axes is given by the triplet of terms corresponding to differentiation of the components in the moving axes, plus the triplet of terms from differentiation of the unit primed vectors, the latter always being of the form $\vec{\omega} \times$ vector. In general we have

$$\left(\frac{d}{dt}\right)_{\text{space}} = \left(\frac{d}{dt}\right)_{\text{moving}} + \vec{\omega} \times \tag{6.21}$$

as an identity which can be applied to any vector. (If the origins of the two frames are in relative motion, one must also be careful to include the $\vec{R}$ terms where needed.)

Thus, we can obtain the acceleration $\vec{a}$ by employing (6.20):

$$\vec{a} \equiv \dot{\vec{v}} = \ddot{\vec{R}} + \dot{\vec{v}}' + \dot{\vec{\omega}} \times \vec{r}' + \vec{\omega} \times \dot{\vec{r}}'.$$

To clarify this expression, note that from (6.21),

$$\vec{\omega} \times \dot{\vec{r}}' = \vec{\omega} \times \vec{v}' + \vec{\omega} \times (\vec{\omega} \times \vec{r}')$$

since in general

$$\dot{\vec{r}}' = \vec{v}' + \vec{\omega} \times \vec{r}',$$

and also
$$\dot{\vec{v}}' = \vec{a}' + \vec{\omega} \times \vec{v}',$$

where
$$\vec{a}' = \ddot{x}'\hat{i}' + \ddot{y}'\hat{j}' + \ddot{z}'\hat{k}'$$

is the apparent acceleration in the moving frame. (Note also that

$$\vec{\alpha} \equiv (\dot{\vec{\omega}})_{\text{space}} = (\dot{\vec{\omega}})_{\text{moving}} + \vec{\omega} \times \vec{\omega} = (\dot{\vec{\omega}})_{\text{moving}};$$

that is, the *angular* acceleration of the rotating axes is the same in either frame.) Putting all these things together, we get

$$\vec{a} = \ddot{\vec{R}} + \vec{a}' + \vec{\omega} \times \vec{v}' + \dot{\vec{\omega}} \times \vec{r}' + \vec{\omega} \times \vec{v}' + \vec{\omega} \times (\vec{\omega} \times \vec{r}')$$

$$= \ddot{\vec{R}} + \dot{\vec{\omega}} \times \vec{r}' + \vec{\omega} \times (\vec{\omega} \times \vec{r}') + 2\vec{\omega} \times \vec{v}' + \vec{a}'. \quad (6.22)$$

By Newton's law, the force on the particle is $\vec{F} = m\vec{a}$, so that if we solve eq. (6.22) for the product $m\vec{a}'$, we find

$$m\vec{a}' = \vec{F} - m\ddot{\vec{R}} - m\dot{\vec{\omega}} \times \vec{r}' - m\vec{\omega} \times (\vec{\omega} \times \vec{r}') - 2m(\vec{\omega} \times \vec{v}'). \quad (6.23)$$

This expression gives the effective force acting on the mass from the standpoint of an observer in the rotating frame. It consists of five terms, the first being the actual physical force $\vec{F}$ that is present, and the second arising from the relative acceleration of the origin O'. The remaining three terms are entirely due to the rotation of the primed axes and need to be discussed further. All the terms subtracted from $\vec{F}$ are often referred to as "fictitious forces," but they are quite real to the primed observer! If the latter wishes to apply Newton's laws in his or her own frame, the effects of these forces must be taken into account; they arise solely because of the attempt to apply Newton's laws in a frame which is moving relative to an inertial frame. The three rotational terms are:

1. The angular acceleration term containing $\dot{\vec{\omega}}$. In many common situations, such as the analysis of the motion of an object on the surface of the rotating Earth (the angular velocity of which is approximately constant), there is no contribution from this term.
2. The centrifugal force term $-m\vec{\omega} \times (\vec{\omega} \times \vec{r}')$. (This is a generalization of the familiar $m\omega^2 r$ force of elementary physics.) Imagine a child whirling a stone on the end of a string, the rock moving around a circle in a horizontal plane;

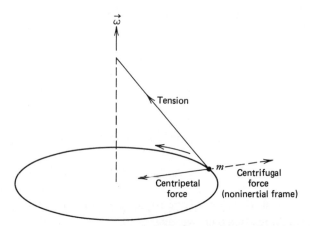

Figure 6.4 Circular motion of stone on string.

for our considerations, gravity can be ignored (Figure 6.4). Taking the angular velocity as directed upward, then the centrifugal force (because of the minus sign) will be outward. In the child's frame of reference, which in this example is assumed to be inertial, the stone is in accelerated motion caused by the inward centripetal force of the tension in the string. Next, consider instead a frame rotating with the rock, in which the rock is stationary. The application of Newton's laws in the moving frame requires that there be an outward force (centrifugal) in order to balance the string tension and allow the stone to remain at rest. The stone is "in equilibrium" in the noninertial frame *only*, a fact which has confused textbook authors in the past and thus has led to numerous erroneous statements.

3. The Coriolis term $-2m\vec{\omega} \times \vec{v}'$. The important effects of the Coriolis force provide a striking illustration of the fictitious forces present in a rotating reference system. As a classic example, consider a physicist who drops a mass from the top of a tall tower of height d, this action taking place at latitude λ in the northern hemisphere (Figure 6.5). As the mass falls, roughly toward the Earth's center, the Coriolis force will be directed eastward. Taking a set of axes attached to the Earth with the x'-axis pointing east, one obtains

$$\frac{d^2x'}{dt^2} = 2\omega v' \cos \lambda = 2\omega gt \cos \lambda$$

for the Coriolis acceleration. We have designated the Earth's angular velocity as ω and have approximated the speed of the falling object by the free-fall formula $v' = gt$. Upon integrating and choosing the constants of integration

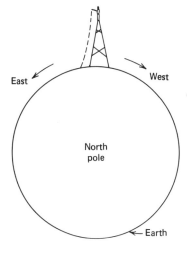

Figure 6.5 Earth observer's view of deflected falling object, looking down on Earth from above.

to be zero, we find

$$x' = \tfrac{1}{3}\omega g t^3 \cos \lambda.$$

The time required to fall a distance equal to the height of the tower is given from the relation

$$d = \tfrac{1}{2} g t^2, \quad \text{or} \quad t = \sqrt{\frac{2d}{g}},$$

whereupon the eastward deflection at the ground will be approximately

$$x' = \tfrac{1}{3}\omega \cos \lambda \sqrt{\frac{8d^3}{g}}. \tag{6.24}$$

This very small amount (less than an inch for a 100-ft tower) is not normally observable, being overwhelmed in significance by the effects of atmospheric turbulence. From the standpoint of an inertial frame observer, the deflection is readily understood: a point at the top of the tower has a slightly greater linear velocity toward the east (because of Earth's rotation and the greater distance of the top from the center of rotation) than a point beneath on the ground. During the time of fall, this greater velocity component causes a slight deflection.

For an object moving across the Earth in a horizontal plane, the Coriolis force tends to cause a deflection to the right in the northern hemisphere and to the left in the southern, thus providing an explanation for commonly observed cyclonic wind patterns. As before, a mass of air moving to the north from the equatorial region will deflect to the east, since it will have a greater eastward component of velocity than those air masses in more northerly latitudes, as far as an inertial observer is concerned. Actually, there is an increase in velocity from conservation of angular momentum as the air mass moves north and the radius of latitude circles shrinks in size.

The rotation of the Earth has been adequately demonstrated ever since the middle of the nineteenth century, when the physicist Foucault used the Coriolis force on a pendulum bob for that purpose. The equation of motion for the pendulum is, from eq. (6.23),

$$m\vec{a}' = m\vec{g} + \vec{T} - 2m\vec{\omega} \times v', \qquad \vec{F} \equiv m\vec{g} + \vec{T}, \tag{6.25}$$

where $\vec{T}$ is the tension in the supporting cord and $\vec{g}$ is the vector acceleration of gravity (which can be considered to include the centrifugal term as a small correction that affects the value of $\vec{g}$ very little). We have taken the primed axes as rigidly fixed on the rotating Earth, with the z'-axis vertically upward and the origin at the point of suspension.

Consider also a double-primed set of axes with the z''- and z'-axes (as well as the origins) coincident, but for which the x''- and y''-axes rotate relative to the primed set with a constant angular velocity $\vec{\omega}_0$. By analogy with (6.22), the acceleration of the bob in the Earth frame is

$$\vec{a}' = \vec{a}'' + \vec{\omega}_0 \times (\vec{\omega}_0 \times \vec{r}'') + 2(\vec{\omega}_0 \times \vec{v}''),$$

where $\vec{R}$ and $\dot{\vec{\omega}}_0$ are both zero and $\vec{v}''$ is the apparent velocity in the "rotating" frame, and so on. We know from (6.20) that

$$\vec{v}' = \vec{v}'' + \vec{\omega}_0 \times \vec{r}'',$$

hence

$$\vec{a}'' = \vec{a}' - 2(\vec{\omega}_0 \times \vec{v}') + \vec{\omega}_0 \times (\vec{\omega}_0 \times \vec{r}'').$$

Upon multiplication by m and insertion of (6.25), this relation becomes

$$m\vec{a}'' = m\vec{g} + \vec{T} - 2m(\vec{\omega} + \vec{\omega}_0) \times \vec{v}' + m\vec{\omega}_0 \times (\vec{\omega}_0 \times \vec{r}''). \tag{6.26}$$

Now, eq. (6.25) gives the effective force on the pendulum as seen by an Earth observer, while the last equation gives the force for an observer attached to the double-primed set of axes; $\vec{\omega}$ represents the Earth's angular velocity, with component $\omega_z \hat{k}'$ along the primed z-axis. Suppose that the quantity $\vec{\omega}_0$, as yet unspecified, is chosen as

$$\vec{\omega}_0 \equiv -\omega_z \hat{k}'.$$

The centrifugal term in (6.26) contains ω_0^2 and is quite small, while the Coriolis term will have vanishing horizontal components for this choice of $\vec{\omega}_0$, if it is assumed that the motion represented by $\vec{v}'$ must lie essentially in the horizontal plane. Therefore, (6.26) is approximately

$$m\vec{a}'' = m\vec{g} + \vec{T} \equiv \vec{F} \quad \text{(horizontal components)}.$$

In the double-primed axes, the equation of motion is identical with that in an inertial frame, the acceleration being determined by the actual force that acts, so in this frame the pendulum oscillates in a plane. Inasmuch as these axes are rotating at a rate ω_z (or $\omega \sin \lambda$) relative to an Earth observer, the plane of oscillation appears to be continuously shifting to such a person, providing proof of the Earth's turning. Due to the motion of the point of suspension at other latitudes, the effect is most easily visualized at the poles, where the Earth rotates "under" the fixed plane of oscillation of the pendulum.

6.4 TIME-DEPENDENT ROTATION AXES; EULER'S EQUATIONS

It is now feasible to attack rigid body problems with the aid of the machinery already assembled. In the applications of Lagrange's equations, the Euler angles can be taken as the generalized rotational coordinates; translational coordinates could be supplied if needed. In the study of rigid body motions with one point fixed, however, it is often more convenient to start from another set of relations, known as **Euler's equations**. These relations are designed so as to involve the principal moments of inertia and the components of torque and angular velocity more directly.

In view of the fact that the Euler relations involve components of the angular velocity along the body axes, it is desirable to express these components in terms of the Euler angles and their time derivatives. The standard textbook procedure for the accomplishment of this goal utilizes the product $\mathbf{R}_3 \mathbf{R}_2 \mathbf{R}_1$ of the partial rotation matrices in Section 5.2 to find the components of $\dot{\phi}$ along the body axes. Since the direction of the angular velocity $\dot{\theta}$ lies along the line of nodes, the matrix $\mathbf{R}_3$ alone gives its body components, while $\dot{\psi}$ is already along the body z'-axis. This method, which we shall refer to as the "projection technique," simply adds the contributions of the three angular velocities and obtains what are, in fact, the correct expressions for the body components of $\vec{\omega}$:

$$\begin{aligned} \omega_{x'} &= \dot{\phi} \sin \theta \sin \psi + \dot{\theta} \cos \psi, \\ \omega_{y'} &= \dot{\phi} \sin \theta \cos \psi - \dot{\theta} \sin \psi, \\ \omega_{z'} &= \dot{\phi} \cos \theta + \dot{\psi}. \end{aligned} \quad (6.27)$$

This procedure seems at first sight to be satisfactory, for it apparently just involves the addition of infinitesimal rotations (as divided by dt). But with rotation axes that are moving in time (such as those of θ or ψ), there are some added complications which will be discussed shortly. Let us therefore develop an alternate method of finding the components (6.27).

Consider a vector $\vec{r}'$ fixed in the body set of axes; it will have instantaneous components in the space set given by

$$\mathbf{r} = \mathbf{R}^{-1}\mathbf{r}'$$

as follows from the inversion of (5.11). Time differentiation (with $\mathbf{r}'$ constant) produces

$$\dot{\mathbf{r}} = \dot{\mathbf{R}}^{-1}\mathbf{r}' = \tilde{\dot{\mathbf{R}}}\mathbf{r}' = \tilde{\dot{\mathbf{R}}}\mathbf{R}\mathbf{r}.$$

But the identity (5.24) or (6.11') tells us that

$$\left(\frac{d\vec{r}}{dt}\right)_{\text{space}} = \vec{\omega} \times \vec{r} = (\vec{\omega} \times \vec{1}) \cdot \vec{r}.$$

Written as a matrix, the dyadic $\vec{\omega} \times \vec{1}$ therefore has components which are identical with those of $\tilde{\dot{\mathbf{R}}}\mathbf{R}$; it is in this way that the angular velocity is related to the Euler angles and their derivatives. To clarify the relation, note from the product (5.18) that

$$\tilde{\dot{\mathbf{R}}}\mathbf{R} = \frac{d}{dt}(\tilde{\mathbf{R}}_1 \tilde{\mathbf{R}}_2 \tilde{\mathbf{R}}_3)(\mathbf{R}_3 \mathbf{R}_2 \mathbf{R}_1)$$

$$= \tilde{\dot{\mathbf{R}}}_1 \mathbf{R}_1 + \tilde{\mathbf{R}}_1 \tilde{\dot{\mathbf{R}}}_2 \mathbf{R}_2 \mathbf{R}_1 + \tilde{\mathbf{R}}_1 \tilde{\mathbf{R}}_2 \tilde{\dot{\mathbf{R}}}_3 \mathbf{R}_3 \mathbf{R}_2 \mathbf{R}_1.$$

These matrix products can be worked out rather simply; for example,

$$\tilde{\dot{\mathbf{R}}}_1 \mathbf{R}_1 = \dot{\phi}\begin{pmatrix} -\sin\phi & -\cos\phi & 0 \\ \cos\phi & -\sin\phi & 0 \\ 0 & 0 & 0 \end{pmatrix}\begin{pmatrix} \cos\phi & \sin\phi & 0 \\ -\sin\phi & \cos\phi & 0 \\ 0 & 0 & 1 \end{pmatrix} = \dot{\phi}\begin{pmatrix} 0 & -1 & 0 \\ 1 & 0 & 0 \\ 0 & 0 & 0 \end{pmatrix}.$$

The final result of the algebra is the antisymmetric matrix

$$\tilde{\dot{\mathbf{R}}}\mathbf{R} = \begin{pmatrix} 0 & -(\dot{\phi} + \dot{\psi}\cos\theta) & (\dot{\theta}\sin\phi - \dot{\psi}\sin\theta\cos\phi) \\ (\dot{\phi} + \dot{\psi}\cos\theta) & 0 & -(\dot{\theta}\cos\phi + \dot{\psi}\sin\theta\sin\phi) \\ -(\dot{\theta}\sin\phi - \dot{\psi}\sin\theta\cos\phi) & (\dot{\theta}\cos\phi + \dot{\psi}\sin\theta\sin\phi) & 0 \end{pmatrix}.$$

Equating this to the matrix of $\vec{\omega} \times \overset{\leftrightarrow}{1}$, which is

$$\begin{pmatrix} 0 & -\omega_z & \omega_y \\ \omega_z & 0 & -\omega_x \\ -\omega_y & \omega_x & 0 \end{pmatrix},$$

we find

$$\omega = \begin{pmatrix} \dot{\theta}\cos\phi + \dot{\psi}\sin\theta\sin\phi \\ \dot{\theta}\sin\phi - \dot{\psi}\sin\theta\cos\phi \\ \dot{\phi} + \dot{\psi}\cos\theta \end{pmatrix} \tag{6.27'}$$

for the space components of the angular velocity. The body components of (6.27) can be obtained by applying the full rotation matrix **R** to the space components (6.27'). (Alternatively, the "angular velocity matrix" $\dot{\mathbf{R}}\mathbf{R}$ can be similarity transformed to the body axes by calculation of $\mathbf{R}(\dot{\mathbf{R}}\mathbf{R})\mathbf{R}^{-1} = \mathbf{R}\dot{\mathbf{R}}$ in much the same way as was just done; the result gives the body components directly.)

There are other aspects of this situation which are more easily investigated from the active point of view. Consider a rotation β about the direction of $\hat{n}$ which carries the initial position vector $\vec{r}(0)$ into that at some later arbitrary time $\vec{r}(t)$. According to eqs. (5.38) and (5.39),

$$\vec{r}(t) = \overset{\leftrightarrow}{R}(\hat{n}, \beta) \cdot \vec{r}(0),$$

where, as always,

$$\overset{\leftrightarrow}{R}(\hat{n}, \beta) = (1 - \cos\beta)\hat{n}\hat{n} + \cos\beta\overset{\leftrightarrow}{1} + \sin\beta(\hat{n} \times \overset{\leftrightarrow}{1})$$

is the familiar dyadic rotation operator. Noting that

$$\frac{d\vec{r}(t)}{dt} = \overset{\leftrightarrow}{\dot{R}} \cdot \vec{r}(0) = \overset{\leftrightarrow}{\dot{R}} \cdot \overset{\leftrightarrow}{R}^T \cdot \vec{r}(t) = \vec{\omega} \times \vec{r}(t) = (\vec{\omega} \times \overset{\leftrightarrow}{1}) \cdot \vec{r}(t),$$

then

$$\vec{\omega} \times \overset{\leftrightarrow}{1} = \overset{\leftrightarrow}{\dot{R}} \cdot \overset{\leftrightarrow}{R}^T \tag{6.28}$$

follows in much the same way as for the matrix form. It is a straightforward, if lengthy, algebraic exercise to evaluate this dyadic product; we leave it for the problem set. Making repeated use of the fact that the derivative $\dot{\hat{n}}$ of the unit vector $\hat{n}$ is perpendicular to $\hat{n}$, one eventually arrives at the expression

$$\vec{\omega} = \dot{\beta}\hat{n} + \sin\beta\dot{\hat{n}} + (1 - \cos\beta)\hat{n} \times \dot{\hat{n}}. \tag{6.29}$$

Relation (6.29) is the finite rotation formula which allows for a variation in time of the direction specified by $\hat{n}$. If that axis is fixed in time or the angle β is infinitesimally small, then the angular velocity $\vec{\omega}$ reduces to the familiar form $\dot{\beta}\hat{n}$.

A glance at the equation is sufficient to indicate why the geometrical derivation of the $\vec{\omega}$-components mentioned earlier may not be a rigorously correct approach. Both the θ- and ψ-rotations are finite and are taken about time-dependent axes. The projection approach properly takes into account the angular velocities $\dot{\phi}, \dot{\theta}, \dot{\psi}$ but does not allow in a proper procedural fashion for those associated with the movement of the θ- and ψ-axes. We have already found the $\vec{\omega}$-components with the analytical matrix considerations, but let us pursue this matter a little further.

Proceeding from the operator $\overset{\leftrightarrow}{R}(\hat{n}, \beta)$ in the product form of eq. (5.45),

$$\overset{\leftrightarrow}{R}(\hat{n}, \beta) = \overset{\leftrightarrow}{R}_3(\hat{k}'', \psi) \cdot \overset{\leftrightarrow}{R}_2(\hat{i}', \theta) \cdot \overset{\leftrightarrow}{R}_1(\hat{k}, \phi),$$

then (6.28) gives

$$\vec{\omega} \times \overset{\leftrightarrow}{1} = \overset{\leftrightarrow}{\dot{R}} \cdot \overset{\leftrightarrow}{R}^T = \frac{d}{dt}[\overset{\leftrightarrow}{R}_3 \cdot \overset{\leftrightarrow}{R}_2 \cdot \overset{\leftrightarrow}{R}_1] \cdot [\overset{\leftrightarrow}{R}_1^T \cdot \overset{\leftrightarrow}{R}_2^T \cdot \overset{\leftrightarrow}{R}_3^T]$$

$$= \overset{\leftrightarrow}{\dot{R}}_3 \cdot \overset{\leftrightarrow}{R}_3^T + \overset{\leftrightarrow}{R}_3 \cdot \overset{\leftrightarrow}{\dot{R}}_2 \cdot \overset{\leftrightarrow}{R}_2^T \cdot \overset{\leftrightarrow}{R}_3^T + \overset{\leftrightarrow}{R}_3 \cdot \overset{\leftrightarrow}{R}_2 \cdot \overset{\leftrightarrow}{\dot{R}}_1 \cdot \overset{\leftrightarrow}{R}_1^T \cdot \overset{\leftrightarrow}{R}_2^T \cdot \overset{\leftrightarrow}{R}_3^T$$

$$= \vec{\omega}_\psi \times \overset{\leftrightarrow}{1} + \overset{\leftrightarrow}{R}_3 \cdot (\vec{\omega}_\theta \times \overset{\leftrightarrow}{1}) \cdot \overset{\leftrightarrow}{R}_3^T + \overset{\leftrightarrow}{R}_3 \cdot \overset{\leftrightarrow}{R}_2 \cdot (\vec{\omega}_\phi \times \overset{\leftrightarrow}{1}) \cdot \overset{\leftrightarrow}{R}_2^T \cdot \overset{\leftrightarrow}{R}_3^T, \quad (6.30)$$

where

$$\vec{\omega}_\phi \times \overset{\leftrightarrow}{1} \equiv \overset{\leftrightarrow}{\dot{R}}_1 \cdot \overset{\leftrightarrow}{R}_1^T, \quad \vec{\omega}_\theta \times \overset{\leftrightarrow}{1} \equiv \overset{\leftrightarrow}{\dot{R}}_2 \cdot \overset{\leftrightarrow}{R}_2^T, \quad \vec{\omega}_\psi \times \overset{\leftrightarrow}{1} \equiv \overset{\leftrightarrow}{\dot{R}}_3 \cdot \overset{\leftrightarrow}{R}_3^T. \quad (6.31)$$

Making use of the previously established identity (5.47), or

$$\overset{\leftrightarrow}{R} \cdot (\vec{V} \times \overset{\leftrightarrow}{1}) \cdot \overset{\leftrightarrow}{R}^T = (\overset{\leftrightarrow}{R} \cdot \vec{V}) \times \overset{\leftrightarrow}{1},$$

eq. (6.30) becomes

$$\vec{\omega} = \vec{\omega}_\psi + \overset{\leftrightarrow}{R}_3 \cdot \vec{\omega}_\theta + \overset{\leftrightarrow}{R}_3 \cdot \overset{\leftrightarrow}{R}_2 \cdot \vec{\omega}_\phi. \quad (6.32)$$

The $\vec{\omega}$ expressions defined in (6.31) are true angular velocities in the sense of (6.29), but the presence of the $\overset{\leftrightarrow}{R}$'s and the time-dependent axes makes the calculation somewhat cumbersome, so the components (6.27') do not follow at once from (6.32).

There is an easier way to proceed, however, through utilization of the "rotation-reversal" theorem embodied in (5.49). It will be recalled that the product $\overset{\leftrightarrow}{R}$ can also be written

$$\overset{\leftrightarrow}{R}(\hat{n}, \beta) = \overset{\leftrightarrow}{R}_1(\hat{k}, \phi) \cdot \overset{\leftrightarrow}{R}_2(\hat{i}, \theta) \cdot \overset{\leftrightarrow}{R}_3(\hat{k}, \psi)$$

Time-Dependent Rotation Axes; Euler's Equations

in terms of rotations about the unprimed axes. But these axes are fixed and not time dependent, hence we need keep only the first term of (6.29) in each of the constituent angular velocities. Repeating the previous derivation, we find now that

$$\vec{\omega} = (\vec{\omega}_\phi)_0 + (\vec{\tilde{R}}_1)_0 \cdot (\vec{\omega}_\theta)_0 + (\vec{\tilde{R}}_1)_0 \cdot (\vec{\tilde{R}}_2)_0 \cdot (\vec{\omega}_\psi)_0, \qquad (6.33)$$

where the angular velocities are defined as in (6.31), except that the $\vec{\tilde{R}}$ expressions now refer solely to the unprimed axes, as indicated by the zero subscripts. Converted into matrix terms, this result agrees with the calculation which eventually led to (6.27′).

Let us make one last comment on these matters. From $(\vec{\omega}_\theta)_0 = \dot{\theta}\hat{i}$, it is easy to show that

$$\vec{\tilde{R}}_1(\hat{k}, \phi) \cdot (\vec{\omega}_\theta)_0 = \dot{\theta}(\cos\phi\, \hat{i} + \sin\phi\, \hat{j}) = \dot{\theta}\hat{i}'.$$

Similarly, one can establish with more effort that

$$\vec{\tilde{R}}_1(\hat{k}, \phi) \cdot \vec{\tilde{R}}_2(\hat{i}, \theta) \cdot (\vec{\omega}_\psi)_0 = \dot{\psi}\hat{k}'',$$

whereupon, from (6.33),

$$\vec{\omega} = \dot{\phi}\hat{k} + \dot{\theta}\hat{i}' + \dot{\psi}\hat{k}''. \qquad (6.34)$$

The scheme of Euler rotations leads to this simple expression for $\vec{\omega}$, which in turn partially justifies the geometrical derivation of the $\vec{\omega}$-components mentioned earlier in connection with (6.27).

In general, however, when moving axes (with their own contribution to the angular momentum) are involved, angular velocities must be combined with caution. A treatment which simply adds the infinitesimal rotations as vector quantities is not adequate in such cases, but instead a lengthy calculation of the type just given will usually be required. (Thus, as an example, even though the Euler rotations can be performed in reverse order about the unprimed axes, it would certainly not be correct to write $\vec{\omega}$ in the form

$$\vec{\omega} = \dot{\phi}\hat{k} + \dot{\theta}\hat{i} + \dot{\psi}\hat{k}.$$

Once the correct form of (6.34) has been properly derived, however, we can take its components along any chosen set of axes.)

Having rigorously obtained the components of $\vec{\omega}$ in terms of the Euler angles and their time derivatives, many kinds of problems can be investigated. First, however, we wish to derive and demonstrate the use of Euler's equations of motion. In an inertial frame of reference such as the space set of axes, eq. (1.11) for the torque is

$$\vec{N} = \left(\frac{d\vec{L}}{dt}\right)_{space} = \left(\frac{d\vec{L}}{dt}\right)_{body} + \vec{\omega} \times \vec{L} \qquad (6.35)$$

when (6.21) has been used. Taking instantaneous components of this vector equation along the body set of *principal* axes, which will simply be labeled x', y', and z', we have, as an example, the x'-component

$$N_{x'} = \frac{dL_{x'}}{dt} + \omega_{y'}L_{z'} - \omega_{z'}L_{y'}.$$

Recalling that in principal axes

$$L_{x'} = I_1\omega_{x'}, \qquad L_{y'} = I_2\omega_{y'}, \qquad L_{z'} = I_3\omega_{z'},$$

one can write

$$N_{x'} = I_1\dot{\omega}_{x'} - \omega_{y'}\omega_{z'}(I_2 - I_3).$$

By cyclic permutation, the full set of Euler equations is obtained:

$$N_{x'} = I_1\dot{\omega}_{x'} - \omega_{y'}\omega_{z'}(I_2 - I_3),$$
$$N_{y'} = I_2\dot{\omega}_{y'} - \omega_{z'}\omega_{x'}(I_3 - I_1), \qquad (6.36)$$
$$N_{z'} = I_3\dot{\omega}_{z'} - \omega_{x'}\omega_{y'}(I_1 - I_2).$$

As a simple example of the use of Euler's equations, consider the problem of finding the angular momentum and torque of a plane disk of mass m and radius a which is rotating so as to keep the angular velocity constant at $\vec{\omega} = \omega(\sin\beta \hat{j}' + \cos\beta \hat{k}')$ in the body set of principal axes (Figure 6.6). Choosing the z'-axis perpendicular to the plane of the disk, we have with the aid of the symmetry that

$$I_1 = I_2 = \tfrac{1}{2}I_3 = \tfrac{1}{4}ma^2.$$

Then, from (5.27), we have

$$L_{x'} = 0,$$

$$L_{y'} = I_2\omega_{y'} = \tfrac{1}{4}ma^2\omega\sin\beta,$$

$$L_{z'} = I_3\omega_{z'} = \tfrac{1}{2}ma^2\omega\cos\beta,$$

Force-Free Motion of a Rigid Body; the Poinsot Construction

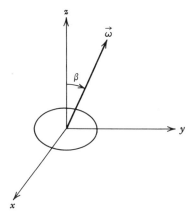

Figure 6.6 Rotating disk and principal axes.

while, from (6.36),

$$N_{x'} = -\omega_{y'}\omega_{z'}(I_2 - I_3) = \tfrac{1}{4}ma^2\omega^2 \sin\beta\cos\beta,$$

$$N_{y'} = N_{z'} = 0.$$

The components of $\vec{L}$ and $\vec{N}$ are constants in the body axes but not in the space axes, which are in effect rotating about the direction of $\vec{\omega}$. Note also that, in general, $\vec{L}$ is not parallel to $\vec{\omega}$.

6.5 FORCE-FREE MOTION OF A RIGID BODY; THE POINSOT CONSTRUCTION

In the case of the motion of a rigid body with one point fixed and no net applied torque acting, the left sides of the Euler equations, eqs. (6.36), must vanish. From the basic torque relation (1.11), we know that $\vec{L}$ is a constant of the motion, and it is not difficult to show that T is also constant when no torques act. Recall that

$$\vec{N} = \left(\frac{d\vec{L}}{dt}\right)_{body} + \vec{\omega} \times \vec{L}.$$

Upon taking the dot product with $\vec{\omega}$, we obtain

$$\vec{\omega} \cdot \vec{N} = \vec{\omega} \cdot \left(\frac{d\vec{L}}{dt}\right)_{body} + \vec{\omega} \cdot (\vec{\omega} \times \vec{L}) = \vec{\omega} \cdot \left(\frac{d\vec{L}}{dt}\right)_{body},$$

the second dot product on the right being zero. According to (5.27), $\vec{L} = \overleftrightarrow{I} \cdot \vec{\omega}$; relative to the body set of axes,

$$\left(\frac{d\vec{L}}{dt}\right)_{body} = \overleftrightarrow{I} \cdot \frac{d\vec{\omega}}{dt},$$

the dyadic $\vec{\vec{I}}$ being constant in that reference frame and the angular acceleration $d\vec{\omega}/dt$ being the same in either body or space axes. Then, we have

$$\vec{\omega} \cdot \vec{N} = \vec{\omega} \cdot \vec{\vec{I}} \cdot \frac{d\vec{\omega}}{dt} = \frac{d\vec{\omega}}{dt} \cdot \vec{\vec{I}} \cdot \vec{\omega} \quad \text{(from the symmetry of } \vec{\vec{I}}\text{)}$$

$$= \frac{1}{2} \frac{d}{dt}(\vec{\omega} \cdot \vec{\vec{I}} \cdot \vec{\omega}) = \frac{dT}{dt},$$

using (5.29). The energy relation is

$$\frac{dT}{dt} = \vec{\omega} \cdot \vec{N}, \tag{6.37}$$

which clearly shows that T is constant whenever $\vec{N}$ is zero. From the constants $\vec{L}$ and T, it is possible to integrate the Euler equations and to solve force-free rotational problems accordingly, but the procedure is often mathematically complex and unrewarding.

A simpler and more elegant geometrical description of the motion was obtained by Poinsot in 1834. At the heart of the Poinsot construction is the figure known as the **inertia ellipsoid**. In order to introduce this concept, let us define a new vector $\vec{r}$ (which is *not* the usual position vector) by

$$\vec{r} = \frac{1}{\sqrt{I}} \hat{n} = \frac{1}{\sqrt{I}} \frac{\vec{\omega}}{\omega} = \frac{\vec{\omega}}{\sqrt{2T}}. \tag{6.38}$$

This vector is along the instantaneous axis of rotation $\hat{n}$, and the use of (5.29) has allowed us to express it as proportional to the angular velocity, I being the moment of inertia about the direction $\hat{n}$. In general, the vector $\vec{\omega}$ undergoes some kind of movement in time, and as it does so, the tip of $\vec{r}$ traces out a geometrical surface, the magnitude of $\vec{r}$ becoming large for those directions in which I is small. Utilizing the usual component notation for $\vec{r}$, we have, from (5.28),

$$\vec{r} \cdot \vec{\vec{I}} \cdot \vec{r} = I_{xx}x^2 + I_{yy}y^2 + I_{zz}z^2 + 2I_{xy}xy + 2I_{xz}xz + 2I_{yz}yz$$

$$= \frac{1}{I}(\hat{n} \cdot \vec{\vec{I}} \cdot \hat{n}) = 1,$$

showing that $\vec{r}$ moves on an ellipsoidal surface. Choosing the axes of $\vec{r}$ so as to coincide with those of the principal body set, the relation takes on the simpler form

$$I_1 x^2 + I_2 y^2 + I_3 z^2 = 1. \tag{6.39}$$

Force-Free Motion of a Rigid Body; the Poinsot Construction

This equation defines the inertia ellipsoid, which is rigidly attached to the body axes and has a shape determined solely by the principal moments. Should all three moments be the same, the ellipsoid becomes a sphere. (There is generally a rough correspondence between the shape of the rigid body and that of the ellipsoid.)

Inasmuch as the figure is characterized by a constant value of $\vec{r} \cdot \ddot{I} \cdot \vec{r}$ equal to unity, the normal to its surface at the position $\vec{r}$ can be found by the gradient operation:

$$\nabla(\vec{r} \cdot \ddot{I} \cdot \vec{r}) = 2I_1 x\hat{i} + 2I_2 y\hat{j} + 2I_3 z\hat{k}.$$

The unit vectors refer to the body axes, although primes have been omitted for simplicity. Now,

$$\vec{L} = \ddot{I} \cdot \vec{\omega} = \sqrt{2T}\, \ddot{I} \cdot \vec{r} = \sqrt{2T}\left[I_1 x\hat{i} + I_2 y\hat{j} + I_3 z\hat{k}\right]$$

$$= \sqrt{T/2}\, \nabla(\vec{r} \cdot \ddot{I} \cdot \vec{r}). \tag{6.40}$$

Expressed in words, this result means that the normal to the ellipsoid is along the fixed direction of the angular momentum, or that the tangent plane to the ellipsoidal surface is perpendicular to $\vec{L}$. Suppose that d is the distance from the origin to this tangent plane, that is,

$$d = \frac{\vec{r} \cdot \vec{L}}{|\vec{L}|} = \frac{1}{\sqrt{2T}|\vec{L}|}(\vec{\omega} \cdot \ddot{I} \cdot \vec{\omega}) = \frac{2T}{\sqrt{2T}|\vec{L}|} = \frac{\sqrt{2T}}{|\vec{L}|}. \tag{6.41}$$

The distance d is then a constant, so that the tangent plane is frequently called the **invariable plane**. The motion is then described as follows (Figure 6.7). The inertia ellipsoid, defined by the tip of the vector $\vec{r}$, rolls without slipping (since the point of contact is along the direction of $\vec{\omega}$, the instantaneous axis of rotation) on the invariable plane. The center of the ellipsoid at the origin always remains a fixed distance above the plane, which is determined completely by the initial specification of T and $|\vec{L}|$. As the ellipsoid rolls, it traces out a curve on its surface called the **polhode**, the corresponding curve traced on the invariable plane being called the **herpolhode**.

If the rigid body is symmetrical about the z-axis, so that $I_1 = I_2$, the motion of the ellipsoid is readily visualized. It rolls such that both the polhode and the herpolhode are circles, the ellipsoid rolling around the fixed direction of $\vec{L}$ while $\vec{\omega}$ precesses in time about the z-axis of the body. That is, the body axes rotate instantaneously at the rate $\vec{\omega}$ while the precession of $\vec{\omega}$ itself just mentioned is an *additional* rotation relative to the body axes, and it is carried out at a different rate. To see this fact, note that in this symmetrical case, the Euler equations, eqs.

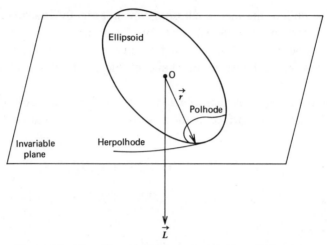

Figure 6.7 The ellipsoid of inertia and the invariable plane.

(6.36), become

$$I_1\dot{\omega}_x - \omega_y\omega_z(I_1 - I_3) = 0,$$

$$I_1\dot{\omega}_y - \omega_z\omega_x(I_3 - I_1) = 0, \qquad (6.42)$$

$$\dot{\omega}_z = 0, \qquad \text{setting } I_2 = I_1 \text{ and } \vec{N} = 0.$$

The last relation tells us that $\omega_z = $ constant. To solve the first two equations, one can either assume a complex exponential solution or introduce a complex variable

$$V \equiv \omega_x + i\omega_y,$$

and let

$$\Omega \equiv \frac{I_3 - I_1}{I_1}\omega_z.$$

In view of the fact that $\dot{\omega}_x + \Omega\omega_y = 0$ and $\dot{\omega}_y - \Omega\omega_x = 0$, V must satisfy $\dot{V} = i\Omega V$, which from its real and imaginary parts gives the first two Euler equations upon substituting the form of V. The solution of the differential equation for V is clearly

$$V = Ae^{i\Omega t}, \qquad A = \text{constant}, \qquad (6.43)$$

which implies $\omega_x = A\cos\Omega t$, $\omega_y = A\sin\Omega t$. These relations along with the con-

Force-Free Motion of a Rigid Body; the Poinsot Construction

stant nature of ω_z demonstrate the precession of $\vec{\omega}$ about the z-axis at the rate Ω as discussed. The magnitude of $\vec{\omega}$ is also constant:

$$|\vec{\omega}| = \sqrt{\omega_x^2 + \omega_y^2 + \omega_z^2} = \sqrt{A^2 + \omega_z^2}.$$

From the relations $T = \frac{1}{2}\vec{\omega} \cdot \vec{I} \cdot \vec{\omega}$ and $\vec{L} = \vec{I} \cdot \vec{\omega}$, A and ω_z can be found in terms of the constants T and $|\vec{L}|$.

Returning to the geometry of the $\vec{\omega}$ precession, we note that the $\vec{\omega}$ vector traces out a cone of half-angle θ_B as it moves around the z-axis in the body set, where

$$\tan \theta_B = \frac{A}{\omega_z}, \qquad (6.44)$$

A being simply the amplitude of the x- and y-components of $\vec{\omega}$. This cone is called the **body cone**. It turns out that, in the space set of axes, $\vec{\omega}$ also precesses on a cone, the **space cone**, about the fixed direction of $\vec{L}$. To see this explicitly, we realize that the half-angle θ_S of the space cone is given by

$$\cos \theta_S = \frac{\vec{\omega} \cdot \vec{L}}{\omega |\vec{L}|} = \frac{\vec{\omega} \cdot \vec{I} \cdot \vec{\omega}}{\omega |\vec{L}|} = \frac{2T}{\omega |\vec{L}|}, \qquad (6.45)$$

the right side being constant. It follows that the body cone and space cone intersect along the instantaneous direction of $\vec{\omega}$; as the motion progresses, the body cone rolls (without slipping!) around the space cone (Figure 6.8). The polhode is the intersection of the body cone and the ellipsoid of inertia.

This force-free analysis can be applied approximately to the Earth's motion, since the torques due to other astronomical bodies are small. The Earth is

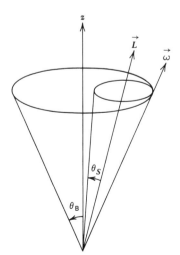

Figure 6.8 The body cone and space cone.

symmetrical about the polar axis, bulging somewhat at the equator so that

$$\frac{I_3 - I_1}{I_1} \simeq \frac{1}{300}, \qquad \Omega \simeq \frac{1}{300}\omega_z,$$

with ω_z being 1 revolution per day. The period of precession of $\vec{\omega}$ about one of the poles then amounts to approximately 300 days. Apparent observations of this phenomenon give a period nearly 50% larger, a difference ascribed to the nonrigidity of the Earth. (The experimental determinations are made by observing the point in the night sky which is stationary by virtue of being on the axis of rotation at that time.)

Let us now examine the case of torque-free motion where all principal moments are different and ordered such that $I_3 > I_2 > I_1$. We know from (6.36) that $\vec{\omega}$ is constant only when it lies along one of the principal axes, for then two of its components must vanish. Suppose that $\vec{\omega}$ is almost but not quite along one of these axes, say, the z-axis. Then, $\omega_z \gg \omega_x$ and $\omega_z \gg \omega_y$; the Euler equations are approximately given by

$$\dot{\omega}_x + \omega_y \Omega_1 = 0,$$

$$\dot{\omega}_y - \omega_x \Omega_2 = 0,$$

$$\dot{\omega}_z \approx 0,$$

with

$$\Omega_1 \equiv \frac{I_3 - I_2}{I_1}\omega_z \quad \text{and} \quad \Omega_2 \equiv \frac{I_3 - I_1}{I_2}\omega_z, \qquad \omega_z \simeq \text{constant}.$$

Assume solutions of the form $\omega_x = Ae^{i\alpha t}$, $\omega_y = Be^{i\alpha t}$, which upon substitution produce $i\alpha A + B\Omega_1 = 0$ and $i\alpha B - A\Omega_2 = 0$. Solving simultaneously, one finds

$$\alpha = \sqrt{\Omega_1 \Omega_2}, \quad B = A\frac{\Omega_2}{i\alpha} = A\sqrt{\frac{\Omega_2}{\Omega_1}}\,e^{-i\pi/2},$$

whereupon

$$\omega_x = A\cos\left(\sqrt{\Omega_1 \Omega_2}\,t\right), \qquad \omega_y = A\sqrt{\frac{\Omega_2}{\Omega_1}}\sin\left(\sqrt{\Omega_1 \Omega_2}\,t\right),$$

from the real parts of the assumed exponentials. (These solutions are generalizations of the ones obtained previously for the symmetrical moment situation.) If $\vec{\omega}$

is directed almost along the x-axis, the formulation and results are similar to those for the z-axis.

But suppose $\vec{\omega}$ is almost along the y-axis, I_2 being the *intermediate* moment in size. The Euler equations then undergo a significant sign change:

$$\dot{\omega}_x + \Omega_4 \omega_z = 0,$$

$$\dot{\omega}_y \simeq 0 \quad (\omega_y \simeq \text{constant}),$$

$$\dot{\omega}_z + \Omega_3 \omega_x = 0,$$

where

$$\Omega_4 \equiv \frac{I_3 - I_2}{I_1} \omega_y, \quad \Omega_3 \equiv \frac{I_2 - I_1}{I_3} \omega_y.$$

The solutions are obtained as above, giving in this case

$$\omega_x = A e^{\mp \sqrt{\Omega_3 \Omega_4}\, t}, \quad \omega_z = \pm A \sqrt{\frac{\Omega_3}{\Omega_4}}\, e^{\mp \sqrt{\Omega_3 \Omega_4}\, t}.$$

In other words, here we get *real* exponential solutions, one of which damps out quickly, while the other increases rapidly in time (the positive exponent). This is associated with the fact that there is instability of the motion around the direction of the intermediate moment, a theorem sometimes referred to as the "tennis racket" theorem. (Figure 6.9 should help in understanding this point; there is a detailed example in the problem set.)

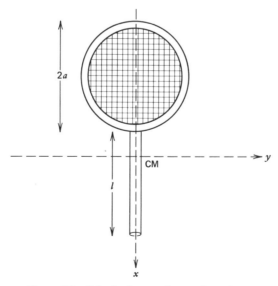

Figure 6.9 Principal axes of a tennis racket.

6.6 MISCELLANEOUS APPLICATIONS

Example 1 Torque on a Satellite in Circular Orbit

In general, an artificial satellite experiences a nonzero torque which tends to align it along the radial gravity gradient vector. In all likelihood, a similar process orients the moon so that the same side always faces Earth. For the geometrical analysis, we find it convenient to choose a set of unprimed coordinates x, y, z with origin at the satellite CM. The axes of these coordinates are not body axes, however, for they will be taken to rotate at the angular velocity $\vec{\omega}$ of the orbital motion of the satellite, so that the body will shift its orientation relative to these axes as it proceeds around the orbit. (In particular, an element of the inertia tensor calculated with these coordinates will *not* be a constant.) The x-axis will be chosen antiparallel to $\vec{\omega}$, the y-axis in the orbit plane, and the z-axis will be chosen along the direction of the vector $\vec{R}$ from the Earth's center to the CM. The vector from Earth's center to an element of mass dm will be denoted as $\vec{\rho}$ (Figure 6.10), while $\vec{r}$ will be taken as the vector from the CM to the element.

Having established the notation, the effective force on dm is then given from (6.23) as

$$d\vec{F} = -GM\,dm\,\frac{\vec{\rho}}{\rho^3} - dm\,\ddot{\vec{R}} - dm\,\dot{\vec{\omega}} \times \vec{r}$$

$$ - dm\,\vec{\omega} \times (\vec{\omega} \times \vec{r}) - 2\,dm(\vec{\omega} \times \vec{v}),$$

where M is the mass of the Earth, G is the gravitational constant, and $\vec{v}$ is the apparent velocity in the xyz-coordinates. The torque about the CM is obtained by integration over the body:

$$\vec{N} = \int \vec{r} \times d\vec{F}.$$

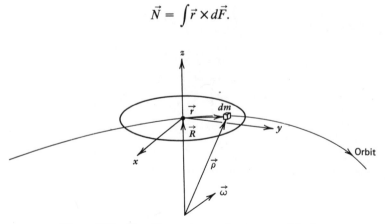

Figure 6.10 Coordinate systems for torque analysis.

Miscellaneous Applications

Making use of the fact that $\vec{R} = R\hat{k}$,

$$\dot{\vec{R}} = R\dot{\hat{k}} = R\vec{\omega} \times \hat{k} = \vec{\omega} \times \vec{R} \quad \text{and} \quad \ddot{\vec{R}} = \vec{\omega} \times (\vec{\omega} \times \vec{R}), \text{ taking } \dot{\vec{\omega}} = 0.$$

In addition, $\vec{\rho} = \vec{R} + \vec{r}$, hence one has

$$\vec{v} = \dot{\vec{r}} - \vec{\omega} \times \vec{r} = \dot{\vec{\rho}} - \dot{\vec{R}} - \vec{\omega} \times \vec{\rho} + \vec{\omega} \times \vec{R} = \dot{\vec{\rho}} - \vec{\omega} \times \vec{\rho} \equiv \dot{\vec{\rho}}_R,$$

the apparent velocity as seen in the rotating frame. Substituting these results into $d\vec{F}$ produces

$$d\vec{F} = -GM\, dm \frac{\vec{\rho}}{\rho^3} - dm\, \vec{\omega} \times (\vec{\omega} \times \vec{\rho}) - 2dm\, \vec{\omega} \times \dot{\vec{\rho}}_R.$$

There are thus three contributions to the torque; we shall examine each term individually.

First, note for the gravity term that

$$\rho^2 = R^2 + r^2 + 2\vec{R} \cdot \vec{r}, \quad \frac{1}{\rho^3} = \frac{1}{R^3}\left(1 + \frac{r^2}{R^2} + \frac{2\vec{R} \cdot \vec{r}}{R^2}\right)^{-3/2}.$$

Neglecting terms of second order and higher in an expansion for $R \gg r$,

$$\frac{1}{\rho^3} \approx \frac{1}{R^3}\left(1 - \frac{3\vec{R} \cdot \vec{r}}{R^2}\right).$$

Kepler's third law requires that $GM/R^3 = \omega^2$, giving for the torque of gravity

$$\vec{N}_g \approx -\omega^2 \int dm(\vec{r} \times \vec{\rho})\left(1 - \frac{3\vec{R} \cdot \vec{r}}{R^2}\right).$$

Using the fact that $\vec{r} = x\hat{i} + y\hat{j} + z\hat{k}$ and $\int \vec{r}\, dm = 0$ by the CM definition, we can simplify this expression to

$$\vec{N}_g = -\omega^2 \int dm(\vec{r} \times \vec{R})\left(1 - \frac{3z}{R}\right) = -\omega^2 \int dm R(-x\hat{j} + y\hat{i})\left(1 - \frac{3z}{R}\right)$$

$$= 3\omega^2 \int dm\, z(-x\hat{j} + y\hat{i}) = 3\omega^2(-I_{yz}\hat{i} + I_{xz}\hat{j}), \tag{6.46}$$

writing the result in terms of products of inertia.

The centrifugal part of the torque is easily found from the fact that $\vec{\omega} = -\omega\hat{i}$, giving

$$\vec{N}_{\text{cen}} = -\int dm\, \vec{r} \times [\vec{\omega} \times (\vec{\omega} \times \vec{\rho})] = -\omega^2 \int dm\, \vec{r} \times [\hat{i} \times (\hat{i} \times \vec{\rho})].$$

Now,

$$\hat{i} \times (\hat{i} \times \vec{\rho}) = \hat{i} \times [\hat{i} \times (R\hat{k} + x\hat{i} + y\hat{j} + z\hat{k})] = -y\hat{j} - (R+z)\hat{k},$$

whereupon we have

$$\vec{N}_{\text{cen}} = \omega^2 \int dm\, \vec{r} \times [y\hat{j} + (R+z)\hat{k}] = \omega^2 \int dm\, x(y\hat{k} - z\hat{j})$$

$$= \omega^2 [I_{xz}\hat{j} - I_{xy}\hat{k}]. \qquad (6.47)$$

Finally, the Coriolis term produces

$$\vec{N}_{\text{cor}} = -2\int dm\, \vec{r} \times (\vec{\omega} \times \dot{\vec{\rho}}_R)$$

$$= 2\omega\left[(\omega_y' I_{xz} - \omega_z' I_{xy})\hat{i} + (\omega_z' I - \omega_x' I_{xz})\hat{j} + (\omega_x' I_{xy} - \omega_y' I)\hat{k}\right], \qquad (6.48)$$

where $I \equiv -\int dm\, x^2$ and $\vec{\omega}'$ is the angular velocity of the satellite relative to the xyz-axes. Since $\vec{R}$ is a constant vector in those axes, $\dot{\vec{\rho}}_R$ is found from $\vec{\omega}' \times \vec{r}$ only.

The addition of the three contributions from (6.46), (6.47), and (6.48) will produce the total torque. Unfortunately, however, these expressions involve time-varying products of inertia, which are not convenient in practice. Thus, we consider a rotational transformation by a matrix **R** which carries **I** from the xyz-axes into the principal body set of axes according to

$$\mathbf{I}' = \mathbf{R}\mathbf{I}\mathbf{R}^{-1},$$

where **I**' is diagonal and constant, the elements of **R** being functions of the time. Simple special cases of this relation apply when the spacecraft is undergoing pure pitch, roll, or yaw motions; these will now be examined.

Pure pitching rotation corresponds to $\omega_y' = \omega_z' = 0$, $\omega_x' = \dot{\theta}_P$, θ_P being the pitch angle. Since the x-axis is the axis of rotation, the rotation matrix **R** is

$$\mathbf{R} = \begin{pmatrix} 1 & 0 & 0 \\ 0 & \cos\theta_P & \sin\theta_P \\ 0 & -\sin\theta_P & \cos\theta_P \end{pmatrix}.$$

The inverse of **R** is found as usual by transposing; calling the principal moments I_1, I_2, and I_3 (diagonal elements of **I**'), we find with a little matrix algebra involving evaluation of $\mathbf{I} = \mathbf{R}^{-1}\mathbf{I}'\mathbf{R}$ that

$$I_{yz} = (I_2 - I_3)\sin\theta_P \cos\theta_P, \qquad I_{xy} = I_{xz} = 0,$$

Miscellaneous Applications   205

and thus

$$\vec{N}_P = -\tfrac{3}{2}\omega^2(I_2 - I_3)\sin(2\theta_P)\hat{i} \tag{6.49}$$

is the total torque for pitching motion [only eq. (6.46) has contributed to it].

Pure rolling takes place about the y-axis and yawing about the z-axis. If θ_R and θ_Y are the roll and yaw angles, respectively, a similar approach in those cases (explored in the problems) yields

$$\vec{N}_{\text{roll}} = -\omega\dot{\theta}_R(I_1 - I_3)\sin(2\theta_R)\hat{i} - 2\omega^2(I_1 - I_3)\sin(2\theta_R)\hat{j}$$
$$+ \omega\dot{\theta}_R[I_2 - (I_1 - I_3)\cos(2\theta_R)]\hat{k} \tag{6.50}$$

and

$$\vec{N}_{\text{yaw}} = -\omega\dot{\theta}_Y(I_1 - I_2)\sin(2\theta_Y)\hat{i} - \omega\dot{\theta}_Y[I_3 - (I_1 - I_2)\cos(2\theta_Y)]\hat{j}$$
$$- \tfrac{1}{2}\omega^2(I_1 - I_2)\sin(2\theta_Y)\hat{k}. \tag{6.51}$$

Example 2 The Heavy Symmetrical Top

It is traditional to employ the methods of rigid body dynamics for the study of the motion of a top; there are many applications, but the analysis of the motion is in itself a striking example of how the powerful mathematical techniques we have developed can be used to extract a great deal of physical information from a minimum amount of calculation.

We choose the symmetry axis of the top as the z-principal axis and fix the bottom (supporting) point of the top at the origin of coordinates. The top is of weight Mg, and this can be considered as concentrated at the CM, a distance l from the origin along the figure (z) axis (Figure 6.11). The Euler angles were especially designed for the treatment of top rotations, and they will prove very convenient. It is found, however, that Lagrangian techniques are simpler to apply to the top than are the Euler equations. With this view in mind, we can write the Lagrangian as

$$L = T - V = \tfrac{1}{2}I_1(\omega_x^2 + \omega_y^2) + \tfrac{1}{2}I_3\omega_z^2 - Mgl\cos\theta,$$

taking the zero of potential at the support point and using the symmetry condition $I_1 = I_2$. It is clear from the geometry of the body axes shown in Figure 6.11, or from eq. (6.27), that the angular velocity is

$$\vec{\omega} = (\dot{\phi}\sin\theta\sin\psi + \dot{\theta}\cos\psi)\hat{i} + (\dot{\phi}\sin\theta\cos\psi - \dot{\theta}\sin\psi)\hat{j} + (\dot{\phi}\cos\theta + \dot{\psi})\hat{k},$$

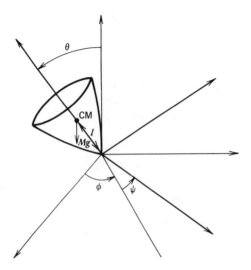

Figure 6.11 Geometry for the study of a top.

whereupon the Lagrangian becomes simply

$$L = \tfrac{1}{2}I_1(\dot{\theta}^2 + \dot{\phi}^2 \sin^2\theta) + \tfrac{1}{2}I_3(\dot{\psi} + \dot{\phi}\cos\theta)^2 - Mgl\cos\theta. \quad (6.52)$$

The conjugate momenta are

$$p_\psi \equiv \frac{\partial L}{\partial \dot{\psi}} = I_3(\dot{\psi} + \dot{\phi}\cos\theta) = I_3\omega_z, \quad (6.53)$$

$$p_\phi \equiv \frac{\partial L}{\partial \dot{\phi}} = I_1(\sin^2\theta)\dot{\phi} + I_3(\dot{\psi} + \dot{\phi}\cos\theta)\cos\theta$$

$$= (I_1\sin^2\theta + I_3\cos^2\theta)\dot{\phi} + I_3\dot{\psi}\cos\theta, \quad (6.54)$$

and

$$p_\theta \equiv \frac{\partial L}{\partial \dot{\theta}} = I_1\dot{\theta}.$$

From the discussion of Chapter 2, the first two angular momenta, p_ψ and p_ϕ, are constants of the motion inasmuch as ψ and ϕ are ignorable coordinates that do not appear in L. There is another constant of the motion available in the total energy $E = T + V$. Subtracting the constant contribution of the ω_z term in T, we find

$$E' \equiv E - \tfrac{1}{2}I_3\omega_z^2 = \tfrac{1}{2}I_1(\dot{\theta}^2 + \dot{\phi}^2\sin^2\theta) + Mgl\cos\theta = \text{constant}.$$

Solving eq. (6.53) for $\dot\psi$ yields

$$\dot\psi = \omega_z - \dot\phi \cos\theta,$$

which upon substitution in (6.54) produces

$$(I_1 \sin^2\theta + I_3 \cos^2\theta)\dot\phi + I_3 \cos\theta(\omega_z - \dot\phi\cos\theta) = I_1(\sin^2\theta)\dot\phi + I_3 \omega_z \cos\theta$$
$$= \text{constant} \equiv I_1 B,$$

or, with $A \equiv (I_3/I_1)\omega_z$,

$$\dot\phi = \frac{B - A\cos\theta}{\sin^2\theta}, \tag{6.55}$$

in terms of new constants A and B which absorb the various constant factors in the relation. Also from (6.53), we have

$$\dot\psi = \omega_z - \frac{\cos\theta}{\sin^2\theta}(B - A\cos\theta). \tag{6.56}$$

Finally, insertion of $\dot\phi$ into the energy equation gives

$$E' = \frac{1}{2}I_1\left[\dot\theta^2 + \frac{(B - A\cos\theta)^2}{\sin^2\theta}\right] + Mgl\cos\theta = \text{constant},$$

or

$$(\sin^2\theta)\dot\theta^2 = (C - D\cos\theta)\sin^2\theta - (B - A\cos\theta)^2, \tag{6.57}$$

where $C \equiv 2E'/I_1$, $D \equiv 2Mgl/I_1$, again introducing new constants C and D to simplify the expression. In principle, the last equation can be integrated to furnish θ as a function of time, which in turn produces solutions for ϕ and ψ by means of eqs. (6.55) and (6.56), the right side of these equations being solely dependent on the angle θ. In practice, however, the integration of (6.57) leads in general to elliptic functions, which are not amenable to an elementary analysis. Nor is any additional insight gained from direct usage of Lagrange's equations, for the first two of them simply give p_ψ and p_ϕ as constants of the motion.

It is possible, nevertheless, to describe the qualitative features of the motion rather simply by means of (6.57). Let us make the change of variable $\mu \equiv \cos\theta$, resulting in

$$\dot\mu^2 = (C - D\mu)(1 - \mu^2) - (B - A\mu)^2 \equiv f(\mu). \tag{6.58}$$

Designating the right side as $f(\mu)$, we note that the dominant term in $f(\mu)$ for large μ is the term $D\mu^3$. Since D as introduced in eq. (6.57) was necessarily

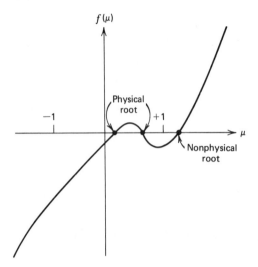

Figure 6.12 Behavior of the top function $f(\mu)$.

positive, $f(\mu) \to +\infty$ as $\mu \to +\infty$ while $f(\mu) \to -\infty$ as $\mu \to -\infty$. For $\mu = \pm 1$, $f(\mu) = -(B - A\mu)^2 \leq 0$. This cubic function has three roots, so it must ordinarily behave somewhat as shown in Figure 6.12; we normally expect two roots between the $+1$ and -1 values of μ. The physical motion of the top takes place entirely in the positive region for $f(\mu)$ between these two roots, or where $\dot\mu^2$ is positive. (Barring the special case of a vertical top where $\sin\theta = 0$, the root $f(\mu) = 0$ implies $\dot\mu = 0$ and hence $\dot\theta = 0$; that is, these roots give turning points in the nutation.) The two extreme values of θ corresponding to these roots of $f(\mu)$ will be designated as $\theta_{\min}$ and $\theta_{\max}$.

It is traditional to investigate the motion of the top through a study of the curve traced by the z-axis on a sphere centered at the fixed support point. The size of the ratio B/A is of paramount importance, since by (6.55)

$$\dot\phi = \frac{A}{\sin^2\theta}(B/A - \cos\theta) = \frac{A}{1-\mu^2}(B/A - \mu).$$

The smallest angle $\theta_{\min}$ corresponds to the largest possible μ value; if B/A is larger than this root of $f(\mu)$, the sign of $\dot\phi$ is fixed and we have regular precession with nutation, as seen in Figure 6.13a. If B/A is equal to the root of $f(\mu)$ corresponding to $\theta_{\min}$, then $\dot\phi = 0$ when $\dot\theta = 0$, giving the cusps of Figure 6.13b; when a spinning top is released it often has initial conditions placing it at the cusp of such a diagram. Finally, if B/A lies *between* the roots of $f(\mu)$, the angular velocity of precession $\dot\phi$ changes sign during the motion, the forward–backward motion causing the loops of Figure 6.13c; the reversal of direction occurs at the instant when $\dot\phi = 0$, or at a point of vertical slope readily identified in the figure.

Miscellaneous Applications

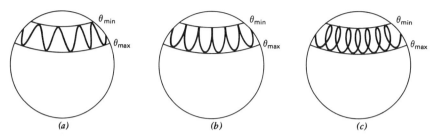

Figure 6.13 Traces of the symmetry axis of a top: (*a*) regular precession with notation; (*b*) motion with cusps; (*c*) loops from precession reversal.

Through analyses of this kind, the general features of the top's motion are easily described; further examples will be found in the problem set.

Thus far, we have said nothing about the effects of friction at the point of contact between the top and the supporting surface. In actual practice, this point is not really fixed but may engage in a rolling or skidding motion across the floor. Frictional forces may damp the skidding motion quickly, then provide the centripetal force for the pure rolling of the point of contact in some circular path, adding another angular velocity to those which have been considered! The initial slipping can also generate frictional torques that are perpendicular to the symmetry axis of the top, along which the angular momentum is mainly directed (for rapid spin); as a consequence, $\vec{L}$ rises toward the vertical. During the rise of the top, some kinetic energy of spin is dissipated and some is converted to potential energy, so the spinning motion is gradually slowed. Upon reaching the vertical, the top is said to "sleep," and the axis will remain vertical until friction reduces the spin below a critical value, at which point the motion becomes unstable.

A similar phenomenon occurs in the event of the flipping of a tippie-top (Figure 6.14). The top, initially rotating with stem upward, very quickly flips over to rest on the stem. The frictional force $\vec{f}$ rotates rapidly with the angular velocity $\vec{\omega}$ of the top, whereupon it and the torque it produces time average to zero. Therefore, the angular momentum is essentially constant and vertical, and $\vec{L} = I\vec{\omega}$ if we take the top as almost spherical with principal moments all equal to I. In a frame rotating with the top, $\vec{N} \neq 0$ in its time average, so we examine instantaneous body components of eq. (6.35):

$$\vec{N} = \left(\frac{d\vec{L}}{dt}\right)_{\text{body}} + \vec{\omega} \times \vec{L} = \left(\frac{d\vec{L}}{dt}\right)_{\text{body}}.$$

Taking *z*-components (along the symmetry axis) on both sides, we have

$$N_z = -N \sin \theta = \left(\frac{dL_z}{dt}\right)_{\text{body}} = \frac{d}{dt}(L \cos \theta)_{\text{body}} = -L(\sin \theta)\dot{\theta},$$

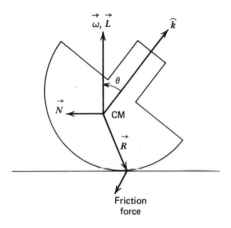

Figure 6.14 Dynamics of a tippie-top.

or

$$\dot{\theta} = \frac{N}{L} \approx \frac{\mu MgR}{I\omega},$$

where μ is the coefficient of friction, M is the mass, and R is the radius of the top. The differential equation shows the increase of θ with time, and its integration reveals that the top will execute a flip within a few seconds (for typical values of the parameters).

Example 3 The Precession of the Equinoxes; Gyroscopic Motion

It is a well-known fact that at present the Earth's axis of daily rotation can be extended into the sky in the approximate direction of the North Star, Polaris. Many centuries in the past or future, however, this extension would lie in another direction. Over a period of several thousand years, the Earth's axis slowly sweeps out a circle in the sky, a phenomenon known as the **precession of the equinoxes**. It is caused by a small asymmetry in the Earth's mass distribution, a bulging at the Equator (from rotation) which makes Earth not quite spherical and leads to a net torque from the gravitational attractions of sun and moon.

To analyze this situation, let us choose a set of $x'y'z'$-axes at the Earth's center (fixed at some instant in time), with the z'-axis along the rotation polar axis and the vector $\vec{R}$ from the sun in the $y'z'$-plane (Figure 6.15). (The vector $\vec{\rho}$, as shown, is not in this plane.) β is the angle between $\vec{R}$ and the z'-axis. Letting $\vec{r} = x'\hat{i}' + y'\hat{j}' + z'\hat{k}'$, the force on a mass element dm is

$$d\vec{F} = -\frac{GM\,dm}{\rho^3}\vec{\rho} = -\frac{GM\,dm}{\rho^3}(\vec{R} + \vec{r})$$

$$= -\frac{GM\,dm}{\rho^3}\left[x'\hat{i}' + (R\sin\beta + y')\hat{j}' + (z' - R\cos\beta)\hat{k}'\right]$$

Miscellaneous Applications

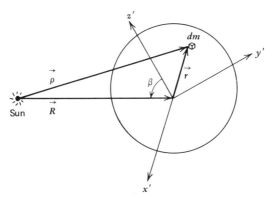

Figure 6.15 Vectors for precession analysis.

where M is the mass of the sun. The torque about the Earth's center due to this force is $\vec{r} \times d\vec{F}$; from the symmetry, it can have only an x-component for the Earth as a whole (after integration over dm). Then,

$$
\begin{aligned}
dN_{x'} &= (\vec{r} \times d\vec{F})_{x'} = y'(d\vec{F})_{z'} - z'(d\vec{F})_{y'} \\
&= \frac{GM\,dm}{\rho^3}[y'(R\cos\beta - z') + z'(R\sin\beta + y')] \\
&= \frac{GMR\,dm}{\rho^3}[y'\cos\beta + z'\sin\beta].
\end{aligned}
$$

Proceeding as with the derivation of eq. (6.46), we have

$$
\frac{1}{\rho^3} \approx \frac{1}{R^3}\left[1 - \frac{3\vec{R}\cdot\vec{r}}{R^2}\right] = \frac{1}{R^3}\left[1 - \frac{3}{R}(y'\sin\beta - z'\cos\beta)\right].
$$

Substitution of $1/\rho^3$ into $dN_{x'}$ followed by integration produces

$$
\begin{aligned}
N_{x'} &= \int \frac{GM\,dm}{R^2}\left(1 - \frac{3y'}{R}\sin\beta + \frac{3z'}{R}\cos\beta\right)(y'\cos\beta + z'\sin\beta) \\
&= \frac{3GM}{R^3}\sin\beta\cos\beta \int dm(z'^2 - y'^2),
\end{aligned}
$$

other terms vanishing from the symmetry of the sphere. The difference of the remaining two integrals is just $I_{y'y'} - I_{z'z'}$, or $I_2 - I_3$, giving

$$
N_{x'} = \frac{3GM}{R^3}(I_2 - I_3)\sin\beta\cos\beta. \tag{6.59}
$$

The torque of the sun tries to align the Earth so that its equatorial bulge is rotated into the plane of revolution, the ecliptic plane. There would be *no* torque if the Earth were a perfect sphere, because the difference of the inertia tensor elements in eq. (6.59) would then be zero. Noting that $\vec{R}\cdot\hat{k}' = -R\cos\beta$ and $\vec{R}\times\hat{k}' = R\sin\beta\hat{i}'$, the torque can be written in vector form as

$$\vec{N} = \frac{3GM}{R^5}(I_3 - I_2)(\vec{R}\cdot\hat{k}')(\vec{R}\times\hat{k}'). \tag{6.59'}$$

As the Earth moves around the sun in its annual journey, the vector $\vec{R}$ sweeps out a circle in the Ecliptic with a roughly constant angular velocity ω_R. In the torque calculation just concluded, we have been considering an instantaneous "snapshot" of this motion where $\vec{R}$ was for convenience taken in the $y'z'$-plane at the instant chosen. Now we shall generalize the situation somewhat by letting the primed axes be the body axes which are rigidly attached to the Earth. In effect, the Earth will be treated as a gigantic top, the center of which is moving around the ecliptic with $\vec{R}$, while the polar axis slowly precesses and nutates. Denoting a fixed set of inertial (space) axes by the unit vectors $\hat{i}$, $\hat{j}$, $\hat{k}$, we have

$$\vec{R} = R[(\cos\omega_R t)\hat{i} + (\sin\omega_R t)\hat{j}],$$

where $\vec{R}$ lies along the x-axis at time zero. Setting $\psi = 0$ and using the inverse (transpose) of the full rotation matrix (5.18), we have

$$\hat{i} = \cos\phi\hat{i}' - \cos\theta\sin\phi\hat{j}' + \sin\theta\sin\phi\hat{k}',$$

$$\hat{j} = \sin\phi\hat{i}' + \cos\theta\cos\phi\hat{j}' - \sin\theta\cos\phi\hat{k}',$$

whereupon $\vec{R}$ in terms of the moving axes is given by

$$\vec{R} = R[\cos(\omega_R t - \phi)\hat{i}' + \cos\theta\sin(\omega_R t - \phi)\hat{j}' - \sin\theta\sin(\omega_R t - \phi)\hat{k}'],$$

after some algebraic simplification. Insertion of this expression into (6.59') gives the torque in terms of the body axes:

$$\begin{aligned}\vec{N} = &-\frac{3GM}{R^3}(I_3 - I_2)\sin\theta\sin(\omega_R t - \phi) \\ &\cdot [\cos\theta\sin(\omega_R t - \phi)\hat{i}' - \cos(\omega_R t - \phi)\hat{j}'] \\ = &\frac{3GM}{2R^3}(I_3 - I_2)\sin\theta\{\cos\theta[\cos 2(\omega_R t - \phi) - 1]\hat{i}' \\ &+ \sin 2(\omega_R t - \phi)\hat{j}'\}. \end{aligned} \tag{6.60}$$

Miscellaneous Applications 213

Note that this time-dependent expression for $\vec{N}$ has both x'- and y'-components, but it is still true that $\vec{R} \cdot \vec{N} = 0$ at every instant of time.

The components of angular velocity in the body axes are, from (6.27),

$$\vec{\omega} = \dot{\theta}\hat{i}' + \dot{\phi}\sin\theta\,\hat{j}' + \dot{\phi}\cos\theta\,\hat{k}',$$

again setting $\psi = 0$; this expression gives the angular velocity for the relatively slow precession and nutation which we are examining. In addition, there is the comparatively rapid daily rotation of the Earth (analogous to the spin of a top) at a rate Ω about $\hat{k}'$. The primed axes themselves are "body axes attached to Earth" for precession and nutation only; they are not undergoing the rotation at Ω as well, so that rotation will simply be added, and the total angular momentum becomes

$$\vec{L} = I_2\dot{\theta}\hat{i}' + I_2\dot{\phi}\sin\theta\,\hat{j}' + I_3(\dot{\phi}\cos\theta + \Omega)\hat{k}',$$

using the fact that $I_1 = I_2$. As in the derivation of the Euler equations, the expression

$$\vec{N} = \left(\frac{d\vec{L}}{dt}\right)_{\text{body}} + \vec{\omega} \times \vec{L}$$

produces

$$N_{x'} = I_2\dot{\omega}_{x'} + \omega_{y'}L_{z'} - \omega_{z'}L_{y'} = I_2\dot{\omega}_{x'} + \omega_{y'}I_3(\omega_{z'} + \Omega) - \omega_{z'}I_2\omega_{y'}$$

$$= I_2\dot{\omega}_{x'} + (I_3 - I_2)\omega_{y'}\omega_{z'} + I_3\omega_{y'}\Omega$$

and

$$N_{y'} = I_2\dot{\omega}_{y'} + \omega_{z'}L_{x'} - \omega_{x'}L_{z'} = I_2\dot{\omega}_{y'} + (I_2 - I_3)\omega_{z'}\omega_{x'} - I_3\omega_{x'}\Omega.$$

Neglecting the small terms in $\dot{\omega}$ and the products of two ω-components (justified by our later results), we have

$$N_{x'} \simeq I_3\omega_{y'}\Omega, \qquad N_{y'} \simeq -I_3\omega_{x'}\Omega,$$

or

$$\dot{\phi} = \frac{\omega_{y'}}{\sin\theta} \simeq \frac{N_{x'}}{I_3\Omega\sin\theta} = -\frac{3GM}{2R^3\Omega}\frac{I_3 - I_2}{I_3}\cos\theta\bigl[1 - \cos 2(\omega_R t - \phi)\bigr],$$

$$(6.61)$$

and
$$\dot{\theta} = \omega_{x'} \simeq -\frac{N_{y'}}{I_3 \Omega} = -\frac{3GM}{2R^3 \Omega} \frac{I_3 - I_2}{I_3} \sin\theta \sin 2(\omega_R t - \phi). \quad (6.62)$$

These are the final expressions for the Earth's motions.

The nutation is quite small, and we shall not be concerned with it further. If the precession of (6.61) is time averaged, one obtains

$$(\dot{\phi})_{av} = -\frac{3GM}{2R^3 \Omega} \frac{I_3 - I_2}{I_2} \cos\theta.$$

Observing that

$$\omega_R^2 = \frac{GM}{R^3}$$

from the earlier discussion of Kepler's laws and planetary motion in Chapter 4, one can write

$$(\dot{\phi})_{av} = -\frac{3\omega_R^2}{2\Omega} \frac{I_3 - I_2}{I_3} \cos\theta,$$

or

$$\frac{\text{precession period}}{\text{revolution period}} = \text{precession period in years}$$

$$= \frac{\omega_R}{(\dot{\phi})_{av}} = -\frac{2\Omega}{3\omega_R} \frac{I_3}{I_3 - I_2} \sec\theta. \quad (6.63)$$

With the angle θ, known to be approximately 23.5°, and the inertia information given earlier in this chapter, one can readily calculate the precession period. It turns out to be about three times larger than the expected result of 26,000 years. The reason is that a computation based on eq. (6.63) has only accounted for the effect of the sun.

While the total (inverse square) gravitational force of the sun is much greater than that of the moon, it has long been known that the moon is primarily responsible for the tides in that it is so much closer to Earth than is the sun, the pertinent differential force law having an inverse cube dependence on distance. A similar phenomenon arises here from the $1/R^3$ factor in eq. (6.61). When the joint effects of moon and sun are considered, the precession period is reduced to the expected value.

Returning to the approximate expressions

$$\omega_{x'} \simeq -\frac{N_{y'}}{I_3 \Omega}, \qquad \omega_{y'} \simeq \frac{N_{x'}}{I_3 \Omega}, \qquad \omega_{z'} \simeq 0,$$

we note that these can be combined in vector form ($\vec{\Omega} = \Omega \hat{k}'$) as

$$\vec{\omega} \simeq \frac{\vec{\Omega} \times \vec{N}}{I_3 \Omega^2}, \qquad |\vec{\omega}| \simeq \frac{|\vec{N}|}{I_3 \Omega}. \tag{6.64}$$

The relation (6.64) is familiar from elementary physics as the equation which governs the precessional rate of a gyroscope. The behavior of the Earth under the torques of other bodies serves as an example of gyroscopic principles.

There are additional examples which could be given. The directional guidance mechanisms of greatly different objects, such as airplanes and torpedoes, are determined by gyroscopes. Often, in such applications the "free gyro" is a heavy spinning metallic disk encased in a gimbal mounting so that it is able to rotate freely about any fixed direction in space. If no torques are applied to the gyro wheel, it will maintain its orientation and thereby serves as a direction indicator. (Of course, if the bearings of the mounting were perfectly frictionless, the gyro would maintain its orientation whether spinning or not, but such ideal conditions are never realized.)

The gyrocompass used on naval vessels is somewhat more sophisticated. Let us choose the x-axis for the direction of the applied torque, the y-axis as vertical, and the z-axis as that of the gyro spin (Figure 6.16). To demonstrate the essential action of the instrument without introducing needless complexities, we shall constrain the gyro axis to the horizontal (xz) plane. This requirement can be accomplished by adjusting the gimbal mounting so that rotation out of the horizontal plane is not permitted. (In actuality, the torque may be provided by attaching weights to the mounting.)

Now, suppose that the gyro is located on Earth at latitude λ, and call α the angle between the spin axis and the horizontal projection of the northern direction. The angular velocity components are then

$$\omega_x = -\Omega \cos \lambda \sin \alpha, \qquad \omega_y = \Omega \sin \lambda + \dot{\alpha}, \qquad \omega_z = \Omega \cos \lambda \cos \alpha. \tag{6.65}$$

Adding the spin angular velocity ω_g for the gyroscope along the z-direction and noting that $I_1 = I_2 =$ constant from the symmetry of the disk, one finds that the angular momentum of the gyroscope is

$$\vec{L} = -I_1 \Omega \cos \lambda \sin \alpha \hat{i} + I_1 (\Omega \sin \lambda + \dot{\alpha}) \hat{j} + I_3 (\omega_g + \Omega \cos \lambda \cos \alpha) \hat{k}.$$

The torque of the suspension system on the gyroscope (which constrains its axis to the horizontal plane) is along the x-axis; an expression for it can be obtained

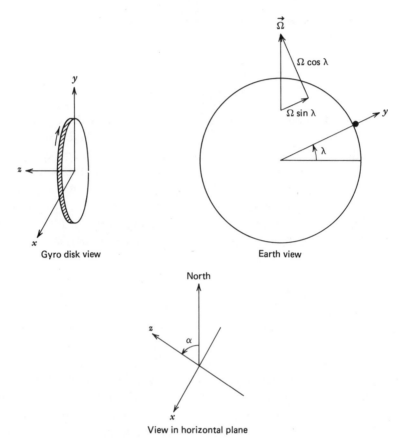

Figure 6.16 Geometry for the gyrocompass.

from relation (6.21):

$$\vec{N} = \left(\frac{d\vec{L}}{dt}\right)_{space} = \left(\frac{d\vec{L}}{dt}\right)_{body} + \vec{\omega} \times \vec{L}$$

$$= -I_1\Omega\dot{\alpha}\cos\lambda\cos\alpha\,\hat{i} + I_1\ddot{\alpha}\,\hat{j} + I_3(\dot{\omega}_g - \Omega\dot{\alpha}\cos\lambda\sin\alpha)\hat{k} + \vec{\omega}\times\vec{L}.$$

Inasmuch as the *y*- and *z*-components of the torque are zero, we find that this expression yields

$$I_1\ddot{\alpha} + (\vec{\omega}\times\vec{L})_y = 0$$

$$= I_1\ddot{\alpha} - I_1\Omega^2\cos^2\lambda\sin\alpha\cos\alpha$$

$$+ I_3\Omega\cos\lambda\sin\alpha(\omega_g + \Omega\cos\lambda\cos\alpha),$$

$$I_3(\dot{\omega}_g - \Omega\dot{\alpha}\cos\lambda\sin\alpha) = 0 = \frac{d}{dt}\left[(\omega_g + \Omega\cos\lambda\cos\alpha)I_3\right].$$

The last equation tells us that the z-component of the angular momentum is constant. Assuming that $\Omega \ll \omega_g$, the y-relation is

$$I_1\ddot{\alpha} + I_3\omega_g\Omega \cos \lambda \sin \alpha \simeq 0. \tag{6.66}$$

For small α, this approximate equation describes simple harmonic oscillations of frequency

$$\frac{1}{2\pi}\sqrt{\frac{I_3\omega_g\Omega \cos \lambda}{I_1}}.$$

The physical meaning of this result is that the gyro axis oscillates about the northern direction in the xz-plane, its frequency being dependent only on the latitude and the angular velocity ω_g in essence. In this manner, a gyrocompass serves to indicate the direction of true north even inside a submarine where a magnetic compass does not function. The actual shipboard apparatus, in order to counteract the effect of friction, is much more complicated than the present discussion has supposed. The basic principles are as outlined, however.

PROBLEMS

6.1 Prove that the quadratic roots of eq. (6.5) both have magnitude unity.

6.2 Show that the elements of the dyadic $\overset{\leftrightarrow}{R}$ of eq. (6.8) can also be obtained by means of relation (5.48).

6.3 Provide the algebraic steps necessary to verify relations (6.9) and (6.10).

6.4 Show that the unit vector $\hat{n}$ given by relation (6.10) has a direction that satisfies the eigenvector eq. (6.6) with unit eigenvalue.

6.5 Two beads of equal mass m are connected to one another by means of a taut inextensible string of length L. The beads are free to slide along a thin, straight rod which rotates in a horizontal plane with a constant angular velocity ω about a vertical axis. (Neglect gravity.) If the beads are initially at rest relative to the rod, the outer one being at a distance D from the axis, find the speed with which they are sliding at a later time t. Why need Coriolis forces not be considered in the solution?

6.6 Coriolis and centrifugal forces can often be used to explain the behavior of fluids, as the following examples will indicate.

(a) Show that the surface contour of the fluid in a rotating pail of water has the shape of a paraboloid of revolution.

(b) A current of water flows northward along a river channel of width d at the northern latitude λ. If v is the flow velocity and ω is the Earth's angular velocity, show that the height of water on the eastern bank is greater than that on the western bank by the amount $2d\omega v(\sin\lambda)/g$.

6.7 Prove that $\vec{\omega}$ is given by expression (6.29) through direct evaluation of $\ddot{\vec{R}}\cdot\vec{R}^T$.

6.8 (a) Work out the general expressions for the angular velocities which appear on the right of expression (6.32):

$$\vec{\omega} = \vec{\omega}_\psi + \ddot{\vec{R}}_3 \cdot \vec{\omega}_\theta + \ddot{\vec{R}}_3 \cdot \ddot{\vec{R}}_2 \cdot \vec{\omega}_\phi$$

(in terms of Euler angles and their derivatives). Notice that $\vec{\omega}_\psi$ depends on $\dot{\theta}$ and $\dot{\phi}$ as well as on $\dot{\psi}$.

(b) Set $\psi \equiv 0$, $\ddot{\vec{R}}_3 \equiv \vec{1}$, and then evaluate $\vec{\omega}$ in detail.

6.9 Consider the rotating disk discussed at the end of Section 6.4. Find the instantaneous components of the angular velocity, the angular momentum, and the torque in the space set of axes. Then show that $\vec{N} = d\vec{L}/dt$ is satisfied.

6.10 (a) Prove that, for a symmetrical rigid body with $I_1 = I_2$, both the polhode and the herpolhode are circles.

(b) For the general rigid body, show that the polhode is a closed curve around the major or minor axis of the ellipsoid but that it is hyperbolic near the intermediate axis.

6.11 For the body and space cones of the symmetrical rigid body, show that the cone half-angles are related by

$$\cos\theta_S = \frac{1 + \left(\dfrac{\Omega}{\omega_z}\right)\cos^2\theta_B}{\sqrt{1 + \left(\dfrac{\Omega}{\omega_z}\right)\left(2 + \dfrac{\Omega}{\omega_z}\right)\cos^2\theta_B}}.$$

6.12 Consider a tennis racket as composed of a thin rod of length l and mass m_l connected to a circular hoop of radius a and mass m_a. Find the moments of inertia about the principal axes, as illustrated in Figure 6.9. Are these axes so ordered that the "intermediate axis" or "tennis racket" theorem will hold?

6.13 Show that the total torques in pitching, rolling, or yawing motions of spacecraft are as given in eqs. (6.49), (6.50), and (6.51), respectively.

6.14 A heavy symmetric top is spun so that its symmetry axis is initially vertical. Find the relation between A and B, and that between C and D. Then find the roots of $f(\mu) = 0$. Show that it is possible to obtain a critical value of angular velocity above which the top must remain in the vertical position.

Problems **219**

6.15 Derive the Lagrange equation for the coordinate θ of a heavy symmetrical top. Then solve this relation for the precessional angular velocity $\dot\phi$ when there is no nutation present. From your result, show that there is a minimum value of ω_z for which pure precession is possible. Finally, for large ω_z (relative to the minimum value), show that there are two permissible values of $\dot\phi$, corresponding to the cases of fast and slow precession.

6.16 Show that the period for the precession of the equinoxes is 26,000 years (as discussed in the text) when the combined effects of sun and moon are considered.

6.17 (a) Prove that

$$\sum_{i,j} \delta_{ijm}\delta_{ijp} = 2\delta_{m,p}.$$

(b) Using this delta relation, show that

$$v'_m = \tfrac{1}{2} \sum_{\substack{i,j,\\k,l}} \delta_{ijm} a_{ik} a_{jl} T_{kl}$$

becomes

$$v'_m = \det \mathbf{A} \sum_n a_{mn} v_n$$

with the aid of eq. (6.16) and (6.14′).

Chapter Seven
OSCILLATIONS AND STABILITY

7.1 STABILITY OF EQUILIBRIUM

A mechanical system that is in an equilibrium configuration must by definition have a resultant force of zero acting on it:

$$\vec{F} = -\nabla V = 0,$$

or

$$\frac{\partial V}{\partial x} = \frac{\partial V}{\partial y} = \frac{\partial V}{\partial z} = 0. \tag{7.1}$$

This fact ensures that the net change of potential energy is

$$\delta V = \frac{\partial V}{\partial x}\delta x + \frac{\partial V}{\partial y}\delta y + \frac{\partial V}{\partial z}\delta z = 0$$

for arbitrary virtual displacements δx, δy, and δz. Since the change of potential energy is the negative of the work done by the forces acting on the system, it follows that our result is in accord with the principle of virtual work, eq. (1.20).

Now, (7.1) tells us that the potential energy has a stationary value at equilibrium, and experience tells us that this value must be a minimum if we are to have a *stable* equilibrium situation. Thus, a ball released from a point slightly up the slope of a rounding valley will oscillate until finally coming to rest at the lowest point. On the other hand, if the ball is released from the top of a hill, it will roll down the slope with no inclination to return to its initial point, a clear example of unstable equilibrium at a position of maximum potential energy.

To amplify these remarks, consider a potential energy which is a function of the single variable x, the origin $x = 0$ being chosen as the equilibrium point. Expanding in a Taylor series about the origin, one obtains

$$V(x) = V_{x=0} + x\left(\frac{dV}{dx}\right)_{x=0} + \frac{x^2}{2!}\left(\frac{d^2V}{dx^2}\right)_{x=0} + \cdots. \tag{7.2}$$

It is customary to take the zero of potential at the equilibrium position, so that $V_{x=0} = 0$. Furthermore, the second term vanishes by virtue of (7.1). We are left with

$$V(x) = \frac{x^2}{2!}\left(\frac{d^2V}{dx^2}\right)_{x=0} + \cdots.$$

In view of the increasing powers of x in the series and the fact that x is small, higher-order terms need not be considered unless the second derivative vanishes in a given problem. In the common situation where it is not zero, note that the sign of the second derivative determines the sign of the potential energy. On either side of a stable minimum, the magnitude of the slope increases as $|x|$ increases and the second derivative is positive, as is $V(x)$, while a position of instability leads to a negative second derivative. The criterion for stability is then as follows: If the second derivative of the potential energy is positive when evaluated at the equilibrium position, the position will be one of stable equilibrium at which $V(x)$ will be a minimum.

As an example, the potential energy of a mass attached to a spring is $V(x) = kx^2/2$, where k is the force constant of the spring. Clearly,

$$\left(\frac{d^2V}{dx^2}\right)_{x=0} = k, \quad k > 0.$$

Thus, in agreement with intuition, we have a stable situation.

7.2 THE ONE-DIMENSIONAL DAMPED HARMONIC OSCILLATOR; VIBRATION OF STRINGS

As in the previous section, we consider a one-dimensional oscillator, with motion along the x-axis about an equilibrium position at the origin. (A detailed review of the linear oscillator is essential to our analyses later in this chapter.) The differential equation describing the motion of the oscillator is

$$m\ddot{x} + R\dot{x} + kx = 0, \tag{7.3}$$

where m is the mass of the moving body and R is the damping constant (giving a retarding force proportional to the velocity for $R > 0$). The last term on the left gives the linear restoring force ($k > 0$) that sustains the oscillations. It is well known that such a differential equation not only depicts the motion of a mass on a spring, but also can be used for the description of electrical oscillations and many other varied phenomena.

One can easily obtain the general solution of the homogeneous equation, (7.3), by assuming it will be of the form $e^{\lambda t}$. Upon substitution,

$$(m\lambda^2 + R\lambda + k)e^{\lambda t} = 0.$$

Equating the quadratic in λ to zero and extracting its roots, we find that

$$\lambda = \frac{-R \pm \sqrt{R^2 - 4mk}}{2m}, \tag{7.4}$$

whereupon

$$x = C_1 \exp\left\{-\frac{Rt}{2m} + \left[\sqrt{\left(\frac{R}{2m}\right)^2 - \frac{k}{m}}\right]t\right\}$$

$$+ C_2 \exp\left\{-\frac{Rt}{2m} - \left[\sqrt{\left(\frac{R}{2m}\right)^2 - \frac{k}{m}}\right]t\right\}. \tag{7.5}$$

C_1 and C_2 are constants that appear in the general solution of the second-order equation and are necessary in order that we may adjust the solution to correspond to appropriate initial values of x and $\dot{x}$ at time zero. The nature of the solution is determined by the radical in the exponents; we shall make this clear by looking at three cases separately.

CASE 1 OVERDAMPED MOTION

In the event that the damping constant R is so large that $[(R/2m)^2 - (k/m)] > 0$, the radical is real and the two exponents are real and negative, so that both terms of (7.5) are decaying exponentials. The motion damps out rapidly, and no oscillations occur.

CASE 2 CRITICALLY DAMPED MOTION

When $(R/2m)^2 = k/m$, so that the radical in the exponents is zero, we have critical damping. In this case, the two terms of (7.5) both coalesce into the same decaying exponential, so that a second solution is needed. With a little algebra, it is easy to show that the general solution is

$$x = (C_1 + C_2 t)e^{-Rt/2m}.$$

Again, however, no sustained oscillations are produced.

CASE 3 UNDERDAMPED MOTION

The final case to be considered is where $[(R/2m)^2 - (k/m)] < 0$. Let us therefore define a positive quantity

$$\omega_0^2 \equiv \frac{k}{m} - \left(\frac{R}{2m}\right)^2, \tag{7.6}$$

whereupon the solution (7.5) becomes

$$x = e^{-Rt/2m}\left[C_1 e^{i\omega_0 t} + C_2 e^{-i\omega_0 t}\right]. \tag{7.7}$$

This case is of the greatest interest, for it corresponds to sustained oscillations which are gradually damped by the real exponential in front of the bracket. The constants C_1 and C_2 in eq. (7.7) are usually complex, while x itself is real. The real nature of x is more readily seen if we adopt real constants,

$$C_1' \equiv C_1 + C_2, \qquad C_2' \equiv i(C_1 - C_2),$$

giving

$$x = e^{-Rt/2m}[C_1' \cos \omega_0 t + C_2' \sin \omega_0 t]. \tag{7.7'}$$

Alternatively, we can identify $C_1' = C \cos \alpha$ and $C_2' = C \sin \alpha$, and obtain

$$x = Ce^{-Rt/2m} \cos(\omega_0 t - \alpha). \tag{7.7''}$$

The form (7.7) is most convenient for algebraic manipulations, but the later forms may offer more physical insight. It is clear from (7.6) and (7.7'') that ω_0 is the "natural frequency" of the damped oscillator and that it is somewhat smaller than it would be if the damping were not present.

Next, suppose that the oscillator is **force driven** by some applied force $f(t)$ which is a function of the time. One example of such a driving force that is frequently considered is sinusoidal in nature, since any periodic force can be decomposed into sinusoidal components. Equation (7.3) now becomes

$$m\ddot{x} + R\dot{x} + kx = f(t), \tag{7.8}$$

where

$$f(t) = F_0 \cos \omega t = \mathrm{Re}\left(F_0 e^{-i\omega t}\right), \qquad F_0 = \text{real constant}.$$

We see that (7.8) is an inhomogeneous differential equation, which can be solved by taking the solution (7.7) of the homogeneous equation (which corresponds to $f(t) = 0$) and adding to it a *particular* solution which will generate the inhomogeneous term $f(t)$ on the right. Thus, we have only to find a particular solution which will give the desired sinusoidal function. From physical intuition, it seems reasonable to assume a solution of the same frequency as the applied force. So let us try x in the complex form $e^{-i\omega t}$ in an attempt to solve the modified equation

$$m\ddot{x} + R\dot{x} + kx = F_0 e^{-i\omega t}. \tag{7.9}$$

Note that there are no imaginary quantities involved in this problem; all physical functions are entirely real. It is merely that the introduction of complex exponentials enables us to do the necessary algebra in a much simpler fashion. Thus, taking the real part of both sides of eq. (7.9), we have

$$m \operatorname{Re}(\ddot{x}) + R \operatorname{Re}(\dot{x}) + k \operatorname{Re}(x) = m \frac{d^2}{dt^2}(\operatorname{Re} x) + R \frac{d}{dt}(\operatorname{Re} x) + k(\operatorname{Re} x)$$

$$= F_0 \operatorname{Re}(e^{-i\omega t}) = F_0 \cos \omega t,$$

showing that the real part of the complex x will satisfy eq. (7.8); the imaginary part of x is of no interest.

Setting $x = Ae^{-i\omega t}$, we find

$$\dot{x} = -iA\omega e^{-i\omega t}, \qquad \ddot{x} = -\omega^2 A e^{-i\omega t},$$

whereupon from eq. (7.9),

$$(-m\omega^2 - iR\omega + k)Ae^{-i\omega t} = F_0 e^{-i\omega t},$$

or

$$A = \frac{F_0/m}{(k/m) - \omega^2 - (iR\omega/m)} = \frac{F_0}{m}\left[\frac{(k/m) - \omega^2 + i(R\omega/m)}{[(k/m) - \omega^2]^2 + (R\omega/m)^2}\right].$$

Letting

$$D \equiv (k/m - \omega^2)^2 + \left(\frac{R\omega}{m}\right)^2,$$

$$\cos \varepsilon \equiv \frac{(k/m) - \omega^2}{\sqrt{D}}, \qquad \sin \varepsilon \equiv \frac{(R\omega/m)}{\sqrt{D}},$$

we have

$$A = \frac{F_0}{m} \frac{(\cos \varepsilon + i \sin \varepsilon)}{\sqrt{D}} = \frac{F_0}{m\sqrt{D}} e^{i\varepsilon},$$

whereupon

$$x = \frac{F_0}{m\sqrt{D}} e^{-i(\omega t - \varepsilon)}.$$

Our physical particular solution is then

$$\operatorname{Re} x = \frac{F_0}{m\sqrt{D}} \cos(\omega t - \varepsilon).$$

There are no arbitrary constants in this solution; F_0 is the amplitude of the driving force, while D and ε are determined by the driving frequency and the system parameters. The general solution of eq. (7.8) will be

$$x(t) = \frac{F_0}{m\sqrt{D}} \cos(\omega t - \varepsilon) + Ce^{-Rt/2m} \cos(\omega_0 t - \alpha), \qquad (7.10)$$

which contains the two arbitrary constants C and α needed for fitting the initial conditions. It is the particular solution that furnishes the steady-state motion, as determined by the driving force; the solution (7.7) of the homogeneous equation has only a transient effect.

Note that the denominator of the steady-state term contains a factor $\sqrt{D}$. In the event that the damping constant R is quite small, we see from eq. (7.6) that the natural frequency ω_0 of the oscillator is approximately $\sqrt{k/m}$ (the exact value for a totally undamped oscillator), which means that $\sqrt{D} \simeq \omega_0^2 - \omega^2$. Thus, whenever the driving frequency ω approaches the natural frequency ω_0, a resonant situation occurs, with the amplitude and the power delivered by the applied force both becoming very large. They are not infinite, because some damping will always be present. (Power considerations are explored in the problem set.)

Finally, let us consider for a moment the superposition principle for linear differential equations. Suppose that there are a number of force functions $f_i(t)$ for which we know the solutions $x_i(t)$. Each such force will satisfy an equation of the type

$$m\ddot{x}_i(t) + R\dot{x}_i(t) + kx_i(t) = f_i(t).$$

Suppose also that the force $f(t)$ in which we are interested can be written as a linear combination of the $f_i(t)$:

$$f(t) = \sum_i C_i f_i(t),$$

where the C_i are constant coefficients in the expansion. Then, the function

$$x(t) \equiv \sum_i C_i x_i(t)$$

is the desired solution for $f(t)$, since

$$m\ddot{x}(t) + R\dot{x}(t) + kx(t) = \sum_i C_i[m\ddot{x}_i(t) + R\dot{x}_i(t) + kx_i(t)]$$

$$= \sum_i C_i f_i(t) = f(t).$$

By means of this principle, we are allowed to express a periodic function $f(t)$ in a Fourier series of sines and cosines; upon evaluation of the coefficients C_i in the series, the functions $x_i(t)$ can be combined to give the general solution $x(t)$, each $x_i(t)$ being of the form (7.10).

As a common example of these concepts, consider transverse wave motion down a long uniform one-dimensional string or wire lying along the x-axis. The transverse displacements, denoted by $y(x, t)$, must satisfy a wave equation

$$\frac{\partial^2 y(x,t)}{\partial x^2} = \frac{1}{v^2} \frac{\partial^2 y(x,t)}{\partial t^2}, \qquad (7.11)$$

as is known from elementary physics or from an analysis like that in Section 5.7.

There is a little trick which can be used to obtain an expression for the wave velocity v in terms of the tension T on the string and the linear density ρ (mass per unit length of string). Consider the string itself as moving at velocity v while being fed from one reel onto another, with the entire string between the reels enclosed in a thin piece of glass tubing which corresponds in shape to the pattern of transverse displacements along the string. A small element ds of the string inside the tubing can be considered as an arc of a circle of some radius R. From Figure 7.1, we see that the vector difference of the tensions at the two ends of this element is directed inward. Equating this difference to the centripetal force on the element gives

$$T\frac{ds}{R} = \rho \, ds \frac{v^2}{R},$$

or
$$v = \sqrt{T/\rho} \qquad (7.12)$$

as the desired expression. The result is independent of R. Therefore, at this speed, each element of the string is supplied (by the tension) with the proper centripetal force and no net force is needed from or on any part of the glass. Then, the tubing could be broken away to leave the wave form standing in place. If a Galilean transformation is made to an inertial frame in which the string is at rest, then the wave form will be seen to move at the velocity v, *unchanged in shape*. This property is a basic one for a uniform string.

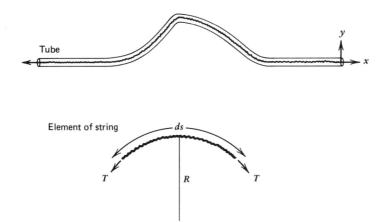

Figure 7.1 Forces on string element in glass tube.

The direction in which the wave moves has not yet been specified. A wave which travels to the right and preserves its form must be given a function $y_1(x - vt)$, while one going left requires some other function $y_2(x + vt)$. The general solution is the superposition

$$y(x, t) = y_1(x - vt) + y_2(x + vt). \tag{7.13}$$

It obviously satisfies the wave equation, (7.11). We shall not give a rigorous mathematical proof that this combination is the most general solution possible, for that is clear on physical grounds from the arguments just presented.

The initial conditions dictate the nature of the particular functions y_1 and y_2 that should be used in a given problem. That is, $y(x, 0)$ and $(\partial y/\partial t)_{t=0}$ must be specified in order to determine y_1 and y_2 at all later times. To be more explicit, we can write

$$y(x, t) = \frac{1}{2}[f(x - vt) + f(x + vt)] + \frac{1}{2v} \int_{x-vt}^{x+vt} g(x') \, dx', \tag{7.14}$$

where

$$y(x, 0) = f(x), \qquad \left(\frac{\partial y}{\partial t}\right)_{t=0} = g(x),$$

as the expression which satisfies the initial conditions. It is known as **D'Alembert's solution.**

Boundary conditions are another matter; these are requirements at the ends of the string which must hold at any and all times. They arise from the fact that a physical string is necessarily finite in length, usually being fastened rigidly at the two ends. As a consequence, severe frequency limitations are imposed on any simple harmonic waves that travel down the string. From a consideration of boundary conditions, we finally return to the Fourier analysis mentioned earlier.

Suppose that our string is of length l and is fixed at its ends, so that

$$y(0, t) = 0 = y(l, t).$$

In general, $y(x, t)$ can be expanded in a Fourier series of sine functions, choosing the sine arguments as

$$y_n(x) \equiv \sin\frac{n\pi x}{l}, \qquad n = 1, 2, 3, \ldots, \infty$$

so that the boundary conditions are both satisfied. Putting in a simple harmonic time dependence leads to the expansion

$$y(x, t) = \sum_{n=1}^{\infty} \sin\frac{n\pi x}{l} [A_n \cos \omega_n t + B_n \sin \omega_n t], \qquad (7.15)$$

where the A_n, B_n, and ω_n are constants to be determined. Now, substitute (7.15) into the wave equation (7.11), producing

$$-\left(\frac{n\pi}{l}\right)^2 = -\frac{\omega_n^2}{v^2},$$

or

$$\omega_n = \frac{n\pi v}{l}. \qquad (7.16)$$

The remaining coefficients A_n and B_n are found in typical Fourier fashion by using the orthogonality of the sine functions:

$$\int_0^l \sin\frac{n\pi x}{l} \sin\frac{m\pi x}{l} dx = \frac{1}{2}\int_0^l \left\{\cos\left[(n-m)\frac{\pi x}{l}\right] - \cos\left[(n+m)\frac{\pi x}{l}\right]\right\} dx$$

$$= \frac{l}{2}\delta_{n,m},$$

whereupon multiplication of $y(x, 0)$ by $\sin(m\pi x/l)\, dx$ and integration yield

$$\int_0^l y(x, 0) \sin\frac{m\pi x}{l} dx = \sum_n A_n \int_0^l \sin\frac{n\pi x}{l} \sin\frac{m\pi x}{l} dx$$

$$= \sum_n A_n \frac{l}{2}\delta_{n,m} = \frac{l}{2}A_m,$$

or

$$A_n = \frac{2}{l}\int_0^l y(x,0)\sin\frac{n\pi x}{l}dx. \qquad (7.16')$$

Similarly, one finds that

$$B_n = \frac{2}{\omega_n l}\int_0^l \left(\frac{\partial y}{\partial t}\right)_{t=0}\sin\frac{n\pi x}{l}dx. \qquad (7.16'')$$

Again, we find that the initial conditions are crucial in the determination of the function $y(x,t)$ at later times. We assume without proof that the Fourier series of (7.15), known as **Bernoulli's solution** to the string problem, is sufficiently general for the treatment of any given set of initial conditions.

The Bernoulli–Fourier expansion is a standard mathematical technique which is often employed, but the traveling wave solution of D'Alembert is more directly connected to physical insight. At this point, we shall try to reconcile these two approaches, understanding that the relationship between them was the object of a great deal of mathematical investigation in the nineteenth century.

First, we note that the D'Alembert solution really applies to an infinitely long string; no boundary conditions are given. On the contrary, the Fourier series repeats periodically and is only intended to apply along the finite string ($0 \leq x \leq l$). Inasmuch as the initial conditions need only be specified in this region, we force agreement between the two methods by insisting that the functions $f(x)$ and $g(x)$ in the D'Alembert solution be periodic and such as to satisfy the boundary conditions for all time. Because of the presence of the sine function in x brought about by these conditions, the series expansion must be odd about $x = 0$ and $x = l$:

$$y(x,0) = f(x) = -f(-x), \qquad f[l+(x-l)] = -f[l-(x-l)],$$

or

$$f(x) = -f(2l-x) = +f(x-2l).$$

Similar results hold for $g(x)$. In other words, extension of the functions $f(x)$ and $g(x)$ beyond the region of the string requires that they be periodic with period $2l$ if the boundary conditions are to be satisfied.

The functions $f(x)$ and $g(x)$ can now be expanded in a Fourier series with the coefficients A_n and B_n as given previously:

$$f(x) = \sum_n A_n \sin\frac{n\pi x}{l}, \qquad g(x) = \sum_n \omega_n B_n \sin\frac{n\pi x}{l}.$$

Substitution of these expansions into D'Alembert's solution (7.14) produces

$$y(x,t) = \frac{1}{2}\sum_n A_n\left[\sin\frac{n\pi}{l}(x-vt) + \sin\frac{n\pi}{l}(x+vt)\right]$$

$$+ (2v)^{-1}\sum_n \omega_n B_n \int_{x-vt}^{x+vt}\sin\frac{n\pi x'}{l}dx'.$$

Elementary integration in the last series and some trigonometric simplification immediately give Bernoulli's form (7.15), demonstrating the equivalence of the two methods.

7.3 TRANSFORM METHODS FOR OSCILLATIONS

We now wish to analyze the motion of an oscillatory system with a driving force $f(t)$ that is nonperiodic, one which in fact may be quite complicated and could even be transient in nature. Such situations may be treated conveniently in general by the transform methods to be discussed. It is assumed that the student is familiar with the basic techniques of contour integration in the complex plane; a review of these procedures may be obtained by consulting Appendix D.

The first step expresses the force $f(t)$ by means of a Fourier integral over the Fourier transform of the force, which will be designated $F(\omega)$. We have

$$f(t) = \int_{-\infty}^{\infty} F(\omega) e^{-i\omega t}\, d\omega, \tag{7.17}$$

where

$$F(\omega) = \frac{1}{2\pi}\int_{-\infty}^{\infty} f(t') e^{i\omega t'}\, dt'. \tag{7.17'}$$

These two functions $f(t)$ and $F(\omega)$ constitute a transform pair in the usual sense. In the same way, the displacement function and its transform are given by

$$x(t) = \int_{-\infty}^{\infty} \chi(\omega) e^{-i\omega t}\, d\omega \tag{7.18}$$

and

$$\chi(\omega) = \frac{1}{2\pi}\int_{-\infty}^{\infty} x(t') e^{i\omega t'}\, dt'. \tag{7.18'}$$

The consistency of this pair of equations can be shown, for example, by substitution of eq. (7.18') into (7.18) and use of the Dirac delta relation from

Appendix B:

$$\delta(t' - t) = \frac{1}{2\pi} \int_{-\infty}^{\infty} e^{i\omega(t'-t)} \, d\omega. \tag{7.19}$$

Furthermore, by differentiation of (7.18) under the integral sign, we have

$$\dot{x}(t) = \int_{-\infty}^{\infty} \chi(\omega)(-i\omega) e^{-i\omega t} \, d\omega; \tag{7.20}$$

upon inversion, the paired equation is

$$-i\omega \chi(\omega) = \frac{1}{2\pi} \int_{-\infty}^{\infty} \dot{x}(t') e^{i\omega t'} \, dt'. \tag{7.20'}$$

In a similar fashion,

$$\ddot{x}(t) = \int_{-\infty}^{\infty} \chi(\omega)(-\omega^2) e^{-i\omega t} \, d\omega, \tag{7.21}$$

which gives upon inversion

$$-\omega^2 \chi(\omega) = \frac{1}{2\pi} \int_{-\infty}^{\infty} \ddot{x}(t') e^{i\omega t'} \, dt'. \tag{7.21'}$$

Equation (7.8) (expressed in the variable t' rather than t) is

$$m\ddot{x}(t') + R\dot{x}(t') + kx(t') = f(t').$$

Multiplying both sides by $(1/2\pi)e^{i\omega t'} \, dt'$ and integrating produces

$$(-m\omega^2 - iR\omega + k) \chi(\omega) = F(\omega)$$

by virtue of eqs. (7.21'), (7.20'), (7.18'), and (7.17'). In other words, we have converted a differential equation into an algebraic equation, which is readily solved for $\chi(\omega)$:

$$\chi(\omega) = \frac{F(\omega)}{k - m\omega^2 - iR\omega}. \tag{7.22}$$

The general procedure for finding $x(t)$ is now clear. Equation (7.17') is employed to find the transform $F(\omega)$ for a given $f(t)$; $\chi(\omega)$ is then obtained from eq. (7.22), and in turn is transformed by (7.18) to arrive at the displacement $x(t)$. Difficulties with the procedure can only arise in the evaluation of the two transform integrals; contour integration techniques can be helpful in that regard.

Substitution of (7.22) into (7.18) leads to

$$x(t) = \int_{-\infty}^{\infty} \frac{F(\omega)e^{-i\omega t}\,d\omega}{k - m\omega^2 - iR\omega}. \tag{7.23}$$

This integral cannot be completely evaluated unless $f(t)$ and hence $F(\omega)$ are known. But at this point, we can calculate in general the contributions from the poles of the quadratic in the denominator (assuming $F(\omega)$ is well behaved). We note that, from (7.6),

$$k - m\omega^2 - iR\omega = -m\left(\omega^2 + \frac{iR}{m}\omega - \frac{k}{m}\right)$$

$$= -m\left(\omega - \omega_0 + \frac{iR}{2m}\right)\left(\omega + \omega_0 + \frac{iR}{2m}\right),$$

so that the poles are located below the real axis of the complex ω-plane at the points $\pm\omega_0 - (iR/2m)$ (see Figure 7.2). Suppose for the present we take t as positive. The path of integration proceeds along the length of the real axis and can be extended into a closed contour by enclosing the lower half-plane with a large semicircle. Because of the factor $e^{-i\omega t}$ in the integrand, the semicircle path will contribute nothing (for $t > 0$). The contribution of the denominator poles is then

$-2\pi i \cdot$ sum of residues at poles

$$= -\frac{2\pi i}{(-m)}\left[\frac{F\left(\omega_0 - \dfrac{iR}{2m}\right)}{2\omega_0}e^{-it(\omega_0 - iR/2m)}\right.$$

$$\left. + \frac{F\left(-\omega_0 - \dfrac{iR}{2m}\right)}{-2\omega_0}e^{-it(-\omega_0 - iR/2m)}\right]$$

$$= \frac{\pi i}{m\omega_0}e^{-Rt/2m}\left[e^{-i\omega_0 t}F\left(\omega_0 - \frac{iR}{2m}\right) - e^{i\omega_0 t}F\left(-\omega_0 - \frac{iR}{2m}\right)\right]. \tag{7.24}$$

This part of the expression for $x(t)$ is transient in nature. For negative t, the contour must be closed with a large semicircle around the upper half-plane; there are no poles in that region, so the integral is zero as far as the contribution from the denominator is concerned.

Transform Methods for Oscillations

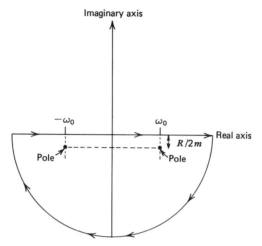

Figure 7.2 Poles of the denominator and contour in the ω-plane.

We can go no further until we specify the form of $f(t)$. Let us choose first the example of the last section with which we are quite familiar, taking $f(t) = F_0 e^{-i\omega' t}$ for $t > 0$ and $f(t) = 0$ for $t < 0$. That is, the driving force of frequency ω' is turned on at time zero and persists thereafter. Again, the exponential form is convenient for calculation; but in the end, we shall take the real part of $x(t)$ as the response to the applied force $F_0 \cos \omega' t$.

The transform $F(\omega)$ is found with the aid of eq. (7.17') as

$$F(\omega) = \frac{1}{2\pi} \int_{-\infty}^{\infty} f(t') e^{i\omega t'} dt'$$

$$= \frac{F_0}{2\pi} \int_0^{\infty} e^{i(\omega - \omega')t'} dt'$$

$$= \lim_{\alpha \to 0} \frac{F_0}{2\pi} \int_0^{\infty} e^{i(\omega - \omega')t' - \alpha t'} dt' = \lim_{\alpha \to 0} \left[\frac{e^{i(\omega - \omega')t' - \alpha t'}}{i(\omega - \omega') - \alpha} \right]_0^{\infty} \frac{F_0}{2\pi}$$

$$= \frac{F_0}{2\pi} \lim_{\alpha \to 0} \left[\frac{-1}{i(\omega - \omega') - \alpha} \right] = \frac{F_0 i}{2\pi(\omega - \omega')}. \tag{7.25}$$

Here, we have employed a convergence factor α in the standard way to eliminate the upper limit term. Note that by (7.19) a delta function would have been obtained had the lower limit been extended over all negative time values; as it is, we find that $F(\omega)$ makes a simple pole contribution to the integral (7.23), giving

$$x(t) = \int_{-\infty}^{\infty} \frac{F_0 i}{2\pi(\omega - \omega')} \frac{e^{-i\omega t} d\omega}{-m\left(\omega - \omega_0 + \dfrac{iR}{2m}\right)\left(\omega + \omega_0 + \dfrac{iR}{2m}\right)}.$$

Unfortunately, the pole of $F(\omega)$ lies on the real axis of the ω-plane. This problem can be avoided by assigning a small imaginary part to ω' (much as the convergence factor was used), so as to displace the pole slightly off the real axis; this imaginary part can be put to zero in the limit after doing the integral. But which way should the sign of the imaginary part be chosen: up or down from the real axis? It is here that the physics must be taken into account. The exponential ensures that we must close the contour in the upper half-plane for negative t; since $x(t) = 0$ for $t < 0$, we must have no poles contributing to the integral in this case, and therefore we insist that the pole be displaced just *below* the real axis. In that event, the pole will contribute its full residue for $t > 0$ (for which the contour is closed in the lower half-plane), adding an amount

$$-\frac{F_0}{m}\frac{e^{-i\omega't}}{[\omega' - \omega_0 + (iR/2m)][\omega' + \omega_0 + (iR/2m)]}$$

$$= -\frac{F_0}{m}\frac{e^{-i\omega't}}{\omega'^2 - \omega_0^2 - (R^2/4m^2) + (iR/m)\omega'}$$

$$= -\frac{F_0}{m}\frac{e^{-i\omega't}}{\omega'^2 - (k/m) + (iR/m)\omega'}$$

$$= -\frac{F_0}{m}\frac{e^{-i\omega't}[\omega'^2 - (k/m) - (iR/m)\omega']}{[\omega'^2 - (k/m)]^2 + (R^2\omega'^2/m^2)}.$$

This is the steady-state part of the solution for $x(t)$, its real part being

$$\frac{F_0}{m}\frac{[-\omega'^2 + (k/m)]\cos\omega't + (R/m)\omega'\sin\omega't}{[\omega'^2 - (k/m)]^2 + (R^2\omega'^2/m^2)}, \quad (7.26)$$

which agrees precisely with the first term of eq. (7.10) when the proper identifications are made.

As for the transient solution (7.24), it becomes in this case

$$-\frac{F_0}{2m\omega_0}e^{-Rt/2m}\left[e^{-i\omega_0 t}\frac{1}{\omega_0 - \omega' - (iR/2m)} - e^{i\omega_0 t}\frac{1}{-\omega_0 - \omega' - (iR/2m)}\right]$$

$$= e^{-Rt/2m}[\text{constant} \cdot \cos\omega_0 t + \text{constant} \cdot \sin\omega_0 t],$$

(7.27)

which agrees with the second term of (7.10) upon appropriate choice of constants. (The "constants" depend on frequencies but not on the time.) It will be left as a problem to show that $x(t)$ and $\dot{x}(t)$ are continuous at time zero with the

Transform Methods for Oscillations

expressions just obtained. That is, $x(t)$ and $\dot{x}(t)$ were assumed to be zero until the application of the driving force—they do not jump instantaneously to finite values.

Next, let us look at the oscillator response to an extreme impulse, the delta function, a force of infinite magnitude applied only for an instant. Letting $f(t) = \delta(t)$, eq. (7.17′) gives

$$F_\delta(\omega) = \frac{1}{2\pi} \int_{-\infty}^{\infty} \delta(t') e^{i\omega t'} dt' = \frac{1}{2\pi}, \tag{7.28}$$

whereupon by (7.23) and (7.24) we have, for $t > 0$,

$$x_\delta(t) = \frac{1}{2\pi} \int_{-\infty}^{\infty} \frac{e^{-i\omega t} d\omega}{k - m\omega^2 - iR\omega}$$

$$= \frac{\pi i}{m\omega_0} e^{-Rt/2m} \left[e^{-i\omega_0 t} \frac{1}{2\pi} - e^{i\omega_0 t} \frac{1}{2\pi} \right]$$

$$= \frac{1}{m\omega_0} e^{-Rt/2m} \sin \omega_0 t. \tag{7.29}$$

The result is what one might expect physically. The oscillator undergoes damped vibrations at its natural frequency.

Closely associated with the delta impulse is the concept of the unit step function force:

$$f(t) \equiv u(t) \equiv \int_{-\infty}^{t} \delta(t') dt' = \begin{Bmatrix} 1 \\ 0 \end{Bmatrix} \quad \begin{matrix} (t > 0) \\ (t < 0) \end{matrix}.$$

Note that $du(t)/dt = \delta(t)$ is a useful identity. The response to the step function from (7.25), (7.26), and (7.27), setting $F_0 = 1$ and $\omega' = 0$, is, for $t > 0$,

$$F_u(\omega) = \frac{i}{2\pi\omega}, \tag{7.30}$$

$$x_u(t) = \frac{1}{k} - \frac{1}{2m\omega_0} e^{-Rt/2m} \left[e^{-i\omega_0 t} \frac{1}{\omega_0 - \dfrac{iR}{2m}} + e^{i\omega_0 t} \frac{1}{\omega_0 + \dfrac{iR}{2m}} \right]$$

$$= \frac{1}{k} - \frac{1}{m\omega_0} e^{-Rt/2m} \left[\frac{\omega_0 \cos \omega_0 t + \dfrac{R}{2m} \sin \omega_0 t}{\omega_0^2 + (R^2/4m^2)} \right]$$

$$= \frac{1}{k} - \frac{1}{k} e^{-Rt/2m} \left(\cos \omega_0 t + \frac{R}{2m\omega_0} \cdot \sin \omega_0 t \right), \tag{7.31}$$

where we have employed eq. (7.6). The constant force simply displaces the system by a constant amount $1/k$ in the steady state.

It turns out that the responses $x(t)$ to the delta and unit step functions are very basic to the theory in that the response to *any* force $f(t)$ can be simply expressed by making use of them. Thus, from (7.29), we note that

$$x_\delta(t-t') = \frac{1}{2\pi}\int_{-\infty}^{\infty}\frac{e^{-i\omega(t-t')}\,d\omega}{k-m\omega^2-iR\omega},$$

which leads to

$$\int_{-\infty}^{\infty}f(t')x_\delta(t-t')\,dt' = \frac{1}{2\pi}\int\int_{-\infty}^{\infty}\frac{e^{-i\omega(t-t')}}{k-m\omega^2-iR\omega}f(t')\,d\omega\,dt'$$

$$= \int_{-\infty}^{\infty}\left[\frac{1}{2\pi}\int_{-\infty}^{\infty}f(t')e^{i\omega t'}\,dt'\right]\frac{e^{-i\omega t}\,d\omega}{k-m\omega^2-iR\omega}$$

$$= \int_{-\infty}^{\infty}F(\omega)\frac{e^{-i\omega t}\,d\omega}{k-m\omega^2-iR\omega} = x(t), \qquad (7.32)$$

by (7.17') and (7.23). This means that $x(t)$ can always be found from the force $f(t)$ by a *single* integration, substituting (7.29) into the integrand of (7.32) along with the force.

Note that (7.32) bears a direct relation to the equation

$$f(t) = \int_{-\infty}^{\infty}f(t')\delta(t-t')\,dt'.$$

An integral of this type is called a **convolution** or **faltung** integral, the latter term being the German word for "folding". This adjective is appropriate if one considers folding a straight line of length t back on itself at the center. The distances from the origin of t' and $t-t'$ will then be superimposed points on the two segments of the folded line; these are the arguments of the integrand.

We can now write the convolution integral for $x(t)$ as

$$x(t) = \int_{-\infty}^{\infty}f(t')x_\delta(t-t')\,dt' = -\int_{-\infty}^{\infty}f(t')\frac{dx_u(t-t')}{dt'}\,dt'$$

$$= -[f(t')x_u(t-t')]_{-\infty}^{\infty} + \int_{-\infty}^{\infty}x_u(t-t')\frac{df(t')}{dt'}\,dt',$$

on integrating by parts and using the fact that

$$x_\delta(t) = \frac{dx_u(t)}{dt}.$$

One factor or the other of the integrated term vanishes at the limits, leaving

$$x(t) = \int_{-\infty}^{\infty} \frac{df(t')}{dt'} x_u(t - t') \, dt'. \tag{7.33}$$

Hence, the response can always be found by differentiation of the force and substitution of eq. (7.31) into (7.33). Any of these approaches may be more convenient in a given problem than the direct calculation of $F(\omega)$, followed by calculation of $x(t)$ through (7.23).

7.4 NORMAL COORDINATES

Let us once more proceed along the lines of Section 7.1, but this time with greater generality. Consider the potential energy as a function of the generalized coordinates q_1 through q_n, and expand it in a Taylor series about the equilibrium position. It will be assumed that the coordinates all take on zero values at the equilibrium positions, so that they are a measure of the displacements from this configuration. We then have

$$V(q_1, \cdots q_n) = V_0 + \sum_{i=1}^{n} \left(\frac{\partial V}{\partial q_i}\right)_0 q_i + \frac{1}{2} \sum_{i,j} \left(\frac{\partial^2 V}{\partial q_i \partial q_j}\right)_0 q_i q_j + \cdots.$$

But all first derivatives $\partial V/\partial q_i$ must vanish at the equilibrium position, for these correspond to generalized forces. If the constant V_0 is chosen as zero, the first approximation to the potential is given by the quadratic terms:

$$V(q_1, \cdots q_n) \simeq \frac{1}{2} \sum_{i,j} \left(\frac{\partial^2 V}{\partial q_i \partial q_j}\right)_0 q_i q_j \equiv \frac{1}{2} \sum_{i,j} V_{ij} q_i q_j, \tag{7.34}$$

where

$$V_{ij} \equiv \left(\frac{\partial^2 V}{\partial q_i \partial q_j}\right)_0 = V_{ji}$$

is a constant symmetric in i and j, the derivative being evaluated at equilibrium.

In essence, an expression similar to (7.34) for the kinetic energy as a quadratic form in the velocities was obtained in Chapter 2 [refer to the lines following eq. (2.15)]. If the coefficients in that expression (which involve the particle masses and the derivatives of the coordinate transformation) are approximated by their values at equilibrium, the kinetic energy becomes

$$T \simeq \frac{1}{2} \sum_{i,j} T_{ij} \dot{q}_i \dot{q}_j, \tag{7.35}$$

the T_{ij} being symmetric constants. Note that the kinetic and potential expressions depend upon the assumption that the oscillations are *small*, for otherwise higher terms in the Taylor expansions would be required.

With the aid of (7.34) and (7.35), the Lagrangian can be written as

$$L = T - V = \tfrac{1}{2} \sum_{i,j} \left(T_{ij} \dot{q}_i \dot{q}_j - V_{ij} q_i q_j \right). \tag{7.36}$$

The equations of motion are then simply

$$\frac{d}{dt}\left(\frac{\partial L}{\partial \dot{q}_i}\right) - \frac{\partial L}{\partial q_i} = 0 = \sum_j \left(T_{ij} \ddot{q}_j + V_{ij} q_j \right). \tag{7.37}$$

The summation on j extends over all n coordinates; there are in addition n equations, one for each value of i. In general, all of the coordinates and their second derivatives appear in each equation, so that we are dealing with a *coupled* system of differential equations which must be solved simultaneously in order to find the q values as a function of time.

Let us consider an oscillatory solution of the type

$$q_j = a_j e^{-i\omega t}, \tag{7.38}$$

the real part of which corresponds to the physical motion. Substitution of eq. (7.38) into (7.37) produces

$$\sum_j \left(V_{ij} - \omega^2 T_{ij} \right) a_j = 0, \tag{7.39}$$

treating the amplitude factors a_j as constants. Here, we have a set of n homogeneous linear equations for the a_j, which have a nontrivial solution only if the coefficient determinant vanishes, a condition which provides us with n roots or values of ω^2 which are the (squared) frequencies allowed in the problem at hand. Constructing square $n \times n$ matrices $\mathbf{T}$ and $\mathbf{V}$ and a column vector $\mathbf{a}$ from the elements T_{ij}, V_{ij}, and a_j, eq. (7.39) becomes a matrix relation

$$\mathbf{Va} = \omega^2 \mathbf{Ta}. \tag{7.40}$$

The whole procedure of solving for the allowed ω^2 is then much like a matrix eigenvalue calculation, and it would be exactly that if it were not for the presence of the **T**-matrix on the right side. It is possible to show from the real symmetric nature of **T** and **V** (much as was done in Chapter 5 with the inertia tensor) that the "eigenvalues" ω^2 are all real and positive or zero. Furthermore, the "eigenvector components" may be chosen entirely real, their ratios being determined in accordance with eq. (7.40).

Normal Coordinates

To carry the eigenvalue concept somewhat further, let us construct a diagonal matrix $\boldsymbol{\lambda}$ with elements $\lambda_{kj} \equiv \omega_k^2 \delta_{k,j}$, the index k on ω^2 being required in order to single out one of the n possible squared frequencies. We shall also construct a square matrix $\mathbf{A}$ with columns which are the n eigenvectors $\mathbf{a}$ of eq. (7.40); the first column of $\mathbf{A}$ satisfies (7.40) for ω_1^2, and so on. Then, the element a_{jk} is the jth component of the kth eigenvector and relates to ω_k^2. By means of these matrices, (7.40) can be written

$$\sum_j V_{ij} a_{jk} = \omega_k^2 \sum_j T_{ij} a_{jk} = \sum_{j,l} T_{ij} a_{jl} \omega_l^2 \delta_{l,k} = \sum_{j,l} T_{ij} a_{jl} \lambda_{lk}$$

or

$$\mathbf{VA} = \mathbf{TA}\boldsymbol{\lambda} \qquad (7.41)$$

in simplest form. To proceed from here, we write (7.40) for the lth eigenvector as

$$\sum_i V_{ji} a_{il} = \sum_i V_{ij} a_{il} = \omega_l^2 \sum_i T_{ji} a_{il} = \omega_l^2 \sum_i T_{ij} a_{il}.$$

Multiply this result by a_{jk} and sum on j, then subtract the relation from that for the kth eigenvector, above (7.41), the latter having been multiplied by a_{il} and summed on i, giving

$$\sum_{i,j} V_{ij} a_{jk} a_{il} - \sum_{i,j} V_{ij} a_{jk} a_{il} \equiv 0 = (\omega_k^2 - \omega_l^2) \sum_{i,j} T_{ij} a_{jk} a_{il}.$$

If the roots ω_k^2 and ω_l^2 are distinct, we have, for $k \neq l$,

$$\sum_{i,j} T_{ij} a_{jk} a_{il} = 0, \qquad (7.42)$$

an orthogonality condition on the elements of $\mathbf{A}$. Now, inasmuch as (7.40) only fixes the ratios of eigenvector components, it is convenient for normalization to require that

$$\sum_{i,j} T_{ij} a_{jk} a_{ik} \equiv 1. \qquad (7.43)$$

Any of the a_{ik} can be expressed in terms of a particular component, which is then normalized through eq. (7.43). The combination of eqs. (7.42) and (7.43) allows us to write the single matrix equation

$$\tilde{\mathbf{A}} \mathbf{T} \mathbf{A} = \mathbf{1}. \qquad (7.44)$$

Thus, the **T**-matrix relates to the unit matrix through a **congruent** transformation, the kind of transformation which is in general appropriate to the metric of a skewed system, as seen in Appendix E. Multiplying (7.41) on the left by $\tilde{\mathbf{A}}$, we have

$$\tilde{\mathbf{A}}\mathbf{V}\mathbf{A} = \tilde{\mathbf{A}}\mathbf{T}\mathbf{A}\lambda = \mathbf{1}\lambda = \lambda, \tag{7.45}$$

so the same transformation also diagonalizes the **V**-matrix.

Let us now define new coordinates Q_i, called **normal coordinates**, which are related to the q_i by a linear transformation $q_i \equiv \sum_j a_{ij} Q_j$, or

$$\mathbf{q} = \mathbf{A}\mathbf{Q}, \tag{7.46}$$

and

$$\mathbf{Q} = \mathbf{A}^{-1}\mathbf{q}, \tag{7.46'}$$

where we have formed column vectors **q** and **Q** in the usual way. Using (7.34), (7.35), and the congruent transformations, it is clear that

$$V = \tfrac{1}{2} \sum_{i,j} V_{ij} q_i q_j = \tfrac{1}{2} \tilde{\mathbf{q}} \mathbf{V} \mathbf{q}$$

and similarly

$$T = \tfrac{1}{2} \tilde{\dot{\mathbf{q}}} \mathbf{T} \dot{\mathbf{q}}.$$

But $\tilde{\mathbf{q}} = \widetilde{(\mathbf{A}\mathbf{Q})} = \tilde{\mathbf{Q}}\tilde{\mathbf{A}}$; hence, one has

$$V = \tfrac{1}{2} \tilde{\mathbf{Q}} \tilde{\mathbf{A}} \mathbf{V} \mathbf{A} \mathbf{Q} = \tfrac{1}{2} \tilde{\mathbf{Q}} \lambda \mathbf{Q} = \tfrac{1}{2} \sum_i \omega_i^2 Q_i^2 \tag{7.47}$$

by (7.45); and from (7.44),

$$T = \tfrac{1}{2} \tilde{\dot{\mathbf{Q}}} \tilde{\mathbf{A}} \mathbf{T} \mathbf{A} \dot{\mathbf{Q}} = \tfrac{1}{2} \tilde{\dot{\mathbf{Q}}} \dot{\mathbf{Q}} = \tfrac{1}{2} \sum_i \dot{Q}_i^2. \tag{7.48}$$

Thus, in the new coordinates, both kinetic and potential energies are simultaneously brought to a sum of squared terms.

The Lagrangian is now

$$L = T - V = \tfrac{1}{2} \sum_i \left(\dot{Q}_i^2 - \omega_i^2 Q_i^2 \right)$$

and the Lagrange equations of motion become of simple harmonic oscillator type

$$\ddot{Q}_i + \omega_i^2 Q_i = 0, \tag{7.49}$$

Normal Coordinates

with solutions

$$Q_i = C_i e^{-i\omega_i t}. \tag{7.50}$$

Combining (7.46) with (7.50), we obtain

$$q_i = \sum_j C_j a_{ij} e^{-i\omega_j t}. \tag{7.51}$$

The real part of eq. (7.51) is the general solution for q_i; it is a linear combination formed from all the normal modes of vibration, one of which is associated with each normal coordinate. If we permit C_j to be complex, each term in the real part of the sum (7.51) will be of form

$$\text{constant} \cdot \cos \omega_j t + \text{constant} \cdot \sin \omega_j t,$$

so it is unnecessary to include an $e^{+i\omega_j t}$ term as a second solution to (7.49).

The introduction of the constants C_j makes it easy to insert the initial conditions on the motion. At time $t = 0$, we have, from (7.51),

$$q_i(0) = \operatorname{Re} \sum_j C_j a_{ij} = \sum_j (\operatorname{Re} C_j) a_{ij},$$

and

$$\dot{q}_i = \operatorname{Re} \frac{d}{dt} \sum_j C_j a_{ij} e^{-i\omega_j t} = \operatorname{Re} \sum_j C_j a_{ij}(-i\omega_j) e^{-i\omega_j t}.$$

Hence, we can write

$$\dot{q}_i(0) = \operatorname{Re} \sum_j C_j a_{ij}(-i\omega_j) = \operatorname{Im} \sum_j C_j a_{ij} \omega_j = \sum_j (\operatorname{Im} C_j) a_{ij} \omega_j.$$

Multiply both sides of these relations by $T_{ik} a_{kl}$ and sum on i and k, using (7.42) and (7.43),

$$\sum_{i,k} q_i(0) T_{ik} a_{kl} = \sum_{i,j,k} (\operatorname{Re} C_j) T_{ik} a_{kl} a_{ij} = \sum_j (\operatorname{Re} C_j) \delta_{j,l} = \operatorname{Re} C_l,$$

$$\sum_{i,k} \dot{q}_i(0) T_{ik} a_{kl} = \sum_{i,j,k} (\operatorname{Im} C_j) \omega_j a_{ij} a_{kl} T_{ik} = \sum_j (\operatorname{Im} C_j) \omega_j \delta_{j,l} = \omega_l \operatorname{Im} C_l,$$

or

$$\operatorname{Re} C_l = \sum_{i,k} q_i(0) T_{ik} a_{kl}, \quad \operatorname{Im} C_l = \frac{1}{\omega_l} \sum_{i,k} \dot{q}_i(0) T_{ik} a_{kl}. \tag{7.52}$$

The constants C_l are determined from the initial conditions, which do not affect

ω_l or a_{ij} in the slightest, the latter being given by the physical nature of the problem.

In summary, an undamped system of n degrees of freedom (the potential and kinetic energies being quadratic functions of the coordinates and their time derivatives) can be most easily described through a transformation to normal coordinates which are linear combinations of the original coordinates. The potential and kinetic energies then become a sum of squared terms in these coordinates and their derivatives, while the equation of motion for each normal coordinate has the form appropriate for a simple harmonic oscillator. The equations for these various coordinates are therefore independent of one another, not coupled, so no energy can be transferred from one normal coordinate to another.

The entire procedure for constructing the normal coordinates can best be clarified with a few examples. First, however, let us see how forced vibrations and resonance can be treated by the matrix formalism. Note that the previous equations of motion (7.37) are given in matrix terms as

$$\mathbf{T\ddot{q} + Vq = 0}.$$

Now suppose that each generalized coordinate q_i has associated with it an applied force F_i. Constructing an $n \times 1$ column matrix $\mathbf{F}$ from the force functions, we have the modified equations for forced vibrations:

$$\mathbf{T\ddot{q} + Vq = F}. \qquad (7.37')$$

Defining $\mathbf{f} \equiv \mathbf{\tilde{A}F}$, multiplication by $\mathbf{\tilde{A}}$ produces

$$\mathbf{\tilde{A}TAA^{-1}\ddot{q} + \tilde{A}VAA^{-1}q = \tilde{A}F = f}.$$

Transforming to normal coordinates (7.46) as before, and utilizing (7.44) and (7.45) as well, there results the modification of (7.49)

$$\mathbf{\ddot{Q} + \lambda Q = f}. \qquad (7.49')$$

To solve (7.49'), we define a diagonal matrix ω such that

$$\omega^2 \equiv \lambda.$$

(The diagonal elements of ω are the allowed frequencies.) The basic relation (7.49') can now be split into two coupled linear equations:

$$(D\mathbf{1} - i\omega)\mathbf{M} = \mathbf{f},$$

$$(D\mathbf{1} + i\omega)\mathbf{Q} \equiv \mathbf{M},$$

where $D \equiv d/dt$. The linear equations are readily solved simultaneously, yielding a single matrix solution for $\mathbf{Q}$ which is equivalent to an entire set of similar equations for the Q_i:

$$\mathbf{Q} = e^{-i\omega t}\mathbf{C}_1 + e^{-i\omega t}\int e^{i\omega t}\mathbf{M}\,dt,$$

$$\mathbf{M} = e^{i\omega t}\mathbf{C}_2' + e^{i\omega t}\int e^{-i\omega t}\mathbf{f}\,dt.$$

On combining, we have

$$\mathbf{Q} = e^{-i\omega t}\mathbf{C}_1 + e^{i\omega t}\mathbf{C}_2 + e^{-i\omega t}\int e^{2i\omega t}\,dt\int e^{-i\omega t'}\mathbf{f}\,dt', \qquad (7.50')$$

where the $\mathbf{C}$ coefficients are arbitrary constant column matrices. The first two terms of $\mathbf{Q}$ are dictated by the initial conditions, and the last by the particular form of $\mathbf{f}$. (Note that $\mathbf{f}$ and $\mathbf{M}$ are usually explicit functions of the time. The exponentials containing matrices are defined by series expansions, as is customary.)

As was done earlier in the chapter, let us apply these results to a cosine function:

$$[\mathbf{f}]_i = B_i \cos \omega' t.$$

Here, the ith component of the column matrix $\mathbf{f}$ has the constant amplitude B_i. Upon substitution of this cosine in (7.50'), it is not difficult to show that one obtains $B_i[\cos \omega' t/(\omega_i^2 - \omega'^2)]$ for the particular integral term in the ith component of $\mathbf{Q}$. This component only contains the ith frequency ω_i, so that typical normal modes are exhibited. The result has the characteristic resonance behavior of (7.26) or (7.10) and agrees with those expressions when the damping terms in them are equated to zero. (It is possible to extend the matrix formalism to include the case of damping, but we shall not do so.)

Example 1 Bead on Ring

A bead of mass m slides without friction on a uniform circular ring of mass M and radius a. If the ring oscillates (under gravity) in its own plane about a fixed point on its circumference, find the frequencies, the eigenvector components a_{ij}, and the normal coordinates (Figure 7.3).

Solution

Taking the y-axis vertically and the x-axis horizontally as usual, with the origin at the fixed point of the ring, we find for the bead the coordinates

$$x = a\sin\theta + a\sin\alpha, \qquad y = -a\cos\theta - a\cos\alpha,$$

whereupon

$$T_{\text{bead}} = \tfrac{1}{2}m(\dot{x}^2 + \dot{y}^2) = \tfrac{1}{2}ma^2[\dot{\theta}^2 + \dot{\alpha}^2 + 2\dot{\theta}\dot{\alpha}\cos(\alpha - \theta)]$$

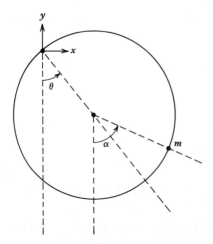

Figure 7.3 Bead sliding on ring.

after some algebraic simplification. According to the parallel-axis theorem, the moment of inertia of the ring about the point of support is $2Ma^2$; hence, the kinetic energy of the ring is

$$T_{\text{ring}} = Ma^2\dot\theta^2,$$

and in total

$$T = T_{\text{bead}} + T_{\text{ring}} = \tfrac{1}{2}a^2\left[(2M+m)\dot\theta^2 + m\dot\alpha^2 + 2m\dot\theta\dot\alpha\cos(\alpha-\theta)\right].$$

Both θ and α are zero at equilibrium. If small oscillations are assumed, we can approximate $\cos(\alpha-\theta)\simeq 1$ so as to obtain a quadratic form for T with constant coefficients, the **T**-matrix being

$$\mathbf{T} = \tfrac{1}{2}a^2\begin{pmatrix} 2(2M+m) & 2m \\ 2m & 2m \end{pmatrix},$$

where we have identified $q_1 \equiv \theta$ and $q_2 \equiv \alpha$, and made use of (7.35). (Factors of 2 appear along the diagonal in the matrix because of the factor $\tfrac{1}{2}$ in (7.35).)

The potential energy, as measured from the equilibrium position, is

$$V = Mga(1-\cos\theta) + mga(2-\cos\theta-\cos\alpha)$$
$$\approx \frac{ga}{2}\left[(M+m)\theta^2 + m\alpha^2\right]$$

for small angles, giving, by (7.34),

$$\mathbf{V} = \tfrac{1}{2}ga\begin{pmatrix} 2(M+m) & 0 \\ 0 & 2m \end{pmatrix}.$$

The determinant for the allowed frequencies is then

$$|\mathbf{V} - \omega^2\mathbf{T}| = 0 = \begin{vmatrix} ga(M+m) - \omega^2 a^2(2M+m) & -ma^2\omega^2 \\ -ma^2\omega^2 & mga - \omega^2 ma^2 \end{vmatrix}.$$

Solving, we find the roots are

$$\omega_1^2 = \frac{1}{2}\frac{g}{a} \quad \text{and} \quad \omega_2^2 = \frac{m+M}{M}\frac{g}{a}.$$

With knowledge of the eigenvalue ω_1^2, (7.40) can be employed to find the first eigenvector:

$$(\mathbf{V} - \omega_1^2 \mathbf{T})\mathbf{a} = 0 = \begin{pmatrix} ga(M+m) - \frac{1}{2}ga(2M+m) & -\frac{1}{2}mga \\ -\frac{1}{2}mga & mga - \frac{1}{2}mga \end{pmatrix} \begin{pmatrix} a_{11} \\ a_{21} \end{pmatrix}$$

$$= \frac{1}{2}mga \begin{pmatrix} 1 & -1 \\ -1 & 1 \end{pmatrix} \begin{pmatrix} a_{11} \\ a_{21} \end{pmatrix},$$

or

$$a_{11} = a_{21}.$$

The ratio of the two components is unity; the normalization is achieved by means of (7.43):

$$T_{11}a_{11}^2 + T_{12}a_{21}a_{11} + T_{21}a_{11}a_{21} + T_{22}a_{21}^2 = 1,$$

or

$$a_{11}^2 = \frac{1}{T_{11} + T_{12} + T_{21} + T_{22}} = \frac{1}{a^2(2M+4m)},$$

$$a_{11} = a_{21} = \frac{1}{\sqrt{2}\,a\sqrt{M+2m}}.$$

In a similar way, the second eigenvalue ω_2^2 leads to

$$a_{22} = -\frac{m+M}{m}a_{12} = \frac{m+M}{a}\frac{1}{\sqrt{mM(M+2m)}},$$

$$a_{12} = -\frac{1}{a}\sqrt{\frac{m}{M(M+2m)}}.$$

Assembling the various results, we find that

$$\mathbf{A} = \frac{1}{a\sqrt{M+2m}} \begin{pmatrix} 1/\sqrt{2} & -\sqrt{m/M} \\ 1/\sqrt{2} & (m+M)/\sqrt{mM} \end{pmatrix},$$

and

$$\mathbf{A}^{-1} = \frac{a\sqrt{2Mm}}{\sqrt{M+2m}} \begin{pmatrix} (m+M)/\sqrt{mM} & \sqrt{m/M} \\ -1/\sqrt{2} & 1/\sqrt{2} \end{pmatrix},$$

as is easily seen by the methods of Appendix C. Equation (7.46′), or $\mathbf{Q} = \mathbf{A}^{-1}\mathbf{q}$, then produces

$$Q_1 = \sqrt{\frac{2}{M+2m}}\, a[(m+M)\theta + m\alpha],$$

$$Q_2 = \sqrt{\frac{Mm}{M+2m}}\, a[\alpha - \theta].$$

These are the normal coordinates. Physically, we note that if $Q_2 = 0$, then $\alpha = \theta$, which means that the bead and ring move together in phase in the *first* normal mode (corresponding to Q_1). On the other hand, setting $Q_1 = 0$ gives

$$\theta = -\frac{m}{m+M}\alpha$$

for the *second* normal mode, the sign showing that the bead and ring would now be moving in opposite directions, out of phase with one another. The determination of the constants C_i from given initial conditions will be left as a problem.

Example 2 Coupled Springs

Consider once more the problem of a mass M suspended by a spring of constant K, with a second mass m attached to the first by a spring of constant k (see Figure 2.1 of Chapter 2). Find the frequencies of the normal modes of vibration.

Solution

The solution to this problem illustrates a technique for obtaining the normal modes directly from the equations of motion, even though the potential energy is not even in quadratic form initially. Some insight into the physics is thereby realized. From Chapter 2, we have the energies

$$T = \tfrac{1}{2}M\dot{y}_1^2 + \tfrac{1}{2}m\dot{y}_2^2,$$

$$V = \tfrac{1}{2}Ky_1^2 + \tfrac{1}{2}k(y_2 - y_1)^2 - Mgy_1 - mgy_2,$$

along with the coupled equations of motion (2.11) and (2.12):

$$\ddot{y}_1 + \frac{K}{M}y_1 - \frac{k}{M}(y_2 - y_1) - g = 0,$$

$$\ddot{y}_2 + \frac{k}{m}(y_2 - y_1) - g = 0.$$

To apply the normal mode theory, the potential energy must be a quadratic function of the coordinates as given in (7.34). This would be the case only if the linear gravitational terms could somehow be erased! To achieve this end, let us make a linear transformation to new coordinates X_1 and X_2 such that

$$X_1 \equiv y_1 - \frac{(M+m)g}{K},$$

$$X_2 \equiv y_2 - \frac{(M+m)g}{K} - \frac{mg}{k}$$

(note the physical significance of the constant terms), in terms of which the potential energy becomes

$$V = \frac{1}{2}K\left[X_1 + \frac{(M+m)g}{K}\right]^2$$
$$+ \frac{1}{2}k\left\{\left[X_2 + \frac{(M+m)g}{K} + \frac{mg}{k}\right] - \left[X_1 + \frac{(M+m)g}{K}\right]\right\}^2$$
$$- Mg\left[X_1 + \frac{(M+m)g}{K}\right] - mg\left[X_2 + \frac{(M+m)g}{K} + \frac{mg}{k}\right]$$
$$= \frac{1}{2}KX_1^2 + \frac{1}{2}k(X_2 - X_1)^2 + \text{constant}.$$

That is, this transformation was so designed as to eliminate the linear terms in V. The constant is irrelevant so far as forces or Lagrangian derivatives are concerned and can simply be omitted. Naturally, the transformation does not affect the kinetic energy:

$$T = \tfrac{1}{2}M\dot{X}_1^2 + \tfrac{1}{2}m\dot{X}_2^2.$$

The standard procedure can now be employed giving the matrices:

$$\mathbf{T} = \begin{pmatrix} M & 0 \\ 0 & m \end{pmatrix}, \quad \mathbf{V} = \begin{pmatrix} K+k & -k \\ -k & k \end{pmatrix},$$

$$\mathbf{V} - \omega^2\mathbf{T} = \begin{pmatrix} K+k-\omega^2 M & -k \\ -k & k-\omega^2 m \end{pmatrix},$$

and therefore

$$\omega^2 = \frac{kM + (K+k)m \pm \sqrt{[kM + (K+k)m]^2 - 4MmkK}}{2Mm}$$

are the two values of frequency squared. From this point on, the components of the eigenvector $\mathbf{a}$ and the normal coordinates are found in the usual way.

But there is an alternative procedure which often is simpler for problems of this type. The Lagrangian in the new coordinates is

$$L = T - V = \tfrac{1}{2}M\dot{X}_1^2 + \tfrac{1}{2}m\dot{X}_2^2 - \tfrac{1}{2}KX_1^2 - \tfrac{1}{2}k(X_2 - X_1)^2,$$

which gives as modified equations of motion (2.11') and (2.12'),

$$\frac{d}{dt}\left(\frac{\partial L}{\partial \dot{X}_1}\right) - \frac{\partial L}{\partial X_1} = 0 = M\ddot{X}_1 + KX_1 - k(X_2 - X_1),$$

$$\frac{d}{dt}\left(\frac{\partial L}{\partial \dot{X}_2}\right) - \frac{\partial L}{\partial X_2} = 0 = m\ddot{X}_2 + k(X_2 - X_1),$$

relations that can also be established by substituting y's in terms of X's into

(2.11) and (2.12). Let us multiply (2.11') by some constant A and (2.12') by another constant B, divide out the masses, and then add the two equations:

$$A\ddot{X}_1 + \frac{AK}{M}X_1 - \frac{Ak}{M}(X_2 - X_1) + B\ddot{X}_2 + \frac{Bk}{m}(X_2 - X_1) = 0$$

$$= A\ddot{X}_1 + B\ddot{X}_2 + \left[\frac{A}{M}(K+k) - \frac{Bk}{m}\right]X_1 + \left[\frac{B}{m} - \frac{A}{M}\right]kX_2.$$

But the normal modes are known to be linear combinations of the form

$$Q = AX_1 + BX_2$$

and so, from (7.49),

$$\ddot{Q} + \omega^2 Q = 0 = A\ddot{X}_1 + B\ddot{X}_2 + \omega^2[AX_1 + BX_2].$$

Identifying the multiplicative factors above with the A and B of the linear combination for Q, we see that the combined equation of motion has exactly this form provided we identify coefficients of the linear terms:

$$\omega^2 A \equiv \frac{A}{M}(K+k) - \frac{Bk}{m}, \qquad \omega^2 B \equiv \left(\frac{B}{m} - \frac{A}{M}\right)k.$$

Dividing one relation by the other (so as to eliminate ω^2) and simplifying the algebra, one obtains

$$\left(\frac{A}{B}\right)^2 + \left(\frac{K+k}{k} - \frac{M}{m}\right)\frac{A}{B} - \frac{M}{m} = 0.$$

It should not be surprising that a quadratic equation is found for the ratio A/B; two separate values of the ratio are needed to determine two distinct normal coordinates. Once the quadratic has been solved for the ratios of A to B, substitution in either of the two coefficient relations will determine the two values of ω^2 required, precisely in agreement with those obtained in the previous method. The normal coordinates, of course, have also been obtained, for

$$Q = (\text{constant } B)\left(\frac{A}{B}X_1 + X_2\right),$$

or

$$Q = (\text{constant } B)\left[\frac{A}{B}y_1 + y_2 - \frac{(M+m)g}{K}\left(\frac{A}{B} + 1\right) - \frac{mg}{k}\right]$$

in terms of the old y-coordinates. The method employed last, which leads directly to ω^2 and the normal coordinates, is physically motivated and easier to follow in simple problems. The matrix method outlined earlier is satisfactory in general, however, even for a complex problem with many coordinates.

Example 3 The One-Dimensional Crystal

Consider small oscillations of N masses m about their equally spaced equilibrium positions (separation a) in a one-dimensional crystal, a very oversimplified model for a solid, but one which yields interesting results. It is difficult to solve even this model problem by the standard methods, however, for the kinetic and potential matrices lead to the construction of a large determinant which must be solved to find the allowed values of ω^2. (In this case, N normal modes are present, and there will be $3N$ modes in a more general three-dimensional example.)

Let us turn instead to the equation of motion for the nth atom. Denote the displacement of this particle relative to its equilibrium position by q_n, and assume that only nearest-neighbor linear forces are acting on it with an "effective spring constant" β. The nth atom is then subjected to a net force

$$\beta(q_{n+1} - q_n) - \beta(q_n - q_{n-1})$$

arising from adjacent atoms. Newton's second law now tells us that

$$m\ddot{q}_n = \beta(q_{n+1} + q_{n-1} - 2q_n).$$

Most physicists are reasonably familiar with this kind of situation and might expect a one-dimensional solution to have the traveling wave form $e^{i(Kx - \omega t)}$. In this discrete problem, however, the continuous variable x must be replaced by the equilibrium position na of the nth atom. In addition, K and ω can only be determined after some consideration of boundary conditions.

There are two kinds of boundary conditions which are often employed. In this discussion, we shall assume a periodic condition such that

$$q_n = q_{n+N},$$

which is equivalent to arranging the masses around a circle so we do not have to deal with effects caused by the two ends of the crystal. We then have the proposed solution

$$q_n \sim e^{i(Kna - \omega t)} = e^{i[K(n+N)a - \omega t]}$$

or

$$e^{iKNa} = 1.$$

The allowed values of K are necessarily discrete and are given by

$$K \rightarrow K_j = \frac{2\pi j}{Na} \quad (j = \text{integer})$$

as a consequence. Taking the usual physical interpretation of K, one can write

$$K_j = \frac{2\pi}{\lambda_j} = \frac{2\pi j}{Na},$$

or
$$\lambda_j = \frac{Na}{j},$$

where λ_j is the wavelength. Substitution in the equation of motion produces

$$-m\omega^2 q_n = \beta q_n [e^{iKa} + e^{-iKa} - 2],$$

$$\omega^2 = -\frac{2\beta}{m}(\cos Ka - 1) = \frac{4\beta}{m}\sin^2\frac{Ka}{2},$$

$$\omega \to \omega_j = \pm 2\sqrt{\beta/m}\,\sin\frac{\pi j}{N}.$$

(We have assigned a subscript to ω, since it depends on the integer j.) The relation between ω and K is called a **dispersion relation**; if it is satisfied, the exponential will be a solution for q_n.

There is another kind of boundary condition which must be considered, namely, that of fixed ends, an analogy to the vibrating string problem. In order to obtain nodal points at both ends for all times, we take q_n to be a linear superposition of waves traveling to right and left:

$$q_n = A_n e^{i(Kna - \omega t)} + B_n e^{-i(Kna + \omega t)};$$

$A_n = -B_n$ results at once from the origin condition $q_n(n = 0) = 0$, giving

$$q_n \sim e^{iKna} - e^{-iKna}.$$

Taking the right end condition as $q_n(n = N) = 0$ for $x = Na$ (which gives only $N - 1$ atoms actually vibrating between the fixed ends), then

$$e^{iKNa} - e^{-iKNa} = 0,$$

or
$$\sin KNa = 0.$$

This relation implies that

$$KNa = j\pi, \qquad K_j = \frac{j\pi}{Na} \qquad (j = \text{integer}).$$

The allowed K values differ by a factor of 2 from those under the periodic condition, but the dispersion relation for ω^2 is still

$$\omega^2 = \frac{4\beta}{m}\sin^2\frac{Ka}{2} \qquad (7.53)$$

as before; substitution of q_n in the equation of motion readily verifies this fact.

Normal Coordinates

Later in this chapter, we shall treat the string again through application of continuum methods; at that point, the fixed end boundary conditions will be needed. For the present we shall apply the periodic conditions only. The latter are really more appropriate for problems of the solid state, for an electron traveling down a lattice sees a very long repetitive path of many atoms in which end (or surface) effects can often be ignored.

Note that it has been possible to find the allowed frequencies ω_j from a simple analysis which has not even mentioned normal coordinates! It seems reasonable to suppose that the latter quantities must be given by an expression such as

$$Q_j = C \sum_n e^{-iKna} q_n \quad (C = \text{constant}),$$

for in this way Q_j will be proportional to $e^{-i\omega_j t}$. If the Q_j relation is multiplied on both sides by $(e^{iKn'a})/CN$ and summed on j, one can invert it to obtain

$$q_n = \frac{1}{CN} \sum_j e^{iKna} Q_j$$

by making use of the geometrical series summation formula

$$\sum_{j=1}^{N} e^{2\pi i j (n'-n)/N} = N \delta_{n,n'}.$$

Evaluation of the energies can now be readily accomplished:

$$T = \frac{1}{2} m \sum_n \dot{q}_n^2 = \frac{1}{2} m \sum_n \left[\frac{1}{CN} \sum_j e^{2\pi i n j/N} \dot{Q}_j \right] \left[\frac{1}{C^*N} \sum_k e^{-2\pi i n k/N} \dot{Q}_k^* \right]$$

$$= \frac{1}{2} \frac{m}{|C|^2 N^2} \sum_{j,k} \dot{Q}_j \dot{Q}_k^* \left[\sum_n e^{2\pi i n (j-k)/N} \right] = \frac{m}{2|C|^2 N^2} \sum_{j,k} \dot{Q}_j \dot{Q}_k^* [N \delta_{k,j}]$$

$$= \frac{m}{2|C|^2 N} \sum_j \dot{Q}_j \dot{Q}_j^*,$$

where the asterisk indicates complex conjugation, which proves convenient in the analysis to follow. The potential energy is found to be

$$V = \frac{1}{2} \beta \sum_n (q_{n+1} - q_n)^2 = \frac{2\beta}{|C|^2 N} \sum_j Q_j Q_j^* \sin^2 \frac{\pi j}{N},$$

after a good bit of algebra similar to that given above. (Note that in differentiating V to find the force on the nth atom, *two* of the squared terms in the sum contribute for a given atom, yielding the force expression given earlier.)

Finally, the Lagrangian is

$$L = T - V = \frac{1}{|C|^2 N} \sum_j \left[\frac{1}{2} m \dot{Q}_j \dot{Q}_j^* - 2\beta Q_j Q_j^* \sin^2 \frac{\pi j}{N} \right].$$

Treating Q_j and Q_j^* as independent, the equation of motion for Q_j^* is

$$\frac{d}{dt} \left(\frac{\partial L}{\partial \dot{Q}_j^*} \right) - \frac{\partial L}{\partial Q_j^*} = 0 = \frac{1}{2} m \ddot{Q}_j + 2\beta Q_j \sin^2 \frac{\pi j}{N}.$$

Since we want the form

$$\ddot{Q}_j + \omega_j^2 Q_j = 0$$

for normal coordinates, one can immediately identify

$$\omega_j^2 = \frac{4\beta}{m} \sin^2 \frac{\pi j}{N},$$

which is just the previously obtained frequency relation. Furthermore, it is clear that the Q_j are the correct normal coordinates, for with their aid it was possible to express T and V as a sum of squares.

Consider now the possibility of assigning the integer j a value of zero. In that event $\omega_j = 0$, whereupon, from (7.49), $\ddot{Q}_j = 0$. Clearly, we are dealing with a situation that calls for translation of the crystal as a whole. Zero frequency results arise often in such calculations; as in the present case, they generally correspond to translation or rotation of the system as a whole, the various parts of the system not being displaced relative to one another. Conditions can be specified in the beginning (such as a requirement that the center of mass be held fixed) which serve to eliminate some of the otherwise necessary coordinates, thereby reducing the number of degrees of freedom and preventing the establishment of zero frequencies from the outset.

As an example, consider the motion of three masses at the vertices of a plane triangle. The requirement that the masses move only within the plane reduces the number of degrees of freedom from 9 to 6. A further reduction to 4 occurs by restricting the CM position, thus eliminating the two possible translation motions of the whole triangle. Insertion of an angular momentum condition for rotation of the triangle about an axis perpendicular to its plane leaves us with only three independent coordinates. Of the normal modes that remain, one is associated

Continuous Media

with an in–out motion which preserves the shape of the triangle. Thus, with a little experience as a guide, very few surprises are in store for us as we investigate the normal modes of simple geometrical configurations.

7.5 CONTINUOUS MEDIA

Transverse oscillations along a one-dimensional string were treated earlier in this chapter by the methods of D'Alembert and Bernoulli. These techniques served as simple examples of solutions to a case where matter is distributed continuously over a medium, rather than located at particular points. The last example of the previous section can also be used to solve the string problem, but in a totally different way.

In that example, we considered the longitudinal motion of atoms along a line, behaving as if they were connected by invisible springs. The transverse motions of many particles lying along a line are exactly the same, for the equation of motion of each atom is the same; furthermore, in the continuous limit, the transverse motions are those of the string problem. To justify these statements, note that in the series of atoms connected so as to form a string under tension T (as shown in Figure 7.4), for the particular atom n, the net transverse force acting is just

$$T \sin \theta - T \sin \alpha \simeq T(\theta - \alpha)$$

for small displacements. Noting that sines and tangents are almost equal,

$$\theta \simeq \frac{q_{n+1} - q_n}{a} \quad \text{and} \quad \alpha \simeq \frac{q_n - q_{n-1}}{a},$$

we can write

$$m\ddot{q}_n = \frac{T}{a}(q_{n+1} + q_{n-1} - 2q_n)$$

from Newton's second law. This is identical with the longitudinal equation derived earlier if we identify the "spring constant" β with T/a. Of course, it is

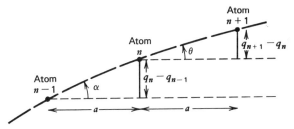

Figure 7.4 Transverse displacements for line of connected atoms under tension.

vital to understand that here the generalized coordinates q_n correspond to transverse displacements and not to the previous longitudinal ones.

The allowed frequencies for the one-dimensional solid with fixed end conditions were given by eq. (7.53):

$$\omega_j = 2\sqrt{\frac{\beta}{m}} \sin\frac{K_j a}{2} = 2\sqrt{\frac{\beta}{m}} \sin\frac{\pi j}{2N}.$$

Consider the limit as the number of atoms N becomes infinite while the spacing a between atoms goes to zero. Let the mass m of an individual atom also go to zero in such a way that the density $\rho(= m/a)$ is held constant. As the sine argument becomes small, it is essentially equal to the sine itself, and we have

$$\omega_j \rightarrow 2\sqrt{\frac{\beta}{m}}\frac{\pi j}{2N} = \sqrt{\frac{T/a}{m}}\frac{\pi j}{N} = \sqrt{\frac{T}{\rho}}\frac{\pi j}{Na} = \frac{\pi j v}{l},$$

where we have utilized $\beta = T/a$, $\rho = m/a$, $v = \sqrt{T/\rho}$, and $Na = l$. The net result is that we have obtained a frequency expression identical with that of the Bernoulli method, as seen in (7.16).

We can go further. The equation of motion for the nth atom is

$$m\ddot{q}_n = \beta(q_{n+1} - q_n) - \beta(q_n - q_{n-1}),$$

or

$$\ddot{q}_n = \frac{T}{\rho a}\left[\left(\frac{q_{n+1} - q_n}{a}\right) - \left(\frac{q_n - q_{n-1}}{a}\right)\right].$$

The difference of the x-coordinates for two adjacent atoms is simply a, hence

$$\frac{q_{n+1} - q_n}{a} \rightarrow \frac{\partial q}{\partial x}$$

in the limit. The expression in the large bracket divided by a in the equation of motion is then a second derivative, giving

$$\ddot{q} = v^2 \frac{\partial^2 q}{\partial x^2},$$

the usual wave equation. (The dots now indicate *partial* differentiation in time.)

The Lagrangian for the line of mass points is

$$L = \tfrac{1}{2} m \sum_n \dot{q}_n^2 - \tfrac{1}{2}\beta \sum_n (q_{n+1} - q_n)^2$$

$$= \tfrac{1}{2}\rho \sum_n \dot{q}_n^2 a - \tfrac{1}{2} T \sum_n \left(\frac{q_{n+1} - q_n}{a}\right)^2 a.$$

In the continuous limit, this expression becomes

$$L \to \frac{1}{2}\rho \int_0^l \dot{q}^2\, dx - \frac{1}{2}T \int_0^l \left(\frac{\partial q}{\partial x}\right)^2 dx. \tag{7.54}$$

This last relationship provides a clue to a better procedure for a treatment of problems involving continuous media. In one dimension, we can define a **Lagrangian density function** $\mathcal{L}$ with the property that

$$L = \int \mathcal{L}\, dx, \tag{7.55}$$

the extension to higher dimensionality being obvious. For the present problem, we then have

$$\mathcal{L} = \tfrac{1}{2}\rho \dot{q}^2 - \tfrac{1}{2}T\left(\frac{\partial q}{\partial x}\right)^2. \tag{7.56}$$

In this density formulation, the continuous variable x has replaced the discrete index n. The generalized coordinate q, its derivatives with respect to x and t, and these two variables themselves may all appear in the Lagrangian density.

The limiting procedure is frequently awkward to pursue, while the earlier methods for the vibrating string are only appropriate for that problem (the Fourier expansion technique can be generalized to other situations, of course). But the Lagrangian density function allows us to construct a variational formulation for problems of continuous media. Note that Hamilton's principle can now be written (in one-dimensional cases) as

$$\delta \int_{t_1}^{t_2} L\, dt = 0 = \delta \int_{t_1}^{t_2} dt \int dx\, \mathcal{L}\left(q, \frac{\partial q}{\partial x}, \frac{\partial q}{\partial t}, x, t\right).$$

The variation of paths is carried out on q and its derivatives (not on x and t) in the same manner as was done in Chapter 3. We saw there that the variation

$$\delta \int_{t_1}^{t_2} L(q, \dot{q}, t)\, dt = 0$$

gave rise to

$$\frac{d}{dt}\left(\frac{\partial L}{\partial \dot{q}}\right) - \frac{\partial L}{\partial q} = 0.$$

The only real difference here is that we are working with $\mathcal{L}$ instead of L and there is an x-derivative involved as well as a t-derivative. A derivation like that of

Chapter 3 yields

$$\frac{\partial}{\partial t}\left(\frac{\partial \mathscr{L}}{\partial(\partial q/\partial t)}\right) + \frac{\partial}{\partial x}\left(\frac{\partial \mathscr{L}}{\partial(\partial q/\partial x)}\right) - \frac{\partial \mathscr{L}}{\partial q} = 0. \tag{7.57}$$

When it is possible to write the Lagrangian density directly, as in the string problem, (7.57) leads at once to the equation of motion.

Consider as an example the string density function (7.56). Since

$$\frac{\partial \mathscr{L}}{\partial \dot{q}} = \rho \dot{q}, \qquad \frac{\partial \mathscr{L}}{\partial(\partial q/\partial x)} = -T\frac{\partial q}{\partial x}, \qquad \frac{\partial \mathscr{L}}{\partial q} = 0,$$

we have

$$\rho \ddot{q} - T\frac{\partial^2 q}{\partial x^2} = 0$$

if the density and tension are fixed quantities. This is the familiar wave equation.

In more specialized texts, the Lagrangian density formulation, when appropriately extended to a greater dimensionality, serves as the basis of a field theoretical description useful in electromagnetism, particle physics, and other areas as well as advanced mechanics. A field description requires the specification of some mathematical function or functions over all space and time. Any convenient form of Lagrangian density that leads to appropriate field equations can be used in a given problem, even though it may not be possible to associate an underlying mechanical system with such a density. The formalism of mechanics has far greater utility than appears at first sight.

7.6 STABILITY OF MOTION; THE ROUTH–HURWITZ CRITERION

We have seen in the preceding sections that stable oscillatory motion occurs whenever the transient terms are such as to damp out in rapid exponential fashion, leaving a steady state motion of fixed amplitude. It also occurs when the amplitude decreases in time. Only in the case where the amplitude grows in time do we have an instability. Let us illustrate this point with a simple example of a type somewhat different from the others in this chapter.

Consider a mass m moving in a circular orbit under an inverse square force of gravity $-k/r^2$, where $k > 0$. The radial equation of motion, from (4.8), is simply

$$m(\ddot{r} - r\dot{\theta}^2) = -\frac{k}{r^2}.$$

Using the constancy of the angular momentum $\mathscr{L}$ to eliminate $\dot{\theta}$, the equation

$$m\left(\ddot{r} - \frac{\mathscr{L}^2}{m^2 r^3}\right) = -\frac{k}{r^2}$$

results. If we suppose that the circular orbit is of radius a and that the particle is slightly disturbed so that it moves outward a small radial distance δ, a displacement in the plane of the orbit yet perpendicular to the initial path, we can substitute
$$r = a + \delta, \qquad \dot{r} = \dot{\delta}, \qquad \ddot{r} = \ddot{\delta}.$$
The equation of motion becomes
$$\ddot{\delta} - \frac{\mathscr{L}^2}{m^2(a+\delta)^3} = \ddot{\delta} - \frac{\mathscr{L}^2}{m^2 a^3 (1+\delta/a)^3} = \frac{-k}{mr^2} = \frac{-k}{m(a+\delta)^2}$$
$$= -\frac{k}{ma^2(1+\delta/a)^2}.$$
Expanding the $(1 + \delta/a)$ factors by means of the binomial theorem,
$$\ddot{\delta} - \frac{\mathscr{L}^2}{m^2 a^3}\left(1 - 3\frac{\delta}{a} + \cdots\right) = -\frac{k}{ma^2}\left(1 - 2\frac{\delta}{a} + \cdots\right).$$
But in the initial circular orbit, $\ddot{r} = 0$, and $\mathscr{L}^2/m^2 a^3 = k/ma^2$ follows from the equation of motion (this is merely the equating of centripetal and gravitational forces). Assuming that the absence of torques keeps the angular momentum $\mathscr{L}$ constant at the value given by this relation, the perturbed equation of motion is now
$$\ddot{\delta} + \left(\frac{\mathscr{L}^2}{m^2 a^3} 3\frac{\delta}{a} - \frac{k}{ma^2} 2\frac{\delta}{a}\right) = 0 = \ddot{\delta} + \frac{k}{ma^3}\delta,$$
an oscillator relationship. The radial oscillations will continue indefinitely (with no damping term) under the linear restoring force, so the motion is stable, the mass moving in and out radially for short distances as it moves around the original circle.

The circular orbit is an example of stability in a marginal sense; if the system is not disturbed too much, the radial oscillations will continue with constant amplitude. A better example of stability calculations is furnished by the damped oscillator of (7.3):
$$m\ddot{x} + R\dot{x} + kx = 0.$$
Assuming a solution $x = e^{\lambda t}$, we found in (7.4) that
$$f(\lambda) \equiv m\lambda^2 + R\lambda + k = 0$$
was satisfied by
$$\lambda = \frac{-R}{2m} \pm \sqrt{\left(\frac{R}{2m}\right)^2 - \frac{k}{m}}.$$

If $R > 0$ and $k > 0$, as is normally true, then the roots of $f(\lambda)$ are either entirely real and negative or at the very least have negative real parts. The solutions for x then contain decaying exponentials, which ensure stability.

Investigations in the late nineteenth century on the stability of a physical system governed by a linear differential equation such as (7.3) led Routh and Hurwitz to the conclusion above: that the roots of the characteristic equation

$$f(\lambda) = a_n \lambda^n + a_{n-1} \lambda^{n-1} + \cdots + a_1 \lambda + a_0 = 0 \qquad (7.58)$$

had to have *negative real parts* for stable motion. In view of the fact that the relation (7.58) cannot always be solved simply for the roots, the **Routh–Hurwitz criterion** was developed as a convenient method of determining, from the coefficients in $f(\lambda)$, when the real part of the roots would indeed be negative. The proof of their criterion is not too difficult, but it makes use of some theorems from the theory of equations[1] with which the reader may not be familiar. We shall therefore simply state the criterion and illustrate its usage without proof.

Assume that the coefficient a_0 is positive. (If not so, multiply $f(\lambda)$ by -1 to make it so.) Construct the set of determinants

$$\Delta_1 \equiv a_1,$$

$$\Delta_2 \equiv \begin{vmatrix} a_1 & a_0 \\ a_3 & a_2 \end{vmatrix},$$

$$\Delta_3 \equiv \begin{vmatrix} a_1 & a_0 & 0 \\ a_3 & a_2 & a_1 \\ a_5 & a_4 & a_3 \end{vmatrix},$$

$$\Delta_4 \equiv \begin{vmatrix} a_1 & a_0 & 0 & 0 \\ a_3 & a_2 & a_1 & a_0 \\ a_5 & a_4 & a_3 & a_2 \\ a_7 & a_6 & a_5 & a_4 \end{vmatrix},$$

$$\vdots$$

$$\Delta_n \equiv \begin{vmatrix} a_1 & a_0 \cdots & & 0 \\ a_3 & a_2 \cdots & & 0 \\ \vdots & & & \\ a_{2n-1} & a_{2n-2} \cdots & & a_n \end{vmatrix}.$$

---

[1] See J. Uspensky, *Theory of Equations*, McGraw-Hill, New York, 1948.

Stability of Motion; the Routh–Hurwitz Criterion

(In this construction, all coefficients with negative subscripts or subscripts greater than the degree of $f(\lambda)$ should automatically be equated to zero, for there is no coefficient in $f(\lambda)$ which corresponds to them.) The Routh–Hurwitz criterion is that all roots of relation (7.58) will have negative real parts if and only if *all* Δ's are positive.

Example 4 Damped Oscillator

Let us apply the criterion to eq. (7.3) as an example. Since $f(\lambda) = m\lambda^2 + R\lambda + k = 0$, $a_2 = m$, $a_1 = R$, and $a_0 = k$,

$$\Delta_1 = R,$$

$$\Delta_2 = \begin{vmatrix} R & k \\ 0 & m \end{vmatrix} = mR.$$

There are no other determinants, since the degree of $f(\lambda)$ is 2. Assuming that both R and k (the constant term) are positive, Δ_1 and Δ_2 are positive, hence the motion is stable. If R had been negative, the criterion would not have been satisfied, and we know from the actual solution (7.4) that increasing exponentials would have resulted in that case.

Example 5 Coupled Springs

We return to the springs of Example 2 in the previous section, setting $K = 2k$ and $M = 2m$. The coupled eqs. (2.11') and (2.12') become

$$\ddot{X}_1 + \frac{k}{m}X_1 - \frac{k}{2m}(X_2 - X_1) = 0,$$

$$\ddot{X}_2 + \frac{k}{m}(X_2 - X_1) = 0,$$

where we have eliminated the constant terms (as suggested earlier) by means of the substitutions

$$y_1 \equiv X_1 + \frac{3}{2}\frac{mg}{k}, \qquad y_2 \equiv X_2 + \frac{5}{2}\frac{mg}{k}.$$

Let us now set $X_1 \equiv Ae^{\lambda t}$, $X_2 \equiv Be^{\lambda t}$, producing

$$A\lambda^2 + A\frac{k}{m} - \frac{k}{2m}B + \frac{k}{2m}A = 0 = A\left(\lambda^2 + \frac{3}{2}\frac{k}{m}\right) - B\frac{k}{2m},$$

$$B\lambda^2 + B\frac{k}{m} - A\frac{k}{m} = 0 = A\left(-\frac{k}{m}\right) + B\left(\lambda^2 + \frac{k}{m}\right),$$

upon substituting in the coupled equations. For this pair of homogeneous equations to have a nontrivial solution for A and B, the determinant of the coefficients must vanish:

$$\begin{vmatrix} \lambda^2 + \dfrac{3}{2}\dfrac{k}{m} & -\dfrac{k}{2m} \\ -\dfrac{k}{m} & \lambda^2 + \dfrac{k}{m} \end{vmatrix} = 0,$$

or

$$\left(\lambda^2 + \frac{3}{2}\frac{k}{m}\right)\left(\lambda^2 + \frac{k}{m}\right) - \frac{1}{2}\frac{k^2}{m^2} = 0 = \lambda^4 + \frac{5}{2}\frac{k}{m}\lambda^2 + \frac{k^2}{m^2} = f(\lambda).$$

Identifying coefficients with (7.58) results in

$$a_4 = 1, \qquad a_3 = a_1 = 0, \qquad a_2 = \frac{5}{2}\frac{k}{m}, \qquad a_0 = \frac{k^2}{m^2}.$$

The set of Δ-determinants is

$$\Delta_1 = 0,$$

$$\Delta_2 = \begin{vmatrix} 0 & \dfrac{k^2}{m^2} \\ 0 & \dfrac{5}{2}\dfrac{k}{m} \end{vmatrix} = 0,$$

$$\Delta_3 = \begin{vmatrix} 0 & \dfrac{k^2}{m^2} & 0 \\ 0 & \dfrac{5}{2}\dfrac{k}{m} & 0 \\ 0 & 1 & 0 \end{vmatrix} = 0,$$

$$\Delta_4 = \begin{vmatrix} 0 & \dfrac{k^2}{m^2} & 0 & 0 \\ 0 & \dfrac{5}{2}\dfrac{k}{m} & 0 & \dfrac{k^2}{m^2} \\ 0 & 1 & 0 & \dfrac{5}{2}\dfrac{k}{m} \\ 0 & 0 & 0 & 1 \end{vmatrix} = 0.$$

All the determinants here are zero, a conclusion which arises from the fact that there are no damping terms present. In fact, direct solution of $f(\lambda) = 0$ gives $\lambda = \pm i\sqrt{k/2m}, \pm i\sqrt{2k/m}$ for the roots, so the solution is (as seen earlier) purely oscillatory in nature. Had damping terms been present, the coefficients a_3 and a_1 would not have been zero, and the determinants would have turned out positive. We have here, as with the circular orbit, another case of borderline stability, but this problem was illuminating in that it showed how to set up $f(\lambda)$ and the Δ-determinants in a more complicated situation than for Example 4.

7.7 OSCILLATIONS IN THE PHASE PLANE

We have seen that for each degree of freedom of a mechanical system, there is a generalized coordinate q_i along with its associated momentum p_i. If there are n degrees of freedom present in a given problem, one can form a $2n$-dimensional space known as **phase space** from all the coordinates and momenta, analyzing the behavior of the system by the study of its representative point in phase space (as is done in statistical mechanics). At a given instant in time, each of the q's and p's will have definite values which determine a point in phase space that completely characterizes the system at that moment; as the motion develops in time, a trajectory will be traced out by this representative point.

With one-dimensional motion, as for the oscillators discussed in this chapter, the phase space reduces to only two dimensions, and we speak of the **phase plane**. As an example, consider an undamped harmonic oscillator, setting $R = 0$ in (7.3). The Hamiltonian is

$$H = \frac{p^2}{2m} + \frac{1}{2}kx^2 = E,$$

or

$$\frac{p^2}{(2mE)} + \frac{x^2}{2E/k} = 1.$$

Since E is a constant of the motion, a phase plane plot of p against x is clearly in the shape of an ellipse (Figure 7.5). As the oscillations continue along the x-axis in the real physical space, the representative point moves repeatedly around the ellipse in the phase plane.

The undamped oscillator is a simple example of a **Hamiltonian system**, defined as one which satisfies the canonical equations (2.17)

$$\dot{x} = \frac{\partial H}{\partial p}, \quad \dot{p} = -\frac{\partial H}{\partial x}.$$

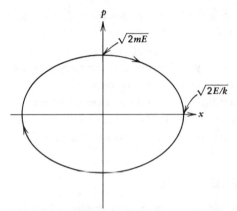

Figure 7.5 The undamped oscillator in the phase plane.

Differentiating the first canonical equation with respect to x and the second with respect to p, we find that

$$\frac{\partial \dot{x}}{\partial x} = \frac{\partial^2 H}{\partial x \partial p} = \frac{\partial^2 H}{\partial p \partial x} = -\frac{\partial \dot{p}}{\partial p},$$

or

$$\frac{\partial \dot{x}}{\partial x} + \frac{\partial \dot{p}}{\partial p} = 0. \tag{7.59}$$

Many simple systems are not Hamiltonian ones and so do not obey the necessary condition (7.59). Thus, the damped oscillator of (7.3) has the differential equation

$$\ddot{x} + \frac{R}{m}\dot{x} + \frac{k}{m}x = 0.$$

Setting

$$y \equiv \dot{x}\left(=\frac{p}{m}\right), \tag{7.60}$$

we have

$$\dot{y} = \ddot{x} = -\frac{R}{m}\dot{x} - \frac{k}{m}x = -\frac{R}{m}y - \frac{k}{m}x,$$

$$\dot{x} = y,$$

and so, for this system,

$$\frac{\partial \dot{x}}{\partial x} = 0 \neq -\frac{\partial \dot{p}}{\partial p} = -\frac{\partial \dot{y}}{\partial y} = \frac{R}{m}.$$

Equation (7.59) will be satisfied only if $R = 0$.

It is customary (as above) to plot the velocity $\dot{x}$ (rather than the momentum) versus x for the description in the phase plane and to represent $\dot{x}$ by the symbol y. The motion can still be depicted readily, even when the system is not a Hamiltonian one. Thus, for the damped oscillator, the second-order relation (7.3) is equivalent to the *pair* of first-order equations

$$\dot{x} = y,$$

$$\dot{y} = -\frac{R}{m}y - \frac{k}{m}x,$$

giving a phase trajectory of slope

$$\frac{dy}{dx} = \frac{\dot{y}}{\dot{x}} = -\frac{1}{m}\frac{Ry + kx}{y}.$$

We shall show shortly that this curve has a spiral nature.

Systems of differential equations of this general type, which do not explicitly involve the time, are known as **autonomous systems**. Much information on the solution in the phase plane can be obtained through investigation of the **singular**, or **equilibrium**, points, where $\dot{x} = \dot{y} = 0$. For these points, dy/dx is indeterminate, so it is possible for many trajectories to pass through them (for ordinary points, only one trajectory can do so). Assume that the origin is an equilibrium point and expand $\dot{y}$ and $\dot{x}$ about it, keeping only the linear terms to get

$$\frac{dy}{dx} = \frac{\dot{y}}{\dot{x}} = \frac{cx + yd}{ax + by}, \tag{7.61}$$

where the quantities a, b, c, and d are real constants. (In the phase plane analysis just given, y was set equal to $\dot{x}$; the present and more general treatment is considering a complicated system of equations in which $\dot{y}$ and $\dot{x}$ are some functions of x and y.)

The system is easily solved if $b = c = 0$, for then

$$\dot{y} = dy, \qquad \dot{x} = ax,$$

hence

$$y = \text{constant } e^{dt}, \qquad x = \text{constant } e^{at}.$$

Under these conditions, the system is said to be in **canonical form**. In order to reach this form, one first needs a matrix **A** which will take the vector with components x, y into one with components $\dot{x}$, $\dot{y}$. This requirement leads to

$$\mathbf{A} = \begin{pmatrix} a & b \\ c & d \end{pmatrix}, \quad \begin{pmatrix} \dot{x} \\ \dot{y} \end{pmatrix} = \mathbf{A} \begin{pmatrix} x \\ y \end{pmatrix}. \tag{7.62}$$

Now let us make a linear transformation through a matrix $\mathbf{T}$ to a new primed set:

$$\begin{pmatrix} x' \\ y' \end{pmatrix} \equiv \mathbf{T} \begin{pmatrix} x \\ y \end{pmatrix}, \qquad \begin{pmatrix} \dot{x}' \\ \dot{y}' \end{pmatrix} = \mathbf{T} \begin{pmatrix} \dot{x} \\ \dot{y} \end{pmatrix}.$$

From (7.62), we have

$$\begin{pmatrix} \dot{x}' \\ \dot{y}' \end{pmatrix} = \mathbf{TA} \begin{pmatrix} x \\ y \end{pmatrix} = \mathbf{TAT}^{-1} \mathbf{T} \begin{pmatrix} x \\ y \end{pmatrix} = \mathbf{TAT}^{-1} \begin{pmatrix} x' \\ y' \end{pmatrix}. \tag{7.62'}$$

If the matrix $\mathbf{T}$ transforms $\mathbf{A}$ so that

$$\mathbf{TAT}^{-1} = \lambda,$$

where λ is a diagonal matrix, then the primed coordinates will be in canonical form. It then follows that

$$|\mathbf{A} - \lambda| = 0,$$

making use of (5.33). The diagonal elements of λ are the eigenvalues of the matrix $\mathbf{A}$; they are found by solving the determinantal equation

$$(a - \lambda)(d - \lambda) - bc = 0 = \lambda^2 - (a + d)\lambda + (ad - bc). \tag{7.63}$$

Let us refer to these characteristic roots as λ_1 and λ_2 and distinguish the following cases:

CASE 1 λ_1 AND λ_2 ARE REAL AND OF THE SAME SIGN

By (7.62'), $\dot{x}' = \lambda_1 x'$ and $\dot{y}' = \lambda_2 y'$, hence

$$\frac{dy'}{dx'} = \frac{\lambda_2}{\lambda_1} \frac{y'}{x'}, \qquad y' = \text{constant} \cdot (x')^{\lambda_2/\lambda_1}. \tag{7.64}$$

The curves in the $x'y'$-plane are similar to parabolas, being tangent to either the x'- or y'-axis according as λ_2/λ_1 is greater or less than unity. For negative λ values, the primed coordinates decrease in time as $e^{\lambda t}$, and the origin (never quite reached by the representative point) is called a **stable node**. If the λ values are both positive, the coordinates increase with time, and the origin is an **unstable node**.

CASE 2 λ_1 AND λ_2 ARE REAL AND OF OPPOSITE SIGN

Exactly as with (7.64), we now obtain

$$y' = \text{constant} \cdot (x')^{-|\lambda_2/\lambda_1|}. \tag{7.65}$$

Oscillations in the Phase Plane

These $x'y'$-curves are similar to hyperbolas; as one primed coordinate increases in time, the other decreases. The origin is unstable and is called a **saddle point**.

CASE 3 λ_1 AND λ_2 ARE COMPLEX CONJUGATES

The reality of the coefficients in (7.63) ensures that complex roots must occur in conjugate pairs. When the common real part of the λ roots is negative, we again have stability at the origin (and instability for a positive real part).

The separation of coordinates achieved by the canonical form is in a sense analogous to the procedure for obtaining independent normal coordinates. Among the many special cases that could be treated, let us once more examine the damped oscillator:

$$\dot{x} = y, \qquad \dot{y} = -\frac{k}{m}x - \frac{R}{m}y. \tag{7.66}$$

Comparison with (7.61) yields $a = 0$, $b = 1$, $c = -k/m$, $d = -R/m$, whereupon (7.63) becomes

$$\lambda^2 + \frac{R}{m}\lambda + \frac{k}{m} = 0. \quad \bullet$$

This eigenvalue equation has a conjugate pair of roots which are given by (7.4), and so (for small R) we have a case 3 situation.

An interesting polar representation can be obtained. Let

$$\omega \equiv \dot{x} + \frac{R}{2m}x = y + \frac{R}{2m}x, \tag{7.67}$$

so that

$$\dot{\omega} = \dot{y} + \frac{R}{2m}\dot{x} = -\frac{R}{m}y - \frac{k}{m}x + \frac{R}{2m}y = -\frac{R}{2m}y - \frac{k}{m}x$$

$$= -\frac{R}{2m}\left(\omega - \frac{R}{2m}x\right) - \frac{k}{m}x = -\frac{R}{2m}\omega - \omega_0^2 x, \tag{7.68}$$

where (7.6) has been employed. Next, let r and θ be defined by

$$\omega \equiv r\sin\theta, \qquad \omega_0 x \equiv r\cos\theta. \tag{7.69}$$

Upon time differentiating, we find that

$$\dot{\omega} = \dot{r}\sin\theta + r\dot{\theta}\cos\theta, \qquad \omega_0\dot{x} = \dot{r}\cos\theta - r\dot{\theta}\sin\theta.$$

Inverting these relations (multiply by $\sin\theta$ or $\cos\theta$ and combine), we get

$$\dot{r} = \dot{\omega}\sin\theta + \omega_0 \dot{x}\cos\theta, \qquad r\dot{\theta} = \dot{\omega}\cos\theta - \omega_0 \dot{x}\sin\theta.$$

Substituting (7.67), (7.68), and (7.69), one obtains

$$\dot{r} = \left(-\frac{R}{2m}\omega - \omega_0^2 x\right)\sin\theta + \omega_0\left(\omega - \frac{R}{2m}x\right)\cos\theta$$

$$= \left(-\frac{R}{2m}\sin\theta + \omega_0\cos\theta\right)\omega + \left(-\omega_0\sin\theta - \frac{R}{2m}\cos\theta\right)\omega_0 x$$

$$= -\frac{R}{2m}r,$$

and similarly

$$\dot{\theta} = -\omega_0.$$

Dividing $\dot{r}$ by $\dot{\theta}$, we have

$$\frac{dr}{d\theta} = \frac{\dot{r}}{\dot{\theta}} = \frac{R}{2m\omega_0}r.$$

Integration now produces

$$r = \text{constant} \cdot e^{R\theta/2m\omega_0}, \qquad (7.70)$$

a simple spiral in the "modified phase plane" with polar coordinates r, θ (Figure 7.6). As time passes (assuming that $R/2m$ and ω_0 are each positive), both θ and r will decrease, the spiral sweeping in clockwise toward the origin, which is called a **stable focus**. Depending on the relation between the parameters R and k, it is also possible that the damped oscillator can correspond to one of the earlier cases as well as case 3.

It follows from Cauchy's existence theorem[2] (which will not be proved) that there is one, and only one, path through each point of the phase plane. An equilibrium point is a kind of "point path," and it is of special interest, as we have seen. An ordinary path may extend toward the singular point, but can never quite reach it in a finite amount of time.

Periodic motions correspond to closed paths in the phase plane, such as the ellipse for the undamped oscillator. If we have as the path such a closed curve C, called a **limit cycle**, then any neighboring path tends either to spiral in toward C

---

[2]A. Andronow and C. Chaikin, *Theory of Oscillations*, Princeton Univ. Press, Princeton, 1949, p. 184.

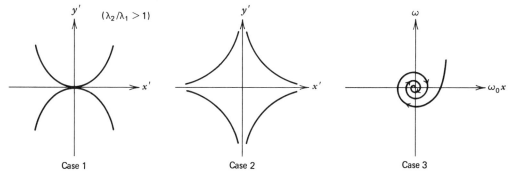

Figure 7.6 Trajectories in the "phase planes" for the three cases.

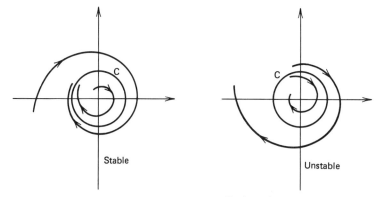

Figure 7.7 Two common limit cycles.

or outward away from C, depending on the stability of the motion (Figure 7.7). A spiral in on one side of the limit cycle and out on the other side does not normally occur and so will not be considered. The work of Poincaré on limit cycles will be further explored in the next section. Together with equilibrium points, these curves are of great value in understanding oscillations as viewed in the phase plane.

7.8 NONLINEAR EQUATIONS; THE METHODS OF POINCARÉ

It is customary in physics to treat oscillation problems as has been described earlier in this chapter, where kinetic and potential energies were quadratic functions and the equation of motion was therefore linear. More complicated situations are often handled through the principle of superposition. The concepts of free and force-driven oscillations are intimately related to this principle, which in fact *no longer holds* when the differential equation for the motion is nonlinear in form.

As an example, a typical nonlinear equation might be represented as

$$m\ddot{x} + g(\dot{x}) + f(x) = F_0 \cos \omega t \tag{7.71}$$

for a one-dimensional force-driven motion. Here, $g(\dot{x})$ is a damping term which depends on the velocity and $f(x)$ is a restoring force function of the displacement. In general, where these functions are *not* proportional to $\dot{x}$ and x, respectively, then the equation is nonlinear and often its solution can only be approximated (as with a series expansion). Attempts at solutions of certain special cases are abundant; but for arbitrary $g(\dot{x})$ and $f(x)$, no solution is possible.

As a typical case, consider a force which is symmetrical about the origin such that the magnitude of the force is the same at equal distances to either side of the origin, while the *direction* is opposite on the two sides. With a situation of this kind, the first correction term to the usual linear restoring force would have to be cubic and not quadratic. The function appearing in eq. (7.71) is then

$$f(x) = kx + \alpha x^3,$$

where α is a small constant.

With a nonlinear differential equation there may appear *qualitative* features of the solution which are entirely unfamiliar. Thus, consider the equation where only the damping term is nonlinear:

$$m\ddot{x} + g(\dot{x}) + kx = F_0 \cos \omega t,$$

with the restrictions

$$\left.\begin{array}{l} g(\dot{x}) < 0 \\ g(\dot{x}) > 0 \end{array}\right\} \quad \begin{array}{l} \text{small } \dot{x} \\ \text{large } \dot{x}. \end{array}$$

The physical system obeying this relation is said to undergo **self-excited** oscillations, even when $F_0 = 0$, because the force $g(\dot{x})$ is in the direction of the velocity for small $\dot{x}$ and even a small disturbance from the rest position will lead to a continuing motion. The oscillations thus begun are eventually limited in amplitude by the fact that the damping term plays its customary role (opposing the motion) for large $\dot{x}$. In the event that $F_0 \neq 0$, it is possible to have solutions that contain both the frequency component of the driving force and that of the self-excited oscillations, giving rise to what are called **combination tones**. Even in an entire book, we could only scratch the surface of this complex field.

Suppose we look more closely at the nonlinear equation

$$\ddot{x} + x = \mu f(x, \dot{x}), \tag{7.71'}$$

where the mass and angular frequency have been chosen as unity for convenience. Thinking in terms of the phase plane as before, eq. (7.71') is clearly equivalent to the coupled system

$$y \equiv \dot{x}, \qquad \dot{y} = \mu f(x, y) - x. \qquad (7.72)$$

If μ is a small constant and $f(x, y)$ can be represented as a polynomial in x and y, Poincaré has suggested a perturbation technique for finding periodic solutions. We want to outline his method as it is typical of the approaches taken in this field. The algebra is a little tedious, however, so the reader not interested in the details might skip at once to the discussion of the conclusions reached.

For $\mu = 0$, eq. (7.71′) has the immediate periodic solution

$$x = R\cos(t - t_0),$$
$$y = -R\sin(t - t_0),$$
$$x^2 + y^2 = R^2. \qquad (7.73)$$

The paths in the phase plane are circles of radius R; if μ is not too large, periodic solutions closely related to the circles would be expected. Now, the circles cross the x-axis at R for some time t_0. Suppose that at t_0 a solution of (7.71′) crosses the x-axis at a neighboring point $(R + \lambda)$, λ being a small positive number that measures the displacement from the circle ($\mu = 0$ solution). Choosing t_0 as zero for simplicity, the (7.71′) solution can be cast into the form

$$x(t) = R\cos t + \delta(R, \lambda, \mu, t),$$
$$y(t) = -R\sin t + n(R, \lambda, \mu, t), \qquad (7.74)$$
$$n = \dot{\delta},$$

where the functions δ and n give the deviations of the coordinates from the circle for the solution in question. It is clear from the zero time conditions that

$$\delta(R, \lambda, \mu, 0) \equiv \lambda, \qquad n(R, \lambda, \mu, 0) \equiv 0. \qquad (7.75)$$

Furthermore, from (7.73), we have that

$$\delta(R, 0, 0, t) \equiv 0, \qquad n(R, 0, 0, t) \equiv 0. \qquad (7.76)$$

The last relations lead us to the conclusion that a series expansion of δ and n in terms of the *two* small variables λ and μ must necessarily be of the form

$$\delta = A\lambda + B\mu + C\lambda\mu + D\mu^2 + \cdots,$$
$$n = \dot{\delta} = \dot{A}\lambda + \dot{B}\mu + \dot{C}\lambda\mu + \dot{D}\mu^2 + \cdots. \qquad (7.77)$$

(The coefficients $A, B, C, \cdots$ are functions of time but not of the other parameters.) Expanding the function $f(x, y)$ in a similar Taylor series gives

$$f(x, y) = f(R\cos t, -R\sin t) + \left[\delta\frac{\partial f}{\partial x} + n\frac{\partial f}{\partial y}\right]_{\substack{x=R\cos t,\\ y=-R\sin t}} + \cdots. \quad (7.78)$$

We now substitute eqs. (7.74) and (7.78) into (7.72), expressing δ and n by means of (7.77):

$$\ddot{n} + \delta = \mu f(R\cos t, -R\sin t) + \mu\left[\delta\frac{\partial f}{\partial x} + n\frac{\partial f}{\partial y}\right] + \cdots,$$

or

$$(\ddot{A} + A)\lambda + (\ddot{B} + B)\mu + (\ddot{C} + C)\lambda\mu + \cdots$$

$$= \mu f(R\cos t, -R\sin t) + \mu\left(A\frac{\partial f}{\partial x} + \dot{A}\frac{\partial f}{\partial y}\right)\lambda + \mu^2\left(B\frac{\partial f}{\partial x} + \dot{B}\frac{\partial f}{\partial y}\right) + \cdots,$$

where the f-derivatives must be evaluated at $(R\cos t, -R\sin t)$. Equating coefficients of λ and μ powers on the two sides of the last relation produces

$$\ddot{A} + A = 0,$$
$$\ddot{B} + B = f(R\cos t, -R\sin t),$$
$$\ddot{C} + C = A\frac{\partial f}{\partial x} + \dot{A}\frac{\partial f}{\partial y}, \quad (7.79)$$
$$\ddot{D} + D = B\frac{\partial f}{\partial x} + \dot{B}\frac{\partial f}{\partial y}, \ldots.$$

The initial conditions which specify the constants in the solutions of (7.79) are given by a combination of (7.75) and (7.77):

$$A(0) = 1,$$
$$B(0) = C(0) = \cdots = 0,$$
$$\dot{A}(0) = \dot{B}(0) = \cdots = 0, \quad (7.80)$$

the parentheses identifying the initial time of zero. The solution of (7.79) for $A(t)$ is trivial, being simply $A(t) = \cos t$ with the aid of (7.80). All other coefficients satisfy equations of form

$$\ddot{F} + F = g(t), \quad F(0) = \dot{F}(0) = 0.$$

By direct inspection or by transform methods, the solution is clearly the convolution integral

$$F(t) = \int_0^t g(t')\sin(t - t')\, dt', \quad (7.81)$$

Nonlinear Equations; the Methods of Poincaré

the $\ddot{F}$ term contributing $g(t)$ from differentiation of the upper limit in the $\dot{F}$ integral. Upon substitution of the relevant quantities into (7.81), we have

$$A(t) = \cos t,$$

$$B(t) = \int_0^t f(R \cos t', -R \sin t') \sin (t - t') \, dt',$$

$$C(t) = \int_0^t \left[\left(\frac{\partial f}{\partial x}\right)_{t=t'} \cos t' - \left(\frac{\partial f}{\partial y}\right)_{t=t'} \sin t' \right] \sin (t - t') \, dt', \ldots . \quad (7.82)$$

Substitution of these results in (7.77) and (7.74) gives the desired periodic solutions for x and y.

Following Poincaré, we suppose that the solutions obtained will have a period almost that of the circular trajectory, or 2π. The period can be written as $2\pi + \tau(\lambda, \mu)$, where τ is a small correction term, such that $\tau(\lambda, 0) = 0$ for any λ. The periodicity will be ensured if we satisfy the relations

$$x(2\pi + \tau) = x(0), \qquad y(2\pi + \tau) = y(0). \quad (7.83)$$

With the aid of (7.74) and (7.75), we have

$$x(2\pi + \tau) - x(0) = 0 = R \cos (2\pi + \tau) + \delta(R, \lambda, \mu, 2\pi + \tau) - R - \lambda,$$

$$y(2\pi + \tau) - y(0) = 0 = -R \sin (2\pi + \tau) + n(R, \lambda, \mu, 2\pi + \tau).$$

Noting that

$$\sin \tau = \tau - \frac{\tau^3}{3!} + \cdots, \qquad \cos \tau = 1 - \frac{\tau^2}{2!} + \cdots,$$

and employing (7.77), the periodicity relations (7.83) become (to first order in τ)

$$R\left(1 - \frac{\tau^2}{2} + \cdots\right) + A(2\pi + \tau)\lambda + B(2\pi + \tau)\mu + C(2\pi + \tau)\lambda\mu$$

$$+ D(2\pi + \tau)\mu^2 + \cdots - R - \lambda = 0$$

$$= R\left(1 - \frac{\tau^2}{2} + \cdots\right) + \lambda \cos (2\pi + \tau) + \left[B(2\pi)\mu + \dot{B}(2\pi)\tau\mu + \cdots\right]$$

$$+ \left[C(2\pi)\lambda\mu + \dot{C}(2\pi)\tau\lambda\mu + \cdots\right] + \cdots - R - \lambda$$

$$\simeq B(2\pi)\mu + \dot{B}(2\pi)\tau\mu + C(2\pi)\lambda\mu + \cdots = 0, \quad (7.84)$$

$$-R \sin (2\pi + \tau) + \dot{A}(2\pi + \tau)\lambda + \dot{B}(2\pi + \tau)\mu + \cdots = 0$$

$$\simeq -R\tau + \ddot{A}(2\pi)\tau\lambda + \dot{B}(2\pi)\mu + \ddot{B}(2\pi)\tau\mu + \cdots, \quad (7.84')$$

where we have used the Taylor expansion

$$F(2\pi + \tau) = F(2\pi) + \dot{F}(2\pi)\tau + \cdots.$$

Recognizing that τ, λ, and μ are all small and keeping only the first-order terms in these quantities, (7.84′) yields

$$\tau \simeq \frac{\mu}{R}\dot{B}(2\pi) = \frac{\mu}{R}\int_0^{2\pi} f\cos t'\, dt' \tag{7.85}$$

as the first approximation to the period increment, while (7.84) gives

$$B(2\pi) = -\int_0^{2\pi} f\sin t'\, dt' \simeq 0, \tag{7.86}$$

this being the only term of first order in the equation. The f in the integrands is a function of $R\cos t'$ and $-R\sin t'$, of course, so that upon integrating, we are left with

$$\Phi(R) = 0, \tag{7.87}$$

where $\Phi(R)$ depends on the form specified for the function f. Actually, this relation is of great interest and importance, for it is like a quantum condition—it specifies the possible radii of the generating circles for a given f. Only those circles with radii which satisfy (7.87) can be used to generate periodic solutions to the nonlinear equation, and these circles then serve as limit cycles for the motion in the phase plane.

Example 6

Find the radii of the generating circles for the equation

$$\ddot{x} + x = \mu(9x^2 - 2x^4)\dot{x}.$$

Solution

Noting that $f(x, y) \equiv (9x^2 - 2x^4)y$, we have, from (7.86),

$$B(2\pi) = -\int_0^{2\pi}(9R^2\cos^2 t' - 2R^4\cos^4 t')(-R\sin t')\sin t'\, dt' = 0,$$

$$9\int_0^{2\pi}\cos^2 t' \sin^2 t'\, dt' - 2R^2\int_0^{2\pi}\cos^4 t' \sin^2 t'\, dt' = 0$$

$$= \frac{9\pi}{4} - \frac{2\pi R^2}{8} = \frac{\pi}{4}(9 - R^2), \qquad R = 3$$

is the desired circle radius.

If the motion is assumed to be stable in this example, there will be periodic solutions in the neighborhood of the circle of radius 3. As time passes, motion which began on the x-axis from either side of this circle will gradually approach the circle as the limit cycle.

The question of stability requires further discussion. Poincaré was responsible for establishing criteria of stability in more general situations than those which have been treated previously. As a simple application of his ideas, we shall calculate the Poincaré **stability index** β for the case of the circular limit cycle just mentioned.

In general, suppose that Hamilton's canonical equations are written as

$$y \equiv \dot{x} \equiv F(x, y), \qquad \dot{y} = \ddot{x} \equiv G(x, y), \qquad (7.88)$$

where F and G are functions that are determined by the Hamiltonian derivatives of a specific problem. By means of a lengthy algebraic derivation which will not be given, it is possible to show[3] that the stability index is determined from the expression

$$\ln \beta = \int_0^T \left(\frac{\partial F}{\partial x} + \frac{\partial G}{\partial y} \right) dt, \qquad (7.89)$$

where T is the period of the motion. The theory of the index tells us that stable motion will occur only when β is less than unity, or $\ln \beta < 0$.

In the case of the generating circles of radius 3, we did not have a Hamiltonian but rather had a nonlinear differential equation

$$\ddot{x} + x = \mu(9x^2 - 2x^4)\dot{x}.$$

The index theory still applies, provided we set

$$y \equiv \dot{x} \equiv F(x, y), \qquad \dot{y} = \ddot{x} \equiv G(x, y) = \mu(9x^2 - 2x^4)y - x.$$

Then,

$$\frac{\partial F}{\partial x} = 0, \qquad \frac{\partial G}{\partial y} = \mu(9x^2 - 2x^4),$$

and with $x = 3 \cos t$,

$$\ln \beta = \mu \int_0^T (9x^2 - 2x^4) \, dt = \mu \int_0^{2\pi} \left[9(3 \cos t)^2 - 2(3 \cos t)^4 \right] dt$$

$$= \mu \left[81\pi - 162 \frac{3\pi}{4} \right] = -\frac{81}{2} \pi \mu,$$

---

[3] See the more detailed discussion in S. McCuskey's *Introduction to Advanced Dynamics*, Addison-Wesley, Reading, Mass., 1959.

evaluating the integral over the limit cycle. If $\mu > 0$, then $\ln \beta < 0$ and the motion is stable. In that event, the trajectories in the phase plane will spiral in toward the generating circle, as described earlier.

The analysis of nonlinear motions by methods such as those of Poincaré is of great physical interest and has numerous applications, but tends to be tedious mathematically. The abundance of techniques which are available for the linear problem are no longer at our disposal. There is no royal road to solutions of nonlinear examples, as is so often true for problems in the real world.

PROBLEMS

7.1 A particle moves along the x-axis with a potential energy
$$V = V_0(e^{-kx} + Cx),$$
where all constants are positive. Find the equilibrium position and show that the equilibrium is stable.

7.2 (a) A force-free underdamped oscillator is initially set into motion. Find the rate at which its total energy decreases in time, and show that this is equal to the rate of dissipation of energy by the damping force.

(b) A damped oscillator is force driven with $f(t) = F_0 \cos \omega t$. Find the time average of the power delivered to the oscillator by the applied force in the steady state. Then find the frequency ω at which the average power is maximized.

7.3 Prove that D'Alembert's solution for the string satisfies both the wave equation and the initial conditions.

7.4 A vibrating string of length l is fixed at both ends. Initially, it is released from rest with its midpoint a distance H from the x-axis, so that the initial shape of the string is triangular. Use Bernoulli's method to find a series expansion for $y(x, t)$.

7.5 Two wires of respective linear densities ρ_1 and ρ_2 are connected to one another and under a uniform tension. A simple harmonic wave moving down one wire is partially reflected and partially transmitted when it reaches the junction. Show that the reflection coefficient R (ratio of reflected to incident wave amplitude) is given by
$$R = \frac{\sqrt{\rho_1} - \sqrt{\rho_2}}{\sqrt{\rho_1} + \sqrt{\rho_2}}.$$

7.6 Show that the contour solution for the sinusoidally driven oscillator, as given in eqs. (7.26) and (7.27), is such that at time zero, both $x(t)$ and $\dot{x}(t)$ must vanish.

7.7 Find the displacement $x(t)$ for an undamped oscillator with a driving force
$$f(t) = \begin{cases} 0 & (t < 0) \\ F_0(t/T) & (0 < t < T) \\ F_0 & (t > T). \end{cases}$$
Assume that the system is at rest until time zero, calculate the transform $F(\omega)$, and employ (7.23).

7.8 Consider again the force $f(t)$ given in Problem 7.7. Calculate $x(t)$ by use of the convolution integral (7.32), which involves the response of the system to a delta function impulse.

7.9 Find the displacement $x(t)$ for an undamped oscillator driven by
$$f(t) = \begin{cases} 0 & (t < 0, t > 2\pi/\omega') \\ F_0 \sin \omega' t & (0 < t < 2\pi/\omega'). \end{cases}$$
Assume that $x(t) = 0$ for $t < 0$.

7.10 Defining an auxiliary complex function z for an undamped force-driven oscillator by
$$z \equiv \dot{x}(t) + i\omega_0 x(t),$$
the equation of motion becomes a simple first-order differential equation in z. (Upon solving it, $x(t)$ is found by dividing $\operatorname{Im}(z)$ by ω_0.)

(a) Derive this equation and solve it by assuming a solution of form $C(t)e^{i\omega_0 t}$, where $C(t)$ is some function of time.

(b) Find the displacement $x(t)$ for the sinusoidal force of Problem 7.9, using the solution found in part (a).

7.11 Consider the normal coordinate example of bead and ring. Show that the expressions (7.47) and (7.48) do indeed give the potential and kinetic energies. Then, find the values of C_1 and C_2 for the initial conditions $\theta = \alpha = \theta_0$, $\dot{\theta} = \dot{\alpha} = 0$ at time zero, obtaining expressions for θ and α as functions of time. Interpret your results.

7.12 Consider again the double pendulum of Chapter 2, Problem 2.6. The pendula are undergoing small oscillations with masses m_1 and m_2, and lengths l_1 and l_2. Find the frequencies of the normal modes of vibration in general. Then, let $m_1 = m_2$, $l_1 = l_2$, and calculate the frequencies, eigenvectors, and normal coordinates for this special case.

7.13 Three identical springs of constant k form the sides of an equilateral triangle, the springs being attached to equal masses m at the vertices of the triangle. Determine the frequencies of the normal modes of vibration. (Simplify the problem by confining the motion to one plane and by assuming that the center of mass is fixed, the latter condition eliminating two of the coordinates.)

7.14 For the equilateral triangle of Problem 7.13, determine the eigenvectors corresponding to each normal frequency, and briefly describe the motion of the masses in each normal mode. (Note: for the repeated frequency root, you must construct two orthogonal eigenvectors. The T-matrix must be included in the orthogonalization process, in accordance with eq. (7.42).)

7.15 Consider again the mass undergoing small radial oscillations about a circular orbit. Prove that stable circular orbits cannot be produced for a central force of form $-k/r^N$ if $N > 3$.

7.16 A simple pendulum with length l and bob mass m oscillates under gravity. Obtain the curve traversed in the phase plane during the motion. Is the pendulum a Hamiltonian system? (You may assume small angles of oscillation.)

7.17 An equation of motion is of the nonlinear form

$$\ddot{x} + \mu(x^2 - 1)\dot{x} + x = 0,$$

in which the damping is dependent on position. Show that the motion is stable for $\mu > 0$ by use of Poincaré's methods, finding the radius of the circular limit cycle. (This is the celebrated van der Pol equation of vacuum tube theory.)

7.18 In the treatment of continuous media, one often defines a **Hamiltonian density** $\mathscr{H}$ such that

$$H = \int \mathscr{H} \, dx$$

in one dimension. Give an explicit relation for $\mathscr{H}$ in the case of the vibrating string fixed at both ends. Then, use it to evaluate H in terms of a sum over an expression in the constants ω_n, A_n, and B_n of the Bernoulli solution.

Chapter Eight
CANONICAL TRANSFORMATIONS

8.1 HAMILTON'S CANONICAL EQUATIONS AND THEIR SYMMETRY

We have seen, in Chapter 2, that the canonical equations of Hamilton, eqs. (2.17),

$$\dot{q}_i = \frac{\partial H}{\partial p_i}, \qquad \dot{p}_i = -\frac{\partial H}{\partial q_i},$$

where $H = H(q, p, t)$, may be used in place of Lagrange's or Newton's equations to generate an alternative formulation of mechanics problems. A fact that has not been emphasized previously is the symmetry between the two equations: except for the sign in the second relation, an interchange of q_i with p_i leads from one equation to the other. That is, a kind of equality of status between coordinates and momenta is exhibited.

This idea can be further examined by the introduction of a new "coordinate" μ_i which takes on a total of $2n$ distinct values:

$$\mu_i \equiv q_i \qquad (i = 1 \text{ to } n),$$

$$\mu_{i+n} \equiv p_i, \tag{8.1}$$

where there are n generalized coordinates in a given problem. With this definition, the first n of the μ-coordinates are the q_i, while the second half of the set comprises the p_i. The μ-index runs all the way from 1 to $2n$, for it must cover the momenta as well as the coordinates. Using this notation, note that the canonical equations become

$$\dot{\mu}_i = \frac{\partial H}{\partial \mu_{i+n}} \qquad \text{and} \qquad \dot{\mu}_{i+n} = -\frac{\partial H}{\partial \mu_i}.$$

The form of the equations now suggests that we construct a $2n \times 2n$ matrix γ

with elements

$$\gamma_{ij} \equiv \begin{cases} 1 & (j = i + n) \\ -1 & (j = i - n) \\ 0 & (\text{all other } j), \end{cases}$$

where i and j must both lie in the range from 1 to $2n$ in order to define an element of the matrix. Alternatively, we can write

$$\gamma_{ij} \equiv \delta_{j,(i+n)} - \delta_{j,(i-n)}. \tag{8.2}$$

With the aid of this matrix, the canonical equations are simply

$$\dot{\mu}_i = \sum_j \gamma_{ij} \frac{\partial H}{\partial \mu_j} \qquad (i, j = 1 \text{ to } 2n). \tag{8.3}$$

In other words, the slight change of notation has allowed us to write both eqs. (2.17) as the single eq. (8.3). This interlocking of coordinates and momenta has become known as the **symplectic** approach, after the pioneering work of Weyl.[1] It can be fully developed in a matrix formalism, but for our purposes, that is not necessary.

In block form, the matrix γ is

$$\gamma = \left(\begin{array}{c|c} 0 & 1 \\ \hline -1 & 0 \end{array} \right)$$

where **0** and **1** refer to $n \times n$ versions of the zero and unit matrices, respectively; γ is both antisymmetric and orthogonal, having unit determinant. The proofs of these properties will be left for the problem set.

As a three-dimensional rectangular coordinate example, let

$$\mu_1 \equiv x, \qquad \mu_2 \equiv y, \qquad \mu_3 \equiv z, \, \mu_4 \equiv p_x, \qquad \mu_5 \equiv p_y, \qquad \mu_6 \equiv p_z.$$

Equations (8.3) are then given in matrix form as

$$\begin{pmatrix} \dot{x} \\ \dot{y} \\ \dot{z} \\ \dot{p}_x \\ \dot{p}_y \\ \dot{p}_z \end{pmatrix} = \begin{pmatrix} 0 & 0 & 0 & 1 & 0 & 0 \\ 0 & 0 & 0 & 0 & 1 & 0 \\ 0 & 0 & 0 & 0 & 0 & 1 \\ -1 & 0 & 0 & 0 & 0 & 0 \\ 0 & -1 & 0 & 0 & 0 & 0 \\ 0 & 0 & -1 & 0 & 0 & 0 \end{pmatrix} \begin{pmatrix} \partial H/\partial x \\ \partial H/\partial y \\ \partial H/\partial z \\ \partial H/\partial p_x \\ \partial H/\partial p_y \\ \partial H/\partial p_z \end{pmatrix}.$$

---

[1] H. Weyl, *The Classical Groups*, Princeton Univ. Press, Princeton, 1939, p. 165.

We have seen in earlier chapters that the solution of a mechanics problem was greatly facilitated when a particular coordinate was ignorable (did not appear in the Lagrangian or Hamiltonian). Thus, in the central field discussions of Chapter 4, the angular coordinate θ was ignorable; as a consequence, the associated momentum p_θ was a constant of the motion. It is clear from the symmetry of the canonical eqs. (2.17) that either the coordinate or its associated momentum will be a constant if the other does not appear in the Hamiltonian.

For the utmost convenience in solving a problem, it would be desirable to have several coordinates ignorable (perhaps even *all* of them!). Yet the generalized coordinates which are suggested by the physical configuration do not often give such a situation. To improve the matter, we note that in past chapters we have been dealing with **point transformations** from an old coordinate set q to a new set Q,

$$Q_i = Q_i(q, t). \tag{8.4}$$

These may depend on time but do not involve momenta (the transformation from spherical to cylindrical coordinates is a simple example). But in the Hamiltonian approach, we have an equality of status between coordinates and momenta, so it is reasonable to transform to a new set both of coordinates *and* momenta, each of which is a function of the old:

$$Q_i = Q_i(q, p, t),$$
$$P_i = P_i(q, p, t). \tag{8.5}$$

A transformation of this type, which preserves the basic form of the canonical equations, provides a new method for the solution of problems; it is called a **canonical transformation**, and will be described more completely in sections to follow.

Note from the outset that such a transformation is independent of the detailed form of the Hamiltonian in a particular situation; that is, other types of problems could also be solved through the same change of variables. It is true, however, that a given transformation may be well suited to only one kind of example, and therefore would not be especially convenient to apply in other cases.

Another concept must be explored in the next section before canonical transformations can be precisely defined and the methodology fully developed.

8.2 POISSON BRACKETS AND THEIR PROPERTIES

In the study of canonical transformations, there is a useful quantity called a **Poisson bracket** which will now be defined. It is expressed in terms of two dynamical variables we shall call f and g; these variables may be functions of either the new Q, P or the old q, p. The definition involves derivatives with respect to the q, p; occasionally we shall indicate differentiation with respect to other variables. The bracket is

$$[f, g] \equiv \sum_{i=1}^{n} \left(\frac{\partial f}{\partial q_i} \frac{\partial g}{\partial p_i} - \frac{\partial f}{\partial p_i} \frac{\partial g}{\partial q_i} \right). \tag{8.6}$$

In the symmetrical μ, γ notation, we have

$$\sum_{i,j=1}^{2n} \frac{\partial f}{\partial \mu_i} \gamma_{ij} \frac{\partial g}{\partial \mu_j} = \sum_{i,j} \left[\frac{\partial f}{\partial \mu_i} (\delta_{j,i+n} - \delta_{j,i-n}) \frac{\partial g}{\partial \mu_j} \right]$$

$$= \sum_{i=1}^{n} \sum_{j=n+1}^{2n} \left(\frac{\partial f}{\partial \mu_i} \delta_{j,i+n} \frac{\partial g}{\partial \mu_j} \right) - \sum_{i=n+1}^{2n} \sum_{j=1}^{n} \left(\frac{\partial f}{\partial \mu_i} \delta_{j,i-n} \frac{\partial g}{\partial \mu_j} \right)$$

$$= \sum_{i=1}^{n} \frac{\partial f}{\partial \mu_i} \frac{\partial g}{\partial \mu_{i+n}} - \sum_{i=n+1}^{2n} \frac{\partial f}{\partial \mu_i} \frac{\partial g}{\partial \mu_{i-n}} \quad \text{(let } k \equiv i - n \text{ in the second sum)}$$

$$= \sum_{i=1}^{n} \frac{\partial f}{\partial q_i} \frac{\partial g}{\partial p_i} - \sum_{k=1}^{n} \frac{\partial f}{\partial \mu_{k+n}} \frac{\partial g}{\partial \mu_k} = \sum_{i=1}^{n} \left(\frac{\partial f}{\partial q_i} \frac{\partial g}{\partial p_i} - \frac{\partial f}{\partial p_i} \frac{\partial g}{\partial q_i} \right),$$

or

$$[f, g] = \sum_{i,j=1}^{2n} \frac{\partial f}{\partial \mu_i} \gamma_{ij} \frac{\partial g}{\partial \mu_j}. \tag{8.7}$$

This is the symmetrical expression for the Poisson bracket.

Now suppose we wish to take the total time derivative of some function $F = F(q, p, t)$. It is clearly

$$\frac{dF}{dt} = \sum_{i=1}^{2n} \frac{\partial F}{\partial \mu_i} \dot{\mu}_i + \frac{\partial F}{\partial t} = \sum_{i,j} \left(\frac{\partial F}{\partial \mu_i} \gamma_{ij} \frac{\partial H}{\partial \mu_j} \right) + \frac{\partial F}{\partial t},$$

or

$$\frac{dF}{dt} = [F, H] + \frac{\partial F}{\partial t}, \tag{8.8}$$

where (8.3) has been employed. Setting $F = \mu_k$, (8.8) gives

$$\dot{\mu}_k = [\mu_k, H]. \tag{8.9}$$

Thus, the time derivative of any of the coordinates or momenta is just its Poisson bracket with the Hamiltonian.

Poisson Brackets and their Properties

Setting F = constant of motion = C in eq. (8.8), one finds that

$$[C, H] = -\frac{\partial C}{\partial t}. \tag{8.10}$$

It must be kept in mind here that the "constant" C is generally some function of the dynamical variables which has a fixed value in a specific problem. Therefore, it could have an explicit time dependence, even though its total time derivative is zero. Usually, however, the right side of (8.10) vanishes, and it is then said that the Poisson bracket of a constant of the motion with H is zero. Thus, in the Kepler problem, the angular momentum is a constant, and its Poisson bracket with the Hamiltonian is readily shown to vanish.

The Poisson brackets satisfy many interesting mathematical properties. From the definition (8.6), they exhibit antisymmetry:

$$[f, g] = -[g, f] \tag{8.11}$$

as well as linearity:

$$[C_1 f + C_2 g, h] = C_1 [f, h] + C_2 [g, h], \tag{8.12}$$

where h is still another function of the dynamical variables. The brackets obey a product rule

$$[f, gh] = [f, g]h + [f, h]g, \tag{8.13}$$

the proof of which is trivial. A more complicated property we shall need is known as **Jacobi's identity**:

$$[f, [g, h]] + [g, [h, f]] + [h, [f, g]] = 0. \tag{8.14}$$

This relation contains brackets within brackets and so is a bit cumbersome algebraically; we leave its proof to the reader (see the problem set).

Next, let us evaluate what are called the **fundamental brackets**, those of the q's and p's themselves. From (8.6), we have

$$[q_j, q_k] = \sum_i \left(\frac{\partial q_j}{\partial q_i} \frac{\partial q_k}{\partial p_i} - \frac{\partial q_j}{\partial p_i} \frac{\partial q_k}{\partial q_i} \right) = 0; \tag{8.15}$$

similarly,

$$[p_i, p_k] = 0, \quad [q_j, p_k] = \delta_{j,k}.$$

In more succinct notation, these are simply

$$[\mu_j, \mu_k] = \sum_{i,l} \frac{\partial \mu_j}{\partial \mu_i} \gamma_{il} \frac{\partial \mu_k}{\partial \mu_l} = \gamma_{jk} \tag{8.16}$$

with the aid of (8.7).

It turns out that the fundamental brackets have their above-stated values regardless of the set of coordinates in which they are expressed, as long as these coordinates are obtained from the original μ_i through a canonical transformation. Thus, it is possible to write

$$[q_j, q_k]_{q,p} = [q_j, q_k]_{Q,P} = 0,$$

$$[p_j, p_k]_{q,p} = [p_j, p_k]_{Q,P} = 0,$$

$$[q_j, p_j]_{q,p} = [q_j, p_k]_{Q,P} = \delta_{j,k}, \tag{8.17}$$

when a transformation of the form (8.5) has been carried out. The subscripts on the Poisson brackets indicate the variables with respect to which the differentiation is performed. (If there are no subscripts, we use the variables q, p as in (8.6).) As an example,

$$[q_j, q_k]_{Q,P} \equiv \sum_i \left(\frac{\partial q_j}{\partial Q_i} \frac{\partial q_k}{\partial P_i} - \frac{\partial q_j}{\partial P_i} \frac{\partial q_k}{\partial Q_i} \right).$$

The right side can only be evaluated when the transformation (8.5) is given and is inverted to provide the q's as a function of Q's and P's. The meaning of (8.17) is that such a procedure is unnecessary; we can evaluate the brackets through use of the "old" q, p, and the result will be the same.

The proof of (8.17) was accomplished by Poincaré through a consideration of area elements in a $2n$-dimensional phase space; but as his argument is rather tedious algebraically, we proceed along more physical lines. First, we define U_i by analogy with (8.1):

$$U_i \equiv Q_i \quad (i = 1 \text{ to } n),$$

$$U_{i+n} \equiv P_i. \tag{8.18}$$

The analogue of the transformation (8.5) is then

$$U_i = U_i(\mu, t).$$

From the chain rule of differentiation and the relation (8.7), we now have

$$[f, g] = \sum_{i,j} \frac{\partial f}{\partial \mu_i} \gamma_{ij} \frac{\partial g}{\partial \mu_j} = \sum_{\substack{i,j, \\ k,l}} \left(\frac{\partial f}{\partial U_k} \frac{\partial U_k}{\partial \mu_i} \right) \gamma_{ij} \left(\frac{\partial g}{\partial U_l} \frac{\partial U_l}{\partial \mu_j} \right)$$

$$= \sum_{k,l} \frac{\partial f}{\partial U_k} \left(\sum_{i,j} \frac{\partial U_k}{\partial \mu_i} \gamma_{ij} \frac{\partial U_l}{\partial \mu_j} \right) \frac{\partial g}{\partial U_l} = \sum_{k,l} \frac{\partial f}{\partial U_k} [U_k, U_l] \frac{\partial g}{\partial U_l}. \tag{8.19}$$

Poisson Brackets and their Properties

But $[U_k, U_l]_{Q, P}$ is a fundamental bracket in the "new" coordinates Q, P and is therefore just γ_{kl} by (8.16). If it has the same value as $[U_k, U_l]_{q, p}$, as stated in (8.17), it follows that

$$[f, g]_{q, p} = \sum_{k, l} \frac{\partial f}{\partial U_k} \gamma_{kl} \frac{\partial g}{\partial U_l} = [f, g]_{Q, P}. \qquad (8.20)$$

That is, as a consequence of the invariance of the fundamental brackets, *any* Poisson bracket turns out to be invariant under the transformation (8.5). (The relationships (8.17) are really just special cases of (8.20).)

Now if we regard g as the Hamiltonian of some hypothetical system and f as a coordinate or momentum, eq. (8.9) tells us that the Poisson bracket $[f, g]$ is simply the derivative df/dt. But such a derivative can depend only on the motion of the system and not on the choice of variables used for describing the motion. Hence, we expect the invariant bracket relation (8.20) to hold from purely physical reasoning.

Now, let us suppose that a system moves along a trajectory in phase space as determined by its Hamiltonian $H(\mu, t)$ in accordance with the equation of motion (8.9); f and g are again two arbitrary dynamical variables. We want to evaluate the total time derivative of the bracket $[f, g]$, which, by (8.8), is

$$\frac{d}{dt}[f, g] = [[f, g], H] + \frac{\partial}{\partial t}[f, g].$$

Using the Jacobi identity in the first term and the relation (8.7) in the second term, we have

$$\frac{d}{dt}[f, g] = [f, [g, H]] + [g, [H, f]] + \sum_{i, j} \frac{\partial^2 f}{\partial t \partial \mu_i} \gamma_{ij} \frac{\partial g}{\partial \mu_j} + \sum_{i, j} \frac{\partial f}{\partial \mu_i} \gamma_{ij} \frac{\partial^2 g}{\partial t \partial \mu_j}$$

$$= [[f, H], g] + [f, [g, H]] + \left[\frac{\partial f}{\partial t}, g\right] + \left[f, \frac{\partial g}{\partial t}\right],$$

interchanging the first two terms and using antisymmetry. But this is

$$\frac{d}{dt}[f, g] = \left[[f, H] + \frac{\partial f}{\partial t}, g\right] + \left[f, [g, H] + \frac{\partial g}{\partial t}\right] = \left[\frac{df}{dt}, g\right] + \left[f, \frac{dg}{dt}\right]. \qquad (8.21)$$

The brackets therefore satisfy a kind of derivative product rule for any two functions f and g.

There is an interesting alternative way of looking at this result. From (8.3) or (8.9), it can be said that the Hamiltonian $H(\mu, t)$ generates or develops the motion of the system in time. One might then ask the question: given a set of $2n$ coupled differential equations for $\dot{\mu}_i = \dot{\mu}_i(\mu, t)$, the solutions of which represent the system motion along some curve in phase space, is it always possible to find a function $H(\mu, t)$ which will produce the various $\dot{\mu}_i$ when substituted in (8.3) or (8.9)? (That is, put more simply, given a solution, can we always find the problem?) The answer is a qualified "yes!" It turns out that the necessary and sufficient condition for the existence of a Hamiltonian is that any two dynamical variables f and g must satisfy the relation (8.21), a result known as the **Poisson bracket theorem**.

It has already been shown that (8.21) is necessary when one assumes the existence of some function $H(\mu, t)$ with the desired properties. The proof of the sufficiency condition takes a little more algebra but leads us to a way of getting at the form of the Hamiltonian. In order to see what it is we are looking for, first consider the relation (8.3). Multiplying by γ_{in} and summing on i produces

$$\sum_i \dot{\mu}_i \gamma_{in} = \sum_{i,j} \gamma_{in} \gamma_{ij} \frac{\partial H}{\partial \mu_j} = \frac{\partial H}{\partial \mu_n} \qquad (8.22)$$

from the γ orthogonality. In the same way one obtains

$$\sum_i \dot{\mu}_i \gamma_{im} = \frac{\partial H}{\partial \mu_m}.$$

The sum on the left can certainly be evaluated at each point along some path in phase space, and it must lead to the H-derivative if the function H exists. Taking second derivatives,

$$\frac{\partial^2 H}{\partial \mu_n \partial \mu_m} = \frac{\partial}{\partial \mu_n}\left(\sum_i \dot{\mu}_i \gamma_{im}\right) = \sum_i \frac{\partial \dot{\mu}_i}{\partial \mu_n} \gamma_{im} = \frac{\partial^2 H}{\partial \mu_m \partial \mu_n} = \frac{\partial}{\partial \mu_m}\left(\sum_i \dot{\mu}_i \gamma_{in}\right) = \sum_i \frac{\partial \dot{\mu}_i}{\partial \mu_m} \gamma_{in}.$$

Changing dummy indices to j and k, this can be written

$$\sum_j \frac{\partial \dot{\mu}_j}{\partial \mu_n} \gamma_{jm} - \sum_k \frac{\partial \dot{\mu}_k}{\partial \mu_m} \gamma_{kn} = 0. \qquad (8.23)$$

This condition must hold so that the Hamiltonian can be twice differentiated and yield the same thing in either order. Put another way, the condition must hold if each one of the set of differential equations given in (8.22) is to have a solution for H in common with the others upon integration. But, in fact, the relation (8.23) follows at once from the bracket derivative rule (8.21), as can now be shown.

Apply (8.21) to the particular bracket $[\mu_j, \mu_k]$, making use of (8.16):

$$\frac{d}{dt}[\mu_j, \mu_k] = \left[\frac{d\mu_j}{dt}, \mu_k\right] + \left[\mu_j, \frac{d\mu_k}{dt}\right] = \frac{d}{dt}\gamma_{jk} = 0$$

$$= \sum_{i,l} \frac{\partial\dot{\mu}_j}{\partial\mu_i}\gamma_{il}\frac{\partial\mu_k}{\partial\mu_l} + \sum_{i,l} \frac{\partial\mu_j}{\partial\mu_i}\gamma_{il}\frac{\partial\dot{\mu}_k}{\partial\mu_l}$$

$$= \sum_{i,l} \frac{\partial\dot{\mu}_j}{\partial\mu_i}\gamma_{il}\delta_{k,l} + \sum_{i,l}\delta_{j,i}\gamma_{il}\frac{\partial\dot{\mu}_k}{\partial\mu_l} = \sum_{i} \frac{\partial\dot{\mu}_j}{\partial\mu_i}\gamma_{ik} + \sum_{l}\gamma_{jl}\frac{\partial\dot{\mu}_k}{\partial\mu_l}.$$

Multiplying by $\gamma_{jm}\gamma_{nk}$ and summing on j and k, we have from the orthogonality of the γ matrix that $\sum_k \gamma_{ik}\gamma_{nk} = \delta_{i,n}$, and so on, or

$$\sum_{i,j,k} \frac{\partial\dot{\mu}_j}{\partial\mu_i}\gamma_{ik}\gamma_{jm}\gamma_{nk} + \sum_{j,k,l} \gamma_{jm}\gamma_{nk}\gamma_{jl}\frac{\partial\dot{\mu}_k}{\partial\mu_l} = 0$$

$$= \sum_j \frac{\partial\dot{\mu}_j}{\partial\mu_n}\gamma_{jm} + \sum_k \gamma_{nk}\frac{\partial\dot{\mu}_k}{\partial\mu_m} = \sum_j \frac{\partial\dot{\mu}_j}{\partial\mu_n}\gamma_{jm} - \sum_k \frac{\partial\dot{\mu}_k}{\partial\mu_m}\gamma_{kn}.$$

Thus, as a direct consequence of (8.21), the relation (8.23) must hold, and therefore (8.22) can be partially integrated with respect to μ_n in order to obtain H. (The $\dot{\mu}_i$ must be expressed as a function of the μ_i and the time.) Some examples will make the procedure clear.

As a simple illustration, suppose we were investigating equations of motion given by

$$\dot{q} = p, \quad \dot{p} = -q \text{ (which imply that } \ddot{q} = -q) \tag{8.24}$$

in one dimension. Then, $\mu_1 = q$, $\mu_2 = p$, and, from (8.22),

$$\frac{\partial H}{\partial \mu_1} = \frac{\partial H}{\partial q} = -\dot{\mu}_2 = -\dot{p} = q,$$

while

$$\frac{\partial H}{\partial \mu_2} = \frac{\partial H}{\partial p} = \dot{\mu}_1 = \dot{q} = p.$$

Since

$$\frac{\partial^2 H}{\partial p \partial q} = 0 = \frac{\partial^2 H}{\partial q \partial p},$$

an H can exist, and by partial integration of $\partial H/\partial q = \dot q$ and $\partial H/\partial p = \dot p$, it must be simply $H = \tfrac{1}{2}p^2 + \tfrac{1}{2}q^2 + $ constant, a function appropriate for a harmonic oscillator of unit constants.

On the other hand, had $\dot\mu_i$ been related to μ_i through equations such as

$$\dot q = p^2 q, \qquad \dot p = -q^2 p, \qquad (8.25)$$

we would have had instead

$$\frac{\partial H}{\partial q} = -\dot p = q^2 p \quad \text{and} \quad \frac{\partial H}{\partial p} = \dot q = p^2 q,$$

so that

$$\frac{\partial^2 H}{\partial p\,\partial q} = q^2, \qquad \frac{\partial^2 H}{\partial q\,\partial p} = p^2.$$

Inasmuch as the second partial derivatives are not equal, no H can possibly exist for these equations of motion. Further details are considered in the problem set; it is easy to show, for example, that the Poisson bracket theorem leads to the same conclusion for eqs. (8.25).

As a final comment, note that if f and g are constants of the motion, then (8.21) tells us that

$$\frac{d}{dt}[f, g] = 0. \qquad (8.26)$$

The bracket of two such constants is yet another constant, a result known as **Poisson's theorem**. It allows us to find new constants of the motion from the Poisson brackets of two such quantities, at least in principle. In practice, however, there may be nothing new or interesting generated in this way.

8.3 CANONICAL TRANSFORMATIONS

It is time that we defined a little more precisely what is meant by a canonical (or **contact**) transformation. From (8.5), we know it is of the form

$$Q_i = Q_i(q, p, t), \quad P_i = P_i(q, p, t),$$

or instead

$$U_i = U_i(\mu, t).$$

In addition, the canonical equations must hold in the new coordinates as well as

Canonical Transformations

the old:

$$\dot{Q}_i = \frac{\partial K}{\partial P_i}, \quad \dot{P}_i = -\frac{\partial K}{\partial Q_i}, \tag{8.27}$$

or

$$\dot{U}_i = \sum_j \gamma_{ij} \frac{\partial K}{\partial U_j} \tag{8.28}$$

by analogy with (8.3). The quantity K which appears in these equations is the "new Hamiltonian." In many cases, as will be shown shortly, it is just the old H expressed in terms of the new Q's and P's.

From the variational techniques of Chapter 3, it was learned that the equations of motion are determined by Hamilton's principle, eq. (3.2):

$$\delta \int_{t_i}^{t_f} L \, dt = 0 = \delta \int_{t_i}^{t_f} \left[\sum_i p_i \dot{q}_i - H(q, p, t) \right] dt,$$

where we have used (2.15) to express the integrand in terms of H. The new canonical coordinates Q, P must likewise satisfy

$$\delta \int_{t_i}^{t_f} \left[\sum_i P_i \dot{Q}_i - K(Q, P, t) \right] dt = 0.$$

Subtracting the two relations produces

$$\delta \int_{t_i}^{t_f} \left[\sum_i p_i \dot{q}_i - H - \sum_i P_i \dot{Q}_i + K \right] dt = 0.$$

The fact that the variation of the integral is zero does not make the integrand zero; rather, it can be identified with the total time derivative of a function F, known as the **generating function** of the transformation, which is clearly a function of both old and new coordinates as well as the time. Letting

$$\sum_i p_i \dot{q}_i - H - \sum_i P_i \dot{Q}_i + K \equiv \frac{dF}{dt}, \tag{8.29}$$

we find that

$$\delta \int_{t_i}^{t_f} \frac{dF}{dt} dt = 0 = \delta \left[F(t_f) - F(t_i) \right],$$

a valid relation in view of the fact that the variation of any function vanishes at the end points of the time interval in Hamilton's principle. Later, it will be seen how the generator and eq. (8.29) allow us to solve many problems more simply through the canonical method.

As an example for clarification, consider the transformation $Q \equiv p$, $P \equiv -q$. This is a very simple transformation, but one which interchanges coordinates and momenta, showing that the transformed variables may not have their customary physical interpretations at all. For convenience, we take K and H as equal in value, though they will naturally be expressed in different coordinates. Then, from (8.29),

$$dF = p\,dq - P\,dQ = p\,dq - (-q)\,dp = p\,dq + q\,dp = d(qp).$$

The result of applying this transformation is that dF turns out to be an exact differential, giving at once that the generator is

$$F = qp + \text{constant},$$

so the transformation is canonical, eqs. (8.27) being trivially satisfied if one assumes that (2.17) is satisfied in the old coordinates. For a valid canonical transformation, we must be able to determine the generator; hence, the dF expression must be an exact differential.

As a further comment, we return to the invariance of the Poisson bracket as stated in (8.20):

$$[f, g]_{Q,P} = [f, g]_{q,p}.$$

Assuming that there is a Hamiltonian $H(q, p, t)$ which generates the motion in phase space, we can differentiate the invariance rule with respect to the time and apply the Poisson bracket relation (8.21) on the right:

$$\frac{d}{dt}[f, g]_{Q,P} = \frac{d}{dt}[f, g]_{q,p} = \left[\frac{df}{dt}, g\right]_{q,p} + \left[f, \frac{dg}{dt}\right]_{q,p}$$

which by the invariance rule gives

$$\left[\frac{df}{dt}, g\right]_{Q,P} + \left[f, \frac{dg}{dt}\right]_{Q,P}$$

on the right side. Realizing that the left side is $(d/dt)[f, g]_{Q,P}$, we see that (8.21) holds in the new coordinates for all f and g. This implies (from the discussion in the previous section) that there is a function $K(Q, P)$ which exists for any Hamiltonian H and which generates the motion in the new coordinates along the lines of eq. (8.28).

Any transformation which preserves the value of Poisson brackets is then canonical, and we may adopt the following definition: A **canonical transformation** is a transformation of both coordinates and momenta which takes the old Hamiltonian $H(q, p)$ into a new function $K(Q, P)$ such that the form of the canonical equations of Hamilton is left intact (with K in the place of H) and such that any Poisson bracket of two dynamical variables is an invariant.

As a consequence of this definition, it is possible to specify a simple test as to whether a given transformation is canonical. If the fundamental brackets for the new coordinates with respect to the old have the values given in (8.15), then all Poisson brackets will be preserved and the transformation must therefore be canonical. There is nothing else which needs to be examined, for a proper choice of the new Hamiltonian K (as from a generator F) will ensure that the canonical equations hold in the new coordinates and momenta.

Unfortunately, the terminology in the literature is not at all uniform in this area. Many writers use the term **contact transformation** for what we have called a canonical transformation, but some use that term in a different sense. Other writers include the adjective **restricted** when they refer to a canonical transformation which preserves Poisson brackets, the omission of the modifier implying that the brackets are equal within a constant factor. Still others mean a transformation that is explicitly independent of time when they discuss a restricted canonical transformation. Adjectives like "extended" are sometimes employed as well. One must be on guard at all times to ensure that the terminology being used is understood correctly.

8.4 GENERATING FUNCTIONS FOR CANONICAL TRANSFORMATIONS

The generating function F, the derivative of which appears in (8.29), is a function of $4n$ variables (both old and new coordinates and momenta) and the time. But owing to the transformation relations (8.5), only $2n$ of these variables are in reality independent. The generator could then be written as a function of any $2n$ of the variables and the time. In practice, however, it would not be convenient in the theory to mix the variables; if we use one of the old coordinates in the generator, we shall use all n of them, along with n of the new coordinates (or n new momenta). There are then possible only four types of generators:

$$F_1(q, Q, t), \quad F_2(q, P, t), \quad F_3(p, Q, t), \quad F_4(p, P, t).$$

Each of these types will satisfy certain simple equations that will be derived shortly and which lead directly to the form of the canonical transformation. For the solution of a given problem, more than one of the four types of generators can often be employed without difficulty. In certain cases, however, relations between some of the variables assure their dependence on one another and rule out the usage of one type of generator. (Thus, with the point transformation between q_i and Q_i, a function of type F_1 could not be employed.)

The necessary equations can be derived most simply for the function of type F_1. Returning to (8.29), we have

$$\frac{dF_1}{dt} = \sum_i p_i \dot{q}_i - \sum_i P_i \dot{Q}_i - H + K.$$

But $F_1 = F_1(q, Q, t)$, hence the chain rule of calculus tells us that

$$\frac{dF_1}{dt} = \sum_i \frac{\partial F_1}{\partial q_i} \dot{q}_i + \sum_i \frac{\partial F_1}{\partial Q_i} \dot{Q}_i + \frac{\partial F_1}{\partial t}.$$

Equating coefficients of $\dot{q}_i$ and $\dot{Q}_i$ in the two expressions produces

$$p_i = \frac{\partial F_1}{\partial q_i}, \qquad P_i = -\frac{\partial F_1}{\partial Q_i}, \qquad K - H = \frac{\partial F_1}{\partial t}. \tag{8.30}$$

After performing the differentiation on the right, the first of the relations (8.30) can be solved for the Q_i as a function of the q's, p's, and t; at least *in principle* the n equations can be solved simultaneously. The results obtained are then substituted into the second (P_i) equation after differentiating, thus giving the P's as functions of the q's, p's, and t. The use of the generator F_1 has therefore led directly to the form of the transformation (8.5). Finally, the third relation of (8.30) shows that K and H have the same numerical value (even though a different functional form in general) as long as F_1 does not depend explicitly on the time, a fact which is also true for the other types of generators. Once K has been determined, it is a straightforward matter to use the canonical eqs. (8.27) in solving the problem at hand. As we shall see later, an adroit choice of K (or of the generator) can aid greatly in finding the solution.

Next, let us consider a generator of the form $F_2(q, P, t)$. As before, the chain rule gives

$$\frac{dF}{dt} = \sum_i \frac{\partial F}{\partial q_i} \dot{q}_i + \sum_i \frac{\partial F}{\partial P_i} \dot{P}_i + \frac{\partial F}{\partial t}.$$

We are no longer able to equate coefficients directly, however, because (8.29) is still expressed in terms of $\dot{q}_i$ and $\dot{Q}_i$, while here we have a relation in $\dot{q}_i$ and $\dot{P}_i$. To circumvent this situation, we note that if q and P are the independent variables, we can express $Q_i = Q_i(q, P, t)$ and write

$$\dot{Q}_i = \sum_j \frac{\partial Q_i}{\partial q_j} \dot{q}_j + \sum_j \frac{\partial Q_i}{\partial P_j} \dot{P}_j + \frac{\partial Q_i}{\partial t}.$$

Substitution in (8.29) yields

$$\frac{dF}{dt} = \sum_i p_i \dot{q}_i - \sum_{i,j} P_i \frac{\partial Q_i}{\partial q_j} \dot{q}_j - \sum_{i,j} P_i \frac{\partial Q_i}{\partial P_j} \dot{P}_j - \sum_i P_i \frac{\partial Q_i}{\partial t} - H + K$$

$$= \sum_i \left(p_i - \sum_j P_j \frac{\partial Q_j}{\partial q_i} \right) \dot{q}_i - \sum_i \left(\sum_j P_j \frac{\partial Q_j}{\partial P_i} \right) \dot{P}_i - \sum_i P_i \frac{\partial Q_i}{\partial t} - H + K$$

upon interchange of dummy indices. Equating coefficients with the chain rule expression for dF/dt produces

$$p_i - \sum_j P_j \frac{\partial Q_j}{\partial q_i} = \frac{\partial F}{\partial q_i},$$

$$\sum_j P_j \frac{\partial Q_j}{\partial P_i} = -\frac{\partial F}{\partial P_i} = \frac{\partial}{\partial P_i}\left[\sum_j P_j Q_j\right] - Q_i,$$

$$K - H - \sum_i P_i \frac{\partial Q_i}{\partial t} = \frac{\partial F}{\partial t}.$$

These equations can be rewritten as

$$p_i = \frac{\partial}{\partial q_i}\left(F + \sum_j P_j Q_j\right), \quad Q_i = \frac{\partial}{\partial P_i}\left(F + \sum_j P_j Q_j\right),$$

$$K - H = \frac{\partial}{\partial t}\left(F + \sum_j P_j Q_j\right),$$

in view of the independence of q's, P's, and t.

Now it is clearly convenient to set

$$F_2(q, P, t) \equiv F + \sum_j P_j Q_j,$$

so that

$$p_i = \frac{\partial F_2}{\partial q_i}, \quad Q_i = \frac{\partial F_2}{\partial P_i}, \quad K - H = \frac{\partial F_2}{\partial t}. \quad (8.31)$$

The function F utilized above is simply an arbitrary generator satisfying (8.29). But we know from (8.30) that $\partial F_1/\partial Q_i = -P_i$, and since $\partial F_2/\partial Q_i = 0$ by definition of F_2, we can identify F with F_1, writing

$$F_2(q, P, t) = F_1(q, Q, t) + \sum_j P_j Q_j.$$

TABLE 8.1 Relationships for Different Types of Generating Functions

Generator Type	Transformation Equations	Relation to F_1
$F_1(q, Q, t)$	$p_i = \dfrac{\partial F_1}{\partial q_i},$ $P_i = -\dfrac{\partial F_1}{\partial Q_i},$ (8.30) $K = H + \dfrac{\partial F_1}{\partial t}.$	$\rightarrow$
$F_2(q, P, t)$	$p_i = \dfrac{\partial F_2}{\partial q_i},$ $Q_i = \dfrac{\partial F_2}{\partial P_i},$ (8.31) $K = H + \dfrac{\partial F_2}{\partial t}.$	$F_2 = F_1 + \sum_j P_j Q_j$
$F_3(p, Q, t)$	$q_i = -\dfrac{\partial F_3}{\partial p_i},$ $P_i = -\dfrac{\partial F_3}{\partial Q_i},$ (8.32) $K = H + \dfrac{\partial F_3}{\partial t}.$	$F_3 = F_1 - \sum_j p_j q_j$
$F_4(p, P, t)$	$q_i = -\dfrac{\partial F_4}{\partial p_i},$ $Q_i = \dfrac{\partial F_4}{\partial P_i},$ (8.33) $K = H + \dfrac{\partial F_4}{\partial t}.$	$F_4 = F_1 + \sum_j P_j Q_j - \sum_j p_j q_j$

That is, one does not obtain F_2 merely by expressing the Q's in F_1 in terms of the q's and P's, but also must add the term $\sum_j P_j Q_j$. The examples will clarify this procedure. Proceeding along similar lines, one can easily develop the analogous relationships for generators of type F_3 and F_4. For convenience, all the needed equations are summarized in Table 8.1.

As usual, we shall acquire a better understanding of the usefulness of these relations by looking at a few examples.

Example 1 Particle Falling Freely in Uniform Gravitational Field

Denote the mass by m and the acceleration of gravity by g for a freely falling particle. The vertical coordinate will be designated as q and its associated momentum as p. The use of such a powerful method as that of canonical

Generating Functions for Canonical Transformations

transformations is scarcely necessary for such a simple problem, the solution of which is well known. However, an example with uncomplicated algebra and familiar physics might help the reader to gain a better understanding of the procedures employed.

Solution

For the generator, we shall select a function of type one,

$$F_1(q, Q, t) = -\frac{1}{3m^2g}[2m^2g(Q - q)]^{3/2},$$

about which more will be said later. To generate the transformation, we find, from eq. (8.30), that

$$p = \frac{\partial F_1}{\partial q} = [2m^2g(Q - q)]^{1/2},$$

or

$$Q = q + \frac{p^2}{2m^2g}.$$

In addition,

$$P = -\frac{\partial F_1}{\partial Q} = [2m^2g(Q - q)]^{1/2} = p.$$

The inverse transformation is clearly given by

$$q = Q - \frac{p^2}{2m^2g}, \qquad p = P.$$

To be convinced that this transformation is indeed canonical, it is an easy matter to check the fundamental Poisson bracket:

$$[Q, P] = \frac{\partial Q}{\partial q}\frac{\partial P}{\partial p} - \frac{\partial Q}{\partial p}\frac{\partial P}{\partial q} = 1 - \frac{p}{m^2g} \cdot 0 = 1,$$

the other fundamental brackets being trivial.

Continuing the application of (8.30), we see that

$$K = H = \frac{p^2}{2m} + mgq = mgQ.$$

The canonical eqs. (8.27) in the new coordinates are now

$$\dot{Q} = \frac{\partial K}{\partial P} = 0, \qquad \dot{P} = -\frac{\partial K}{\partial Q} = -mg,$$

whereupon the solutions are obtained at once:

$$Q = \text{constant} = C_1, \qquad P = -mgt + \text{constant} = -mgt + C_2.$$

The desired solution in the original coordinates is then

$$q = Q - \frac{P^2}{2m^2g} = C_1 - \frac{(-mgt + C_2)^2}{2m^2g} = -\frac{1}{2}gt^2 + \text{constant} \cdot t + \text{constant},$$

$$p = P = -mgt + C_2,$$

the constants being fixed by the initial conditions; this final solution is the one expected.

At this point, the reader who has been carefully following the details of the method of solution might be wondering where we obtained the generator F_1. That is the crucial issue; it is not always easy to find a generator which leads to a convenient solution, and there is no standard procedure for doing so. (The author obtained the above F_1 by recognizing that the Hamiltonian was a constant of the motion, setting it proportional to Q and working "backward" with the relations (8.30) until F_1 was reached.)

Now let us approach the problem with a generator of type 2, using Table 8.1:

$$F_2(q, P, t) = F_1 + PQ = -\frac{P^3}{3m^2g} + P\left(q + \frac{P^2}{2m^2g}\right)$$

$$= qP + \frac{1}{6}\frac{P^3}{m^2g},$$

expressing the generator in terms of the variables q and P. From (8.31), we have

$$p = \frac{\partial F_2}{\partial q} = P, \qquad Q = \frac{\partial F_2}{\partial P} = q + \frac{1}{2}\frac{P^2}{m^2g},$$

$$K = H = mgQ,$$

the remainder of the solution proceeding as before. For type 3, one obtains

$$F_3(p, Q, t) = F_1 - pq = -\frac{p^3}{3m^2g} - p\left(Q - \frac{p^2}{2m^2g}\right) = \frac{1}{6}\frac{p^3}{m^2g} - pQ.$$

Equation (8.32) yields

$$q = -\frac{\partial F_3}{\partial p} = -\frac{p^2}{2m^2g} + Q, \qquad P = -\frac{\partial F_3}{\partial Q} = p, \qquad K = H.$$

The transformation equations have been employed in order to express each generator in the appropriate variables, so all we have done here is to show consistency. Starting from F_2 or F_3 in the form obtained, however, we could have solved the problem exactly as was done with F_1. (A function of type F_4 cannot be used with this transformation, which sets $P = p$, for the variables p and P would not be independent as required by the theory.)

Example 2 Orthogonal and Point Transformations

Suppose we are given a generating function such as

$$F_3 = -\sum_j p_j Q_j.$$

Application of (8.32) at once produces

$$q_i = -\frac{\partial F_3}{\partial p_i} = Q_i, \quad P_i = -\frac{\partial F_3}{\partial Q_i} = p_i, \quad K = H.$$

In words, this F_3 is a generator of the identity transformation. (It is naturally not unique in this regard; $F_2 = \sum_j q_j P_j$ would have done just as well, for example.)

Generalizing this idea, consider a generator of the form

$$F_3 = -\sum_{j,k} a_{jk} p_k Q_j,$$

where the a_{jk} are the elements of an orthogonal matrix as discussed in Chapter 5. Since

$$q_i = -\frac{\partial F_3}{\partial p_i} = \sum_j a_{ji} Q_j \quad \text{and} \quad P_i = -\frac{\partial F_3}{\partial Q_i} = \sum_k a_{ik} p_k,$$

we can employ the orthogonality relation (5.16) to obtain

$$\sum_i a_{ki} q_i = \sum_i a_{ki} \left(\sum_j a_{ji} Q_j \right) = \sum_j Q_j \left(\sum_i a_{ki} a_{ji} \right) = \sum_j Q_j \delta_{j,k} = Q_k,$$

showing that F_3 is the generator of an arbitrary orthogonal transformation. (The momenta transform similarly, as is apparent from the P_i relation.) In such an example, the new coordinates are linear functions of the old, transforming independently of the momenta.

A general point transformation is reminiscent of an orthogonal transformation, except that the a_{ij} may be functions of the coordinates rather than constants and the simple linear form of the transformation may be abandoned. Generating functions can always be found, however, no matter how complex the transformation. All orthogonal and all point transformations are then canonical, so perhaps the latter concept is not unfamiliar after all.

To clarify these points, consider the conversion from rectangular to plane polar coordinates. The familiar relations

$$x = r\cos\theta, \quad y = r\sin\theta$$

become

$$q_1 = Q_1 \cos Q_2, \qquad q_2 = Q_1 \sin Q_2,$$

if the "old" coordinates are chosen as the rectangular ones and the "new" as the polar. Inverting the transformation leads to

$$Q_1 = \sqrt{q_1^2 + q_2^2}, \qquad Q_2 = \arctan \frac{q_2}{q_1}.$$

Taking as the generator a function of type 2,

$$F_2 = \sqrt{q_1^2 + q_2^2} \cdot P_1 + \arctan\left(\frac{q_2}{q_1}\right) P_2,$$

we have from (8.31)

$$Q_1 = \frac{\partial F_2}{\partial P_1} = \sqrt{q_1^2 + q_2^2}, \qquad Q_2 = \frac{\partial F_2}{\partial P_2} = \arctan \frac{q_2}{q_1},$$

as above. The momenta p_1 and p_2 are given by

$$p_1 = \frac{\partial F_2}{\partial q_1} = \frac{q_1 P_1}{\sqrt{q_1^2 + q_2^2}} - \frac{(q_2 P_2/q_1^2)}{1 + (q_2/q_1)^2} = \frac{q_1 P_1}{\sqrt{q_1^2 + q_2^2}} - \frac{q_2 P_2}{q_1^2 + q_2^2},$$

$$p_2 = \frac{\partial F_2}{\partial q_2} = \frac{q_2 P_1}{\sqrt{q_1^2 + q_2^2}} + \frac{q_1 P_2}{q_1^2 + q_2^2}.$$

Expressing the q's in terms of the Q's, the momenta become

$$p_1 = P_1 \cos Q_2 - \frac{\sin Q_2}{Q_1} P_2, \qquad p_2 = P_1 \sin Q_2 + \frac{\cos Q_2}{Q_1} P_2.$$

Inverting the momentum transformation yields

$$P_1 = \frac{q_1 p_1 + q_2 p_2}{\sqrt{q_1^2 + q_2^2}}, \qquad P_2 = q_1 p_2 - q_2 p_1.$$

Now take a simple problem like that of a free particle with $H = (p_1^2 + p_2^2)/2m$. The "old" canonical equations give

$$\dot{q}_1 = \frac{\partial H}{\partial p_1} = \frac{p_1}{m}, \qquad \dot{q}_2 = \frac{\partial H}{\partial p_2} = \frac{p_2}{m}, \qquad \dot{p}_1 = -\frac{\partial H}{\partial q_1} = 0, \qquad \dot{p}_2 = 0,$$

while with $K = H = P_1^2/2m + P_2^2/2mQ_1^2$, the "new" canonical equations produce

$$\dot{Q}_1 = \frac{\partial K}{\partial P_1} = \frac{P_1}{m}, \qquad \dot{Q}_2 = \frac{\partial K}{\partial P_2} = \frac{P_2}{mQ_1^2},$$

$$\dot{P}_1 = -\frac{\partial K}{\partial Q_1} = \frac{P_2^2}{mQ_1^3}, \qquad \dot{P}_2 = -\frac{\partial K}{\partial Q_2} = 0.$$

Stated in more familiar terms, these equations have produced

$$P_1 = m\dot{r}, \qquad P_2 = mr^2\dot{\theta} = \text{constant}, \qquad \dot{P}_1 = \frac{P_2^2}{mr^3} = m\ddot{r},$$

which are in agreement with (2.8) and (2.9), P_1 being the radial component and P_2 the angular momentum for the plane motion. (One must set $\theta = 0$ in the equations of Chapter 2, identifying axes appropriately.)

8.5 THE HAMILTON – JACOBI EQUATION

Suppose that the original Hamiltonian H of a given problem does not involve the time explicitly, and is therefore a constant of the motion which is designated as α_1 in accordance with customary practice:

$$H(q_1, q_2, \ldots q_n; p_1, \ldots p_n) = \alpha_1. \tag{8.34}$$

The integral that is required in the use of Hamilton's principle, eq. (3.2), is

$$\int_{t_0}^{t} L \, dt = \int_{t_0}^{t} \left(\sum_i p_i \dot{q}_i - H \right) dt = \int_{t_0}^{t} \sum_i p_i \, dq_i - \int_{t_0}^{t} H \, dt, \tag{8.35}$$

where we have employed the definition of H. The integrand of the first term is merely

$$dS = \sum_i p_i \, dq_i,$$

$$p_i = \frac{\partial S}{\partial q_i}, \tag{8.36}$$

$S = S(q)$ being an indefinite form of the integral called the **action** in Section 3.3. The second term can also be integrated at once upon insertion of (8.34), but we shall not proceed in that way. Instead, substituting (8.36) into (8.34), one obtains

$$H\left(q_1, \ldots q_n; \frac{\partial S}{\partial q_1}, \frac{\partial S}{\partial q_2}, \ldots \frac{\partial S}{\partial q_n}\right) = \alpha_1, \tag{8.37}$$

a partial differential equation for the function S known as the **Hamilton–Jacobi** (or simply HJ) **equation**.

Historically, the HJ equation had its heyday in the early part of this century as the early quantum theory was developed. It does have some advantages in the analysis of periodic motions, but generally one of the methods given earlier would be more practical for problem solving, in the author's opinion. The HJ method is mainly of interest here as an illustration of canonical transformations, and because the technique is frequently employed in the literature.

The equation (8.37) can only be used when it can be separated into n ordinary differential equations. Since the relation is first order in the n variables q_i, its solution must contain n arbitrary constants of integration. One of these constants should be a final overall additive constant, for only derivatives of S appear in the HJ equation. Such a constant can be ignored as it is of no interest, and the remaining $n-1$ constants denoted by $\alpha_2, \alpha_3, \ldots \alpha_n$. Upon solving (8.37), the action can be expressed through the procedure described below as

$$S = S(q_1, q_2, \ldots q_n; \alpha_1, \alpha_2, \ldots \alpha_n).$$

To solve a given problem by means of (8.37), it is first necessary to find the action S (sometimes referred to as **Hamilton's characteristic function** in this context) as a function of q's and α's through the separation of variables just mentioned. As a start, all terms containing q_1 or $\partial S/\partial q_1$ are put on one side of (8.37), the other side being a different function of the remaining variables. If the equality is to hold with all the variables independent of one another, each side must be equal to a separation constant which will be called α_2:

$$\alpha_2 \equiv f\left(q_1, \frac{\partial S}{\partial q_1}, \alpha_1\right) = g\left(q_2, q_3, \ldots q_n; \frac{\partial S}{\partial q_2}, \ldots \frac{\partial S}{\partial q_n}; \alpha_1\right).$$

An ordinary differential equation has therefore been produced for that part of S which depends only on q_1, and it can be called $S_1(q_1)$. The separation process is continued to find each S_i in turn, a procedure which will become clear in the examples. The full action S is thus a sum of the form

$$S = \sum_i S_i(q_i), \tag{8.38}$$

where each S_i may also contain some of the α's.

Having found the action function, consider a canonical transformation in which the new momenta are chosen as the separation constants α_i:

$$P_i \equiv \alpha_i. \tag{8.39}$$

The Hamilton–Jacobi Equation

The action is then $S(q, \alpha) \equiv S(q, P)$, so S itself can be taken as a generating function of type 2:

$$F_2(q, P) \equiv S(q, P). \tag{8.40}$$

Then it follows from (8.31), that

$$p_i = \frac{\partial F_2}{\partial q_i} = \frac{\partial S}{\partial q_i}, \quad Q_i = \frac{\partial F_2}{\partial P_i} = \frac{\partial S}{\partial \alpha_i}, \tag{8.41}$$

the first of which agrees with (8.36). The second relation, of course, gives a way of determining the "new" coordinates from the action. Furthermore, since S (or F_2) does not depend explicitly on the time,

$$K = H = \alpha_1 \equiv P_1.$$

This means that the "new" canonical equations are

$$\dot{Q}_i = \frac{\partial K}{\partial P_i} = \delta_{i,1}, \quad \dot{P}_i = -\frac{\partial K}{\partial Q_i} = 0,$$

or

$$Q_1 = t + \beta_1, \quad Q_i(i \neq 1) = \beta_i, \quad P_i = \text{constant} = \alpha_i. \tag{8.42}$$

The β_i are merely integration constants which can be specified by the initial conditions. Thus, all new coordinates and momenta are constants except Q_1, which is a simple linear function of the time. (Note that the momentum conjugate to Q_1 is just $P_1 = \alpha_1$, which is the Hamiltonian itself, Q_1 being in essence the time.) Inverting the transformation to the original coordinates by means of $Q_i = \partial S/\partial \alpha_i$ (a set of equations which are solved simultaneously, each $\partial S/\partial \alpha_i$ being dependent on the various q_i), the problem has been solved. The elegance and power of this canonical approach is indeed striking!

The same results can be obtained with more labor in the traditional (if somewhat old-fashioned) approach through a variation of paths in Hamilton's principle. This cumbersome, obscure older method suffers badly by comparison with the canonical technique, yet in the end each road leads to the same destination.

As a final comment, let the Hamiltonian, and the action (which in this context is often called **Hamilton's principal function**) be time dependent. Make a canonical transformation such that the new Hamiltonian K is identically zero (rather than the constant α_1) which will ensure that *all* the Q's and P's are constants. The HJ equation will become, using eq. (8.31),

$$K = H\left(q_1, q_2, \ldots q_n; \frac{\partial F_2}{\partial q_1}, \ldots \frac{\partial F_2}{\partial q_n}; t\right) + \frac{\partial F_2}{\partial t} \equiv 0, \tag{8.43}$$

where the equations (8.39) through (8.41) are still in effect. In view of the fact that $\dot{Q}_i = \partial K/\partial P_i \equiv 0$ and $\dot{P}_i = -\partial K/\partial Q_i \equiv 0$,

$$Q_i = \beta_i, \qquad P_i = \alpha_i, \qquad (8.44)$$

much as before. The transformation can again be inverted at this point to obtain an expression for each q_i as a function of time. In fact, the time-independent formalism can be recovered quite easily from this approach. Simply subtract a term $\alpha_1 t$ from the generator F_2, and (8.43) will reduce to (8.34) immediately by separation of the time variable.

It might be wondered why the new Hamiltonian K is set equal to zero in the time-dependent approach, where previously it had been the constant α_1. The answer is that the time entered before through Q_1 in (8.42), which allowed a solution for the old coordinates as functions of time. But in the time-dependent approach, the time is already contained in H and S and therefore need not be present in Q_i or P_i, each of which is taken as a constant for simplicity by putting $K \equiv 0$. (An example of this method is given in the last section of this chapter.)

Example 3 Projectile Motion

Apply the HJ equation to the motion of a projectile of mass m in the xy-plane. Assume that $x = y = 0$ and $\dot{x} = \dot{y} = v_0$ at time zero. The Hamiltonian is

$$H = \frac{1}{2m}(p_x^2 + p_y^2) + mgy = \alpha_1.$$

The HJ equation is then

$$\frac{1}{2m}\left[\left(\frac{\partial S_x}{\partial x}\right)^2 + \left(\frac{\partial S_y}{\partial y}\right)^2\right] + mgy = \alpha_1, \qquad S \equiv S_x(x) + S_y(y).$$

Separating the x-term, then the y-terms, we find

$$\frac{\partial S_x}{\partial x} = \sqrt{2m(\alpha_1 - mgy) - \left(\frac{\partial S_y}{\partial y}\right)^2} \equiv \alpha_2, \qquad S_x = \alpha_2 x,$$

and

$$\frac{\partial S_y}{\partial y} = \sqrt{2m(\alpha_1 - mgy) - \alpha_2^2}, \qquad S_y = -\frac{1}{3m^2 g}[2m(\alpha_1 - mgy) - \alpha_2^2]^{3/2},$$

whereupon

$$S = \alpha_2 x - \frac{1}{3m^2 g}[2m(\alpha_1 - mgy) - \alpha_2^2]^{3/2}$$

is the action integral. (The new momenta are not needed here, but note that $P_1 = \alpha_1$ is the energy, and $P_2 = \alpha_2 = p_x$.)

Now, by (8.42), one has

$$\frac{\partial S}{\partial \alpha_1} = t + \beta_1 = -\frac{1}{mg}\sqrt{2m(\alpha_1 - mgy) - \alpha_2^2},$$

and

$$\frac{\partial S}{\partial \alpha_2} = \beta_2 = x + \frac{\alpha_2}{m^2 g}\sqrt{2m(\alpha_1 - mgy) - \alpha_2^2}.$$

Solving these two equations simultaneously, we have for x

$$\beta_2 = x - \frac{\alpha_2}{m}(t + \beta_1),$$

or

$$x = \frac{\alpha_2}{m}(t + \beta_1) + \beta_2, \qquad \dot{x} = \frac{\alpha_2}{m}.$$

From the conditions at zero time, $\alpha_2 = mv_0$, $0 = (\alpha_2/m)\beta_1 + \beta_2$, hence $x = v_0 t$. In addition, for y, one has

$$m^2 g^2 (t + \beta_1)^2 = 2m(\alpha_1 - mgy) - \alpha_2^2,$$

$$y = -\frac{g(t + \beta_1)^2}{2} + \frac{(2m\alpha_1 - \alpha_2^2)}{2m^2 g},$$

$$\dot{y} = -g(t + \beta_1).$$

The initial conditions yield

$$-g\beta_1 = v_0, \qquad (y)_{t=0} = 0 = -\frac{g}{2}\left(\frac{-v_0}{g}\right)^2 + \frac{\alpha_1}{mg} - \frac{(mv_0)^2}{2m^2 g}, \qquad \alpha_1 = mv_0^2;$$

therefore,

$$y = -\frac{g}{2}t^2 + v_0 t.$$

The equation of the trajectory is found by eliminating the time:

$$y = -\frac{g}{2}\frac{x^2}{v_0^2} + x.$$

Thus, the problem is solved, the physical results being what one would expect.

Example 4 Damped Harmonic Oscillator

Some problems can be greatly simplified through a prior canonical transformation before the Hamilton–Jacobi techniques are applied. Consider again the damped oscillator of eq. (7.3):

$$m\ddot{x} + R\dot{x} + kx = 0.$$

The system is dissipative so there is no potential energy function in the usual sense. Nevertheless, the equation of motion can be obtained easily from Lagrange's equation if we *define* the Lagrangian as

$$L \equiv e^{Rt/m}\left(\tfrac{1}{2}m\dot{x}^2 - \tfrac{1}{2}kx^2\right).$$

(This expression for L, however, is not unique in that regard.) The momentum is

$$p_x = \frac{\partial L}{\partial \dot{x}} = m\dot{x}e^{Rt/m}$$

and the Hamiltonian is

$$H = p_x\dot{x} - L = e^{Rt/m}\left(\frac{1}{2}m\dot{x}^2 + \frac{1}{2}kx^2\right) = \frac{p_x^2}{2m}e^{-Rt/m} + \frac{kx^2}{2}e^{Rt/m}.$$

Let us now make a canonical transformation with

$$q \equiv xe^{Rt/2m}, \qquad p \equiv p_x e^{-Rt/2m}.$$

(The generator is $F_2 = e^{Rt/2m}xp$.) The transformed Hamiltonian becomes

$$K = H + \frac{\partial F_2}{\partial t} = \frac{p^2}{2m} + \frac{kq^2}{2} + \frac{R}{2m}qp = \alpha_1.$$

(Note that K is independent of time explicitly and so is a constant of the motion.) Applying the Hamilton–Jacobi technique to the new Hamiltonian K, the HJ equation is

$$\frac{1}{2m}\left(\frac{\partial S}{\partial q}\right)^2 + \frac{kq^2}{2} + \frac{R}{2m}q\frac{\partial S}{\partial q} = \alpha_1.$$

This relation is a quadratic in $\partial S/\partial q$, having the solution

$$\frac{\partial S}{\partial q} = -\frac{R}{2}q \pm \frac{1}{2}\sqrt{(R^2 - 4mk)q^2 + 8m\alpha_1},$$

whereupon

$$S = -\frac{R}{4}q^2 \pm \frac{1}{2}\int \sqrt{(R^2 - 4mk)q^2 + 8m\alpha_1}\, dq.$$

Without evaluating the integral of the second term, (8.42) gives

$$\frac{\partial S}{\partial \alpha_1} = t + \beta_1 = \pm 2m \int \frac{dq}{\sqrt{(R^2 - 4mk)q^2 + 8m\alpha_1}}.$$

This becomes upon integration

$$t + \beta_1 = \pm \frac{2m}{\sqrt{4mk - R^2}} \arcsin\left[\sqrt{\frac{4mk - R^2}{8m\alpha_1}}\, q\right]$$

$$= \pm \frac{1}{\omega_0} \arcsin\left(\sqrt{\frac{m}{2\alpha_1}}\,\omega_0 q\right),$$

$$q = \pm \frac{\sqrt{2\alpha_1/m}}{\omega_0} \sin\omega_0(t + \beta_1),$$

where we have taken $4mk > R^2$ (the underdamped case) and defined ω_0 as in (7.6). The sign factor is irrelevant; so for the original variable x, we obtain, as expected from (7.9),

$$x = qe^{-Rt/2m} = \frac{\sqrt{2\alpha_1/m}}{\omega_0} e^{-Rt/2m} \sin\omega_0(t + \beta_1),$$

the constants α_1 and β_1 being fixed by the initial conditions.

8.6 ACTION AND ANGLE VARIABLES

When the motion of a physical system is periodic in nature, it is possible to apply the Hamilton–Jacobi techniques in a very elegant yet convenient fashion. (The damped oscillator treated above is, however, *not* an example of a periodic system because of the presence of the attenuating exponential.) We refer to the method of **action and angle variables**,[2] which was historically important in the development of quantum theory and is still of great interest to astronomers.

---

[2] K. Schwarzschild, *Sitzungsber. der Kgl. Akad. der Wiss.*, 548 (1916).

The essence of the method can be outlined simply enough. From eqs. (8.41) and (8.38), we have

$$p_i = \frac{\partial S}{\partial q_i} = \frac{\partial S_i(q_i, \alpha_1, \ldots \alpha_n)}{\partial q_i} = p_i(q_i, \alpha_1, \ldots \alpha_n).$$

Integrating over one full cycle or period of the coordinate q_i (as is indicated by the circle on the integral sign), we define the **action variable** as

$$J_i \equiv \oint p_i dq_i = J_i(\alpha_1, \ldots \alpha_n). \tag{8.45}$$

This quantity is very much like S_i in that $p_i dq_i$ must be integrated in order to obtain it, but with S_i the integral is indefinite, while here the definite full cycle integral leaves J_i as a function of the α's only. When the integral (8.45) has been calculated for each J_i, the n relations are inverted to yield

$$\alpha_i = \alpha_i(J_1, \ldots J_n). \tag{8.46}$$

The Hamiltonian and the characteristic function of Hamilton now become

$$H = \alpha_1 = H(J_1, \ldots J_n), \qquad S = S(q_1, \ldots q_n; J_1, \ldots J_n). \tag{8.47}$$

Next, the J_i themselves are chosen as the new momenta (rather than the α_i) for the canonical transformation:

$$P_i \equiv J_i. \tag{8.48}$$

As before, the generator S is of type 2, and from (8.31), we have

$$\delta_i \equiv Q_i = \frac{\partial F_2}{\partial P_i} = \frac{\partial S}{\partial J_i}, \qquad K = H. \tag{8.49}$$

The symbol δ_i, called the **angle variable**, is used for the new coordinate (it is a dimensionless quantity). The "new" canonical equations are, from the fact that $K = H = \alpha_1(J)$,

$$\dot{P}_i = -\frac{\partial K}{\partial Q_i} = -\frac{\partial K}{\partial \delta_i} = 0, \qquad (K \text{ is a function of the } J\text{'s only})$$

$$\dot{Q}_i = \dot{\delta}_i = \frac{\partial K}{\partial P_i} = \frac{\partial K}{\partial J_i} \equiv \nu_i(J_1, \ldots J_n), \tag{8.50}$$

or

$$\delta_i = \nu_i t + \beta_i, \tag{8.51}$$

Action and Angle Variables

upon integrating the $\dot{\delta}_i$ equation, with β_i as an integration constant. That such an integration is possible follows from the knowledge that each ν_i depends on the constants J_i but not on the time.

The physical meaning of the ν's is not difficult to determine. If we express the angle variable δ_i as a function of the original coordinates and new momenta, and let one of the coordinates q_j go through a full period of the motion while the others are held fixed, we can write (with the aid of the transformation relations above) that the change in δ_i thus produced is

$$\Delta \delta_i = \oint \frac{\partial \delta_i}{\partial q_j} dq_j = \oint \frac{\partial}{\partial q_j} \frac{\partial S}{\partial J_i} dq_j = \frac{\partial}{\partial J_i} \oint \frac{\partial S}{\partial q_j} dq_j = \frac{\partial}{\partial J_i} \oint p_j dq_j$$

$$= \frac{\partial J_j}{\partial J_i} = \delta_{i,j}.$$

(The last symbol is a Kronecker delta function.) Thus, δ_i is unaffected by a full period motion of q_j unless $j = i$, in which case δ_i changes by unity. Now, for the q_i motion, we have

$$\Delta \delta_i = \nu_i \Delta t = 1, \qquad \nu_i = \frac{1}{\Delta t}$$

from (8.51), where Δt is the period. Therefore, ν_i is clearly the *frequency* of this motion.

The action-angle method then gives a very direct procedure for finding all the allowed frequencies. (Those frequencies associated with different q_i need not all be the same. But, unless one gets a rational fraction for each frequency ratio, the system will not execute motion along a closed curve in space.) First, the action integrals (8.45) are worked out as a function of the α's, and then are inverted to give $\alpha_i(J)$ as in (8.46). Finally, the Hamiltonian $\alpha_1(J)$ is differentiated with respect to J_i by (8.50), thus yielding the frequencies ν_i. The procedure is both elegant and straightforward, the difficulties usually arising from evaluation of the action integrals.

Example 5 Inverse Square Central Force

For a particle moving in an attractive potential $-k/r$, we have seen in Chapter 4 that the Hamiltonian is

$$H = \frac{p_r^2}{2m} + \frac{p_\theta^2}{2mr^2} - \frac{k}{r} = \alpha_1,$$

where p_θ is the constant angular momentum. The characteristic function has the form

$$S = S_r(r) + S_\theta(\theta),$$

and therefore

$$\frac{1}{2m}\left(\frac{\partial S_r}{\partial r}\right)^2 + \frac{1}{2mr^2}\left(\frac{\partial S_\theta}{\partial \theta}\right)^2 - \frac{k}{r} = \alpha_1.$$

Only the term in $\partial S_\theta/\partial \theta$ involves θ, so, clearly, $S_\theta = \alpha_\theta \theta$, and

$$\frac{\partial S_r}{\partial r} = \sqrt{2m\alpha_1 + \frac{2mk}{r} - \frac{\alpha_\theta^2}{r^2}}.$$

The action integrals then become

$$J_\theta = \oint p_\theta \, d\theta = \oint \frac{\partial S_\theta}{\partial \theta} \, d\theta = \oint \alpha_\theta \, d\theta = 2\pi\alpha_\theta,$$

$$J_r = \oint p_r \, dr = \oint \frac{\partial S_r}{\partial r} \, dr = \oint \sqrt{2m\alpha_1 + \frac{2mk}{r} - \frac{\alpha_\theta^2}{r^2}} \, dr.$$

As we shall show below,

$$J_r = 2\pi\left[-\alpha_\theta + \frac{mk}{\sqrt{-2m\alpha_1}}\right].$$

Inverting the J relations,

$$\alpha_\theta = \frac{J_\theta}{2\pi}, \qquad \alpha_r \equiv \alpha_1 = -\frac{2\pi^2 mk^2}{(J_r + J_\theta)^2} = K.$$

Then, by (8.50),

$$\nu_r = \frac{\partial K}{\partial J_r} = +\frac{4\pi^2 mk^2}{(J_r + J_\theta)^3} = \frac{\partial K}{\partial J_\theta} = \nu_\theta.$$

The only two frequencies involved are equal to one another, a situation known as **frequency degeneracy**. As we have indicated, such degeneracy occurs when a particle undergoes motion around a closed orbit (here, an ellipse for negative total

Action and Angle Variables

energy). The period T is the reciprocal of the frequency, so from the α_1 and J_r relations,

$$T = \frac{(J_r + J_\theta)^3}{4\pi^2 mk^2} = \frac{(2\pi^2 mk^2)^{3/2}}{4\pi^2 mk^2(-\alpha_1)^{3/2}} = \pi k \sqrt{\frac{m}{2(-\alpha_1)^3}},$$

$$T^2 = \frac{\pi^2 k^2 m}{2(-\alpha_1)^3} = \left(\frac{4\pi^2 m}{k}\right) a^3,$$

where we have used the expression $\alpha_1 = -k/2a$ derived in Chapter 4; in this way, Kepler's third law has been obtained once more in a new and interesting way. The extension of the action angle techniques to perturbation problems in astronomy is straightforward but algebraically laborious, so we shall not pursue it further.

The only detail still remaining is the evaluation of the integral $J_r = \oint p_r \, dr$. Now, $p_r = m\dot{r}$ is positive as the mass moves from $r_{\min}$ to $r_{\max}$, and is negative as it returns to $r_{\min}$. At these two extreme values of r (the turning points on the orbit), $\dot{r} = 0$, and the energy equation (4.9) gives a quadratic in r,

$$\alpha_1 = \frac{\alpha_\theta^2}{2mr^2} - \frac{k}{r},$$

or

$$\alpha_1 r^2 + kr - \frac{\alpha_\theta^2}{2m} = 0,$$

the two roots of which are the desired values of $r_{\min}$ and $r_{\max}$. These roots are

$$-\frac{k}{2\alpha_1} \pm \frac{1}{2\alpha_1} \sqrt{k^2 + \frac{2\alpha_1 \alpha_\theta^2}{m}} = \frac{-k}{2\alpha_1}\left(1 \mp \sqrt{1 + \frac{2\alpha_1 \alpha_\theta^2}{mk^2}}\right)$$

$$= a(1 \mp \varepsilon),$$

making use of the eccentricity relation (4.12) (with $p_\theta = \partial S/\partial \theta = \alpha_\theta \equiv \mathscr{L}$ and $r_{\max} + r_{\min} = 2a$). Exploiting the orbit symmetry, J_r can then be evaluated in direct fashion by using standard integral tables to evaluate the expression

$$J_r = 2\sqrt{2m} \int_{r_{\min}}^{r_{\max}} \frac{dr}{r} \sqrt{\alpha_1 r^2 + kr - \frac{\alpha_\theta^2}{2m}},$$

the roots being required for the limits of integration.

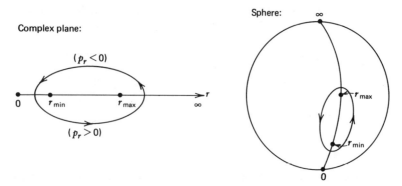

Figure 8.1 The contour of integration for J_r.

Alternatively, Born and Sommerfeld popularized a more elegant way of doing this integral by the residue techniques outlined in Appendix D. They kept J_r as a contour integral, and used the turning point scheme above to factor the quadratic under the square root of the integrand:

$$J_r = \sqrt{2m\alpha_1} \oint \frac{dr}{r} \sqrt{(r - r_{\min})(r - r_{\max})} \quad \left(= \oint p_r\, dr \right).$$

Since a full cycle involves both signs of p_r, we are dealing in the complex r-plane with a double-valued function having branch points at $r_{\min}$ and $r_{\max}$ (Figure 8.1). A branch cut can be taken between these points and the contour of integration chosen to encircle the cut so that the sign of the function will change as we pass around one of the branch points on the contour.

Unfortunately, the residue method cannot be applied to such a contour, but a little trick (often useful) can be performed to make it applicable. Imagine wrapping the r-plane about the surface of a sphere, with origin at the south pole and *any* infinity at the north pole. The real r-axis becomes a meridian of longitude, and the contour can now be considered as *clockwise* around the rest of the sphere away from the cut (where the function is single valued). There are poles of the integrand only at the origin and infinity, and their contributions can now be found.

Residue at origin: $\qquad \sqrt{2m\alpha_1 r_{\min} r_{\max}}$.

(This is the coefficient of $1/r$ for $r \approx 0$.)

Residue at infinity: Setting $u \equiv 1/r$, the integrand is

$$-\frac{du}{u^2}\left[1 - (r_{\min} + r_{\max})u + r_{\min}r_{\max}u^2\right]^{1/2} \simeq -\frac{du}{u^2}\left[1 - \tfrac{1}{2}(r_{\min} + r_{\max})u\right]$$

upon expanding about $u = 0$, giving residue $\sqrt{2m\alpha_1}\,\frac{1}{2}(r_{\min} + r_{\max})$. Adding these contributions,

$$J_r = -2\pi i\sqrt{2m\alpha_1}\left[\sqrt{r_{\min}r_{\max}} + \frac{1}{2}(r_{\min} + r_{\max})\right]$$

$$= -2\pi i\sqrt{2m\alpha_1}\left[\sqrt{-\frac{\alpha_\theta^2}{2m\alpha_1}} + \frac{1}{2}\left(-\frac{k}{\alpha_1}\right)\right] = 2\pi\left[-\alpha_\theta + \frac{mk}{\sqrt{2m(-\alpha_1)}}\right],$$

as desired.

8.7 INFINITESIMAL CANONICAL TRANSFORMATIONS AND THE TIME DEVELOPMENT OPERATOR

Suppose we make a canonical transformation in which the new coordinates and momenta differ only by infinitesimal amounts from the original ones:

$$Q_i \equiv q_i + dq_i, \qquad P_i \equiv p_i + dp_i. \tag{8.52}$$

Earlier in the chapter, we noted that the generator $-\sum p_j Q_j$ would produce the identity transformation, and since we are here dealing with a transformation differing by only a tiny amount from the identity, we take as the generator

$$F_3 = -\sum_j p_j Q_j + \varepsilon G(Q, p), \tag{8.53}$$

where ε is a small parameter that specifies the transformation in some sense. (As ε goes to zero, dq_i and dp_i must do likewise.) From (8.32) and (8.52), we have

$$q_i = -\frac{\partial F_3}{\partial p_i} = Q_i - \varepsilon\frac{\partial G}{\partial p_i} = Q_i - dq_i, \qquad dq_i = \varepsilon\frac{\partial G}{\partial p_i},$$

$$P_i = -\frac{\partial F_3}{\partial Q_i} = p_i - \varepsilon\frac{\partial G}{\partial Q_i} = p_i + dp_i, \qquad dp_i = -\varepsilon\frac{\partial G}{\partial Q_i}.$$

In the last relation, we are still correct to first order if we replace Q_i by q_i in the G factor, considering G as a function of q and p:

$$dq_i = \varepsilon\frac{\partial G}{\partial p_i}, \qquad dp_i = -\varepsilon\frac{\partial G}{\partial q_i}. \tag{8.54}$$

Now, suppose that the physical system evolves in time through an interval dt, giving with the aid of the canonical equations

$$dq_i = \dot{q}_i\, dt = \frac{\partial H}{\partial p_i} dt,$$

$$dp_i = \dot{p}_i\, dt = -\frac{\partial H}{\partial q_i} dt. \tag{8.55}$$

A quick comparison of eqs. (8.54) and (8.55) shows that the parameter ε can be identified with dt, and G with the Hamiltonian. Regarding G as the "effective generator" of the infinitesimal transformation, it is then often stated that H generates the motion of the system in time. Since the motion at time $t + dt$ is determined from that at time t, a further succession of such transformations will determine the motion at a *finite* time interval after the initial moment. The motion evolves from a canonical transformation unfolding continuously in time and is generated by the Hamiltonian!

In this light, the result (8.8),

$$\frac{dF}{dt} = [F, H] + \frac{\partial F}{\partial t},$$

is not surprising. The evolution of the system motion is determined by H, and this relation shows how a particular F develops in time through the influence of H; as a consequence, (8.8) is sometimes called **the unfolding theorem**. It turns out that there is a neat practical way of solving dynamical problems by means of this theorem. Expanding $q(t)$ and $p(t)$ as functions of time about their initial values q_0 and p_0, we have

$$q(t) = q_0 + \left(\frac{dq}{dt}\right)_{t=0} t + \left(\frac{d^2q}{dt^2}\right)_{t=0} \frac{t^2}{2!} + \cdots, \tag{8.56}$$

$$p(t) = p_0 + \left(\frac{dp}{dt}\right)_{t=0} t + \cdots.$$

According to (8.8),

$$\frac{dq}{dt} = [q, H], \qquad \frac{d^2q}{dt^2} = \frac{d}{dt}\frac{dq}{dt} = [[q, H], H],$$

and so on, with similar results for the p-derivatives. It follows that

$$q(t) = q_0 + [q, H]_{t=0} t + [[q, H], H]_{t=0} \frac{t^2}{2!} + \cdots. \tag{8.57}$$

Defining the *classical operator* T by

$$T \equiv \frac{d}{dt} \equiv [\, , H] + \frac{\partial}{\partial t} = \sum_i \left(\frac{\partial H}{\partial p_i} \frac{\partial}{\partial q_i} - \frac{\partial H}{\partial q_i} \frac{\partial}{\partial p_i} \right) + \frac{\partial}{\partial t}, \qquad (8.58)$$

it is clear that

$$q(t) = q_0 + t(T_0 q_0) + \frac{t^2}{2!}(T_0^2 q_0) + \cdots \equiv (e^{tT_0}) q_0, \qquad (8.59)$$

where the subscripts zero indicate that the derivatives must finally be evaluated at time zero [$p(t)$ can be treated in the same way]. The exponential e^{tT_0} is known as the **time development operator**; it guides the evolution of $q(t)$ in time from its initial value q_0. All of this theory may seem more abstract and complicated than it really is, as the following application should make clear.

Example 6 Harmonic Oscillator
Given a Hamiltonian $H = p^2/2m + \frac{1}{2}m\omega^2 q^2$. Find the solution $q(t)$ by means of the time development operator.

Solution
From (8.58), we have

$$T = \frac{\partial H}{\partial p} \frac{\partial}{\partial q} - \frac{\partial H}{\partial q} \frac{\partial}{\partial p} = \frac{p}{m} \frac{\partial}{\partial q} - m\omega^2 q \frac{\partial}{\partial p},$$

hence $Tq = p/m$, $T^2 q = -\omega^2 q$, $T^3 q = -\omega^2 p/m$, $T^4 q = \omega^4 q$, and so on. These expressions are all to be evaluated at time zero, then substituted into (8.59) to yield

$$q(t) = q_0 + t\frac{p_0}{m} + \frac{t^2}{2!}(-\omega^2 q_0) + \frac{t^3}{3!}\left(-\omega^2 \frac{p_0}{m}\right) + \cdots$$

$$= q_0 \left[1 - \frac{\omega^2 t^2}{2!} + \cdots \right] + \frac{p_0}{m\omega}\left[\omega t - \frac{\omega^3 t^3}{3!} + \cdots \right]$$

$$= q_0 \cos \omega t + \frac{p_0}{m\omega} \sin \omega t,$$

with a similar development for $p(t)$. Where the series expansions are familiar ones (as these), or terminate rapidly, the solutions can be given in a simple closed form.

8.8 TIME-DEPENDENT PERTURBATION THEORY

The methods outlined so far in this book have provided a number of ways in which *exact* solutions for mechanics problems can be obtained, even where the problems exhibit a considerable degree of complexity. In basic research and in practical applications, however, life is rarely so simple—most realistic problems can only be solved in some approximate sense. From Newton's pioneering work in celestial mechanics to the modern space physicist's computerized orbital calculations, a form of perturbation theory has often been used in an attempt to find approximate solutions to complicated equations of motion.

It is beyond the scope of the present text to consider modern perturbation approaches in detail, but we should like to consider classical time-dependent perturbation theory in the Hamilton–Jacobi formulation as one example which can be worked out rather simply in theory. The classical development proceeds along much the same lines as for the corresponding treatment in quantum mechanics, with which the reader may be familiar.

The Hamiltonian for the problem is split into two parts:

$$H(q, p, t) = H_0(q, p, t) + H'(q, p, t). \quad (8.60)$$

The decomposition must be such that the first term H_0 corresponds to a motion which can be solved exactly, perhaps by means of the HJ technique specified in (8.43) and (8.44). The remainder H' is known as the **perturbing term**, and its inclusion in the full H makes a complete solution difficult (if not impossible) to obtain.

In accordance with the prescription of the HJ method, consider a generator F_2 or $S(q, \alpha, t)$ which determines a canonical transformation that is applied to the unperturbed Hamiltonian H_0, taking it into a new Hamiltonian K_0 that is identically zero by definition:

$$K_0 = H_0 + \frac{\partial S}{\partial t} \equiv 0.$$

The transformed canonical equations are therefore given by

$$\dot{Q}_i = 0, \quad \dot{P}_i = 0,$$

or

$$Q_i = \beta_i, \quad P_i = \alpha_i,$$

where, as before,

$$Q_i = \frac{\partial S}{\partial \alpha_i}, \quad p_i = \frac{\partial S}{\partial q_i}.$$

Direct calculation of S (which includes the term $-\alpha_1 t$) from separation of variables is followed by evaluation of $\partial S/\partial \alpha_i$, which when equated to the

Time-Dependent Perturbation Theory 313

constant β_i eventually leads to the solution of the problem. All of this is familiar: the new coordinates and momenta are the constants of the motion β_i, α_i.

If now the full H is formed by adding the perturbation H' to H_0, the new Hamiltonian K will *no longer vanish* under the effect of the *same* canonical transformation applied to H_0. (Recall that a particular transformation is either canonical or it isn't! The form of the Hamiltonian is irrelevant, so we can again transform through the same generator as before.) That is,

$$K(\alpha, \beta, t) = H_0 + H' + \frac{\partial S}{\partial t} = \left(H_0 + \frac{\partial S}{\partial t}\right) + H' = H', \qquad (8.61)$$

using the K_0 relation. Then, the canonical equations for the perturbed problem become

$$\dot{\alpha}_i = -\frac{\partial H'}{\partial \beta_i}, \qquad \dot{\beta}_i = \frac{\partial H'}{\partial \alpha_i} \qquad (8.62)$$

$$\left(\text{using } \dot{P}_i = \frac{-\partial K}{\partial Q_i}, \qquad \dot{Q}_i = \frac{\partial K}{\partial P_i}\right).$$

To employ (8.62), one must first express H' in terms of the α_i, β_i; this is achieved by means of the transformation equations *already derived* in solving the H_0 problem. In the full perturbed problem, however, eq. (8.62) tells us that, in general, α_i, β_i are *no longer constant* (except in those cases where β_i, α_i turn out to be ignorable in H')! The technique is therefore often called the **variation of constants** method. But nothing may have been gained as yet, for the exact eqs. (8.62) can often be as hard to solve as the original canonical equations were in the old coordinates.

The perturbation treatment now makes the assumption that H' is small, so that (hopefully) α_i, β_i will vary slowly in time. Therefore, it seems reasonable to take the initial unperturbed constant values of α_i, β_i at the time zero on the right of (8.62) as a zeroth-order approximation to α_i, β_i:

$$(\dot{\alpha}_i)_1 = \left(-\frac{\partial H'}{\partial \beta_i}\right)_0, \qquad (\dot{\beta}_i)_1 = \left(\frac{\partial H'}{\partial \alpha_i}\right)_0. \qquad (8.63)$$

(The outer subscripts indicate the order of the approximation; clearly, the differentiation must be performed before the constant values are inserted on the right.) The functions of time can often be integrated without difficulty, leading to α_i, β_i in first order. Substitution of these expressions on the right of (8.62) then gives α_i, β_i to second order, and the process can be continued as far as one wishes to go. Upon reaching satisfactory values for α_i, β_i to some order, the inverse canonical transformation will yield a solution in the original variables as a function of time. An example should greatly clarify the procedure.

Example 7 Perturbation Approach in One-Dimensional Motion

Consider a mass moving along the x-axis and subject to a force proportional to its x-coordinate:

$$m\ddot{x} = kx.$$

The equation is not difficult to solve exactly by any of several methods; the solution, in fact, is just $x = \sinh\sqrt{k/m}\, t$ if the initial conditions $(x)_0 = 0$, $(\dot{x})_0 = \sqrt{k/m}$ are selected. But let us attempt to solve it by the perturbation approach.

Solution

The full Hamiltonian is

$$H = \frac{p^2}{2m} - \frac{1}{2}kx^2.$$

Take $H_0 = p^2/2m$ for the free particle as the unperturbed part of H, and $H' = -\frac{1}{2}kx^2$ as the perturbation. First, the free particle problem must be solved by the HJ method, which of course is not difficult. The HJ equation is

$$H_0 + \frac{\partial S}{\partial t} = 0 = \frac{1}{2m}\left(\frac{\partial S}{\partial x}\right)^2 + \frac{\partial S}{\partial t}.$$

With $\partial S/\partial t = -\alpha$, this relation becomes

$$\frac{\partial S}{\partial x} = \sqrt{2m\alpha},$$

so

$$S = \sqrt{2m\alpha}\, x - \alpha t.$$

Then

$$\frac{\partial S}{\partial \alpha} = \beta = \sqrt{\frac{m}{2\alpha}}\, x - t,$$

or

$$x = \sqrt{\frac{2\alpha}{m}}\,(t + \beta) \tag{8.64}$$

is the free particle solution. The initial conditions dictate that $(\alpha)_0 = k/2$, $(\beta)_0 = 0$ are the unperturbed values of the constants for the free particle or the initial values for the perturbed problem.

Making use of these results, the perturbation Hamiltonian is

$$H' = -\frac{1}{2}kx^2 = -\frac{k\alpha}{m}(t + \beta)^2$$

when expressed as a function of the transformed canonical variables. From (8.62), we now have

$$\dot{\alpha} = -\frac{\partial H'}{\partial \beta} = \frac{2k\alpha}{m}(t+\beta),$$

$$\dot{\beta} = \frac{\partial H'}{\partial \alpha} = -\frac{k}{m}(t+\beta)^2. \qquad (8.65)$$

It is, in fact, possible to solve these equations exactly for this relatively simple problem.

But since we are really trying to present a demonstration of the perturbation approach, let us attack this problem once more in that way. Substituting $(\alpha)_0$ and $(\beta)_0$ on the right of (8.65), we obtain the first-order approximations

$$(\dot{\alpha})_1 = \frac{k^2}{m}t, \qquad (\dot{\beta})_1 = -\frac{k}{m}t^2.$$

By straightforward integration,

$$(\alpha)_1 = \frac{k^2}{2m}t^2 + (\alpha)_0 = \frac{k}{2}\left(1 + \frac{k}{m}t^2\right),$$

$$(\beta)_1 = -\frac{k}{3m}t^3 + (\beta)_0 = -\frac{k}{3m}t^3.$$

When these expressions are substituted on the right of (8.65), second-order approximations $(\alpha)_2$ and $(\beta)_2$ can be found at once, and so on. The solution to first order is

$$(x)_1 = \sqrt{\frac{2(\alpha)_1}{m}}(t + (\beta)_1) = \sqrt{\frac{k}{m}}\sqrt{1 + \frac{k}{m}t^2}\left(t - \frac{k}{3m}t^3\right)$$

$$\approx \sqrt{\frac{k}{m}}\left(1 + \frac{k}{2m}t^2\right)\left(t - \frac{k}{3m}t^3\right) \approx \sqrt{\frac{k}{m}}\,t + \frac{1}{3!}\left(\sqrt{\frac{k}{m}}\,t\right)^3.$$

As expected, the first two terms of the expansion are the famililar ones from the sinh expansion; perturbation theory would have to be carried out to higher orders in order for further terms in the series to have the correct coefficients, however.

In concluding, we stress that we have been discussing time-dependent perturbation theory, which examines the way that the unperturbed constants α and β vary with time under a perturbation. On the other hand, there is also a time-independent theory which looks at those parameters that are constants of the motion for the perturbed Hamiltonian. The treatment of the latter is algebraically lengthy and complicated, so we can only refer the interested reader to a more specialized reference.[3]

---

[3] See the second edition of H. Goldstein's *Classical Mechanics*, Addison-Wesley, Reading, Mass., 1980, p. 515.

PROBLEMS

8.1 Show in general that the γ matrix, as defined in eq. (8.2), is antisymmetric and orthogonal and has unit determinant.

8.2 Prove the Jacobi identity of (8.14) by direct calculation of the Poisson brackets.

8.3 Show that the Poisson bracket theorem, when applied to the variables q and p, will hold for the oscillator eqs. (8.24) but not for eqs. (8.25). Then, find in each case the differential equation satisfied by q as a function of time.

8.4 Show that the transformation given by

$$Q_1 = q_1^2 + \lambda^2 p_1^2,$$

$$Q_2 = \frac{1}{2\lambda^2}[q_1^2 + q_2^2 + \lambda^2(p_1^2 + p_2^2)],$$

$$P_1 = \frac{1}{2\lambda}\left[\arctan\left(\frac{q_2}{\lambda p_2}\right) - \arctan\left(\frac{q_1}{\lambda p_1}\right)\right],$$

$$P_2 = \lambda \arctan\left(-\frac{q_2}{\lambda p_2}\right),$$

where λ = constant, is canonical.

8.5 An inverse canonical transformation is given by

$$q_1 = \sqrt{\frac{2Q_1}{A}}\cos P_1 + \sqrt{\frac{2Q_2}{B}}\cos P_2,$$

$$q_2 = -\sqrt{\frac{2Q_1}{A}}\cos P_1 + \sqrt{\frac{2Q_2}{B}}\cos P_2,$$

$$p_1 = \sqrt{\frac{AQ_1}{2}}\sin P_1 + \sqrt{\frac{BQ_2}{2}}\sin P_2,$$

$$p_2 = -\sqrt{\frac{AQ_1}{2}}\sin P_1 + \sqrt{\frac{BQ_2}{2}}\sin P_2,$$

with A and B as constants. If the Hamiltonian is

$$H = p_1^2 + p_2^2 + \frac{1}{8}A^2(q_1 - q_2)^2 + \frac{1}{8}B^2(q_1 + q_2)^2,$$

find K. Once this has been obtained, integrate the "new" canonical equations to find the solutions as functions of time for the "old" coordinates and momenta.

8.6 A Hamiltonian is of the form

$$H = C_2 q_2^2 - C_1 q_1^2 + q_1 p_1 - q_2 p_2,$$

where C_1 and C_2 are constants. Using Poisson bracket methods, show that

$$f_1 \equiv q_1 q_2 \quad \text{and} \quad f_2 \equiv \frac{1}{q_1}(p_2 - C_2 q_2)$$

are constants of the motion. Show also that $[f_1, f_2]$ is a constant of the motion. Is H itself a constant? Check your results by finding $q_1, q_2, p_1,$ and p_2 as explicit functions of time.

8.7 Consider an oscillator with Hamiltonian

$$H = \frac{p^2}{2m} + \frac{1}{2}m\omega^2 q^2$$

and a generating function

$$F_1 = \frac{1}{2}m\omega q^2 \cot Q.$$

Derive the equations of the transformation and show that the expected solution can be obtained easily from them. Then express the generator in turn as types F_2, F_3, and F_4 and show that for each of these types the correct transformation relations are readily derived.

8.8 Consider again the inverse cube central field problem discussed in Chapter 4. Show that the problem can be solved in a straightforward manner by making a canonical transformation such that

$$Q \equiv (2mE - p^2)q^2,$$

$$P \equiv \frac{1}{2q\sqrt{2mE - p^2}} \arcsin\left(\frac{p}{\sqrt{2mE}}\right).$$

Also find a generator of type 3 and the new Hamiltonian K.

8.9 In the HJ treatment of the damped oscillator, it was stated that the quantity

$$K = \frac{p^2}{2m} + \frac{kq^2}{2} + \frac{R}{2m}qp$$

was a constant of the motion. Express K in terms of the original coordinates x and p_x, then show directly that it is constant.

8.10 Three springs, each of constant k, are connected by two masses m in a straight line between supporting walls. (See Figure 8.2) Letting q_1 and q_2 be the displacements of the masses from equilibrium, write the Hamiltonian and show that the HJ equation is not separable in these coordinates. Then, make an appropriate transformation and solve the problem by the HJ techniques.

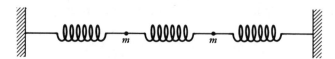

Figure 8.2 Coupled masses.

8.11 Solve the HJ equation for the problem of a mass m moving in an attractive inverse cube central field of force.

8.12 Work out the action and angle variables as well as the frequency for the simple harmonic oscillator with Hamiltonian

$$H = \frac{p^2}{2m} + \frac{1}{2}kq^2.$$

8.13 Derive the action variables and also the frequencies of motion for the coupled masses of Problem 8.10.

8.14 A projectile of weight mg moves through a trajectory in the xy-plane. Assuming arbitrary initial conditions, use the time development operator to find x, y, p_x, and p_y as functions of time.

8.15 A straight wire rotates with uniform angular velocity ω in a vertical plane, one end being fixed at the origin. If a bead of weight mg slides on the wire without friction, use the Lagrangian to find the radial position of the bead as a function of time, and compare your result with that of the time development operator. (Assume that $r = r_0$, $\dot{r} = v_0$ at time zero.)

8.16 Consider again the familiar problem of a particle falling under gravity. Treating the gravitational potential as a perturbation on free particle motion, show that the eqs. (8.62) of time-dependent perturbation theory can be solved exactly to yield the usual solution. What do these equations give in first-order approximation?

8.17 Solve eqs. (8.65) exactly for the given initial conditions, showing that the expressions obtained for $\alpha(t)$ and $\beta(t)$ lead to the hyperbolic solution for x in Example 7.

Chapter Nine
THE THEORY OF RELATIVITY

9.1 INTRODUCTION

A modern textbook on mechanics would be incomplete without a discussion of the modifications in the classical theory brought about by the development of relativity early in this century. We outline the essential parts of that theory in this chapter; the reader who needs to have further details should consult one of the specialized books listed in the reference section. (Note that the excellent literature now available, and the greater mathematical sophistication of the modern student, have dispelled much of the mystery once inherent in the theory of relativity.)

We begin with what is now called the **special theory of relativity**, proposed by Einstein in 1905. The traditional approach considers two systems of coordinates in uniform motion relative to one another; each frame will be designated as a CS for short (Figure 9.1). As an added convenience, we shall choose the primed CS to be moving with the relative velocity v along the x-axis of the unprimed frame. By this statement, one should *not* infer that the xyz-system is "stationary" with respect to the Earth, the sun, the distant fixed stars, or anything else, and that the primed CS is *really* the one which is moving. Instead, the situation is simply that the primed frame is in motion *relative* to the unprimed, and that is where the theory derives its name. The two coordinate systems are the relativistic analogues of inertial frames in the Newtonian scheme.

If it is assumed that the two frames have coordinate axes that remain parallel and are exactly coincident at time zero, then the classical relation between the two sets of coordinates is given by the Galilean transformation (1.5) of Chapter 1:

$$x' = x - vt, \quad y' = y, \quad z' = z, \quad t' = t,$$

and

$$\dot{x}' = \dot{x} - v, \quad \ddot{x}' = \ddot{x},$$

where v is the constant relative velocity along the x-axis.

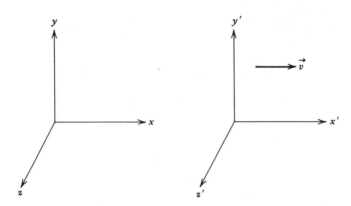

Figure 9.1 Relative motion of two coordinate frames.

Letting these coordinates represent the position of a particle of mass m as seen in the two frames, it is apparent that an observer in the primed and one in the unprimed frame will differ (by v) in their estimates of the particle velocity, but they will agree on the acceleration. In the Newtonian sense, it then follows that the force and the laws of mechanics in general will appear the same to either observer.

Unfortunately, eqs. (1.5) also imply that the velocity of a spreading wave from a light source will differ in the two frames, at least as far as the x-direction is concerned. But such a conclusion contradicts the results of the famous Michelson–Morley experiment, which showed that the velocity of light always has the constant value denoted by c in vacuum, regardless of the relative motions of source, observer, or medium. Thus, there is an inconsistency in the use of the Galilean transformation, however plausible it may seem. (Historically, Einstein may have known only vaguely of the Michelson–Morley result which is so emphasized in texts today. He was concerned instead with problems in the electrodynamics of moving bodies—the appearance of the fields in a light wave to an observer moving at speed c.)

Einstein's solution of these difficulties was bold and innovative. He chose to modify the usual concept of time, making it (as well as x) dependent upon the particular CS in question, rather than retaining it as something which flowed absolutely, the same for an observer in any system. Basically, only two assumptions were needed:

1. The laws of mechanics, indeed of *all* physics, must have the same form in two systems moving uniformly relative to one another.
2. The speed of light in vacuum must have the *same* value c in either system.

The second assumption implies that a wave equation of the form given by Maxwell's electromagnetic equations,

$$\nabla^2 f - \frac{1}{c^2} \frac{\partial^2 f}{\partial t^2} = 0,$$

where f is a component of an electric or magnetic field in a light wave, must be invariant under the desired transformation; the details are left to the problem set. The above assumptions lead at once to a set of equations known as the **Lorentz transformation**, which will be derived in the next section. Strangely enough, it turns out that the mechanical laws of Newton, invariant under the Galilean transformation, are *not* preserved under the Lorentz transformation. At speeds small compared with that of light, however, the Newtonian laws are recovered in their usual form from the equations of relativity.

It is the fact that the speeds of objects perceived in everyday life are so small relative to c that often makes relativity run counter to our common sense; the Galilean transformation "feels" correct intuitively, and indeed it is roughly correct for the speeds of our experience. Paradoxes arise easily in the theory as a result of this limited perception; we deal with some of them later.

9.2 THE LORENTZ TRANSFORMATION

When an occurrence of physical significance takes place at a given location in space (ideally at a point), one speaks in relativity of an **event**; such a happening must also be characterized by a particular value of the time coordinate, for other events may occur at that location before or after the one in question. Thus it is that a particle or a person moves continuously along a trajectory, called a **world line**, in a "four-dimensional space–time continuum," which merely means that three spatial coordinates and one time coordinate are required to specify the location of an event. But time is still unidirectional; it flows toward the future, not the past. In no sense does relativity put time on an equal footing with a spatial coordinate along which one can move either forward or backward.

Nevertheless, our knowledge of orthogonal transformation theory makes it very advantageous *mathematically* to define a fourth coordinate x_4 with dimensions of length but proportional to time:

$$x_1 \equiv x, \quad x_2 \equiv y, \quad x_3 \equiv z, \quad x_4 \equiv ict. \tag{9.1}$$

The imaginary unit in the definition of x_4 is suggested by the fact that the equation of a spherical light wave front, emanating from a point source at the origin at time zero, is just

$$x^2 + y^2 + z^2 = c^2 t^2,$$

or

$$\sum_{\alpha=1}^{4} x_\alpha^2 = 0.$$

(The fourfold summation of relativity is traditionally indicated by a Greek subscript.) Furthermore, if we let a primed system move along the x-axis as in Figure 9.1 and have its origin initially coincident with the unprimed origin, postulate 2 assures us that

$$\sum_\alpha x_\alpha^2 = 0 = \sum_\alpha (x'_\alpha)^2. \tag{9.2}$$

In the complex four-dimensional space time, known as **Minkowski space**, the squared length of the position vector is zero in either frame for points on the wave front. Alternatively, in particle language we might say that the position vector for a photon has zero magnitude. However, the important point is not that the magnitude is zero but rather that it is an invariant under transformation to the primed frame. Later, the notion of invariance will be used repeatedly in connection with several different vectors.

With the choice which we have adopted for x_4, we have produced a situation formally analogous to that of Chapter 5, which lies at the heart of the rotation theory developed there. Indeed, all the machinery of Chapter 5 is still intact, the only difference being that here we must sum over four indices instead of three, the vectors here being designated as **four-vectors** because of their four components (and transformation laws to be discussed later). In particular, a linear transformation in four-space of the form (5.4)

$$x'_\alpha = \sum_{\beta=1}^{4} a_{\alpha\beta} x_\beta \tag{9.3}$$

leads to orthogonality conditions on the a-coefficients exactly as in eq. (5.5).

We assume that the transverse coordinates y and z are unaffected by the transformation (with relative motion along x), as was the case for the Galilean relations (1.5); this assumption is warranted by the isotropy of space. For example, marks traced by the moving body with a separation of 1 cm along the y-direction in the unprimed frame must also be 1 cm apart in the primed CS. If this were not true, the right–left symmetry of space would be violated.

The Lorentz Transformation

It then follows that the rotation matrix must have the form

$$\mathbf{A} = \begin{pmatrix} a_{11} & 0 & 0 & a_{14} \\ 0 & 1 & 0 & 0 \\ 0 & 0 & 1 & 0 \\ a_{41} & 0 & 0 & a_{44} \end{pmatrix}. \tag{9.4}$$

The orthogonality conditions now produce

$$a_{11}^2 + a_{14}^2 = 1, \qquad a_{41}^2 + a_{44}^2 = 1, \qquad a_{11}a_{41} + a_{14}a_{44} = 0.$$

In addition, as the primed origin moves along the x-axis, its position is given by

$$x_1 = vt = -\frac{iv}{c}(ict) \equiv -i\beta x_4, \tag{9.5}$$

where $\beta \equiv v/c$. But for this origin, we can also write, from (9.3),

$$x_1' = 0 = a_{11}x_1 + a_{14}x_4 = (-i\beta a_{11} + a_{14})x_4.$$

Hence we have

$$a_{14} = i\beta a_{11},$$

and from the first of the orthogonality relations,

$$a_{11}^2(1 - \beta^2) = 1, \qquad \text{or} \qquad a_{11} = \gamma,$$

with

$$\gamma \equiv \frac{1}{\sqrt{1 - \beta^2}}. \tag{9.6}$$

It then follows immediately that

$$a_{14} = i\beta\gamma, \qquad a_{41} = -i\beta a_{44}, \qquad a_{44} = \gamma, \qquad a_{41} = -i\beta\gamma,$$

where we have employed the second and third orthogonality conditions. The Lorentz matrix (9.4) is now

$$\mathbf{A} = \begin{pmatrix} \gamma & 0 & 0 & i\beta\gamma \\ 0 & 1 & 0 & 0 \\ 0 & 0 & 1 & 0 \\ -i\beta\gamma & 0 & 0 & \gamma \end{pmatrix}, \tag{9.4'}$$

and the transformation (9.3) becomes

$$x_1' = \gamma(x_1 + i\beta x_4),$$

or

$$x' = \gamma(x - vt),$$
$$y' = y, \quad z' = z, \quad (9.7)$$

and

$$t' = \gamma\left(t - \frac{v}{c^2}x\right),$$

from

$$x_4' = \gamma(-i\beta x_1 + x_4),$$

as desired. When $v \ll c$, $\beta \simeq 0$, $\gamma \simeq 1$, the Lorentz transformation reduces to that of the Galilean equations, (1.5).

Furthermore, we can invert the set (9.7) as an application of postulate 1 by viewing the situation from the standpoint of a primed observer for whom the unprimed axes are moving at velocity $-v$ along the x'-axis. It follows that

$$x = \gamma(x' + vt'), \quad y = y', \quad z = z', \quad t = \gamma\left(t' + \frac{v}{c^2}x'\right) \quad (9.8)$$

are the inverse relations.

Certain simple examples of the use of (9.7) or (9.8) now follow easily. Consider a rigid rod of length $l = x_2 - x_1$ at rest along the unprimed x-axis. To an observer in the primed frame, the rod is in motion at speed v along the negative x'-axis; therefore, to measure its length accurately, the observer and a cohort must agree to note the two end coordinates of the rod at a common time t'. Such simultaneous measurements assume that all primed clocks are properly synchronized with one another. Equations (9.8) yield

$$l = x_2 - x_1 = \gamma(x_2' + vt') - \gamma(x_1' + vt') = \gamma(x_2' - x_1'),$$

or

$$x_2' - x_1' \equiv l' = \sqrt{1 - \beta^2}\, l. \quad (9.9)$$

The length l' appears shortened to the primed observers, for whom the rod is in motion. This is a real physical effect, not an illusion of some sort; it is known as

The Lorentz Transformation

the **Lorentz–Fitzgerald contraction**. The unprimed observer would argue that the shortened result was obtained because the two moving experimentalists did not *simultaneously* record the end positions of the rod. Since time is no longer absolute but relative to the CS, observers in different frames cannot even agree on the simultaneity of events, a point to which we shall return.

Next, consider a clock fixed at the primed origin. It has x-coordinates in the two frames of

$$x' = 0, \quad x = vt.$$

A number of synchronized clocks are placed along the x-axis, and are at rest in the unprimed CS. As each in turn is passed by the primed clock, readings of the two instantaneously coincident clocks are taken. From (9.7), the difference of two successive readings of the "moving" clock is

$$t'_2 - t'_1 = \gamma\left[(t_2 - t_1) - \frac{v}{c^2}(x_2 - x_1)\right]$$

$$= \gamma(t_2 - t_1)\left[1 - \frac{v}{c^2}v\right] = \sqrt{1 - \beta^2}\,(t_2 - t_1), \qquad (9.10)$$

using the fact that $x = vt$. When a time interval of 1 hour elapses on the unprimed set of clocks, only $\sqrt{1 - \beta^2}$ hour has passed on the primed. Thus, we see the basis for the statement that a "moving clock runs slow."

Before proceeding with further examples, we wish to comment somewhat more fully on the mathematical aspects of the Lorentz transformation. The xt-submatrix of the Lorentz matrix (9.4') is

$$\begin{pmatrix} \gamma & i\beta\gamma \\ -i\beta\gamma & \gamma \end{pmatrix},$$

which is to be compared with the xy-submatrix of (5.17):

$$\begin{pmatrix} \cos\theta & \sin\theta \\ -\sin\theta & \cos\theta \end{pmatrix}.$$

Upon identification of submatrix elements, producing

$$\cos\theta = \gamma, \quad \sin\theta = i\beta\gamma, \qquad (9.11)$$

we conclude that the rotation angle θ (for the rotation in the xt-plane) must be purely imaginary.

This result is a natural consequence of our rotational analogy in the complex Minkowski space. Some writers prefer instead to use a real angle of rotation χ, defining $\chi \equiv -i\theta$. Note (see Appendix D) that $\cos\theta = \cos i\chi = \cosh\chi$, and $\sin\theta = \sin i\chi = i\sinh\chi$, whereupon

$$\cosh\chi = \gamma, \qquad \sinh\chi = \beta\gamma, \qquad (9.12)$$

and the submatrix becomes

$$\begin{pmatrix} \cosh\chi & i\sinh\chi \\ -i\sinh\chi & \cosh\chi \end{pmatrix}.$$

Both these approaches are frequently seen in the literature.

In addition, one often encounters the notation

$$x_0 \equiv ct \qquad (9.13)$$

used as an alternative to our coordinate x_4. The motivation is much the same as for the angle χ; with x_0, all the coordinates are real, and no complex space is needed. We prefer to employ x_4, however, so that the analogy to Chapter 5 is maintained, as seen in (9.2).

Suppose we now consider a matrix for a rotation in the xy-plane by angle θ, that is, a matrix such as (5.17), but write it in the four-space form with $\beta = 0$, $\gamma = 1$, giving

$$\begin{pmatrix} \cos\theta & \sin\theta & 0 & 0 \\ -\sin\theta & \cos\theta & 0 & 0 \\ 0 & 0 & 1 & 0 \\ 0 & 0 & 0 & 1 \end{pmatrix}.$$

Letting this matrix act on the column four-vector **x** so as to produce **x**′, we find once more that the invariance property is satisfied:

$$\sum_{\alpha=1}^{4} x_\alpha^2 = \sum_{\alpha=1}^{4} x_\alpha'^2.$$

The matrix is orthogonal in Minkowski space, meeting all the mathematical conditions needed for a Lorentz transformation. An ordinary rotation is therefore a special case of the Lorentz transformation. In performing classical calculations, we do not normally think of rotations in this sense, of course, but here the framework is four-dimensional (even though the time coordinate does not enter significantly when β is zero), and many different special cases are possible.

By contrast, the Lorentz matrix (9.4′) corresponds to a pure axial translation with no rotation. In the literature, this traditional form of the Lorentz transformation has recently been called a **boost**. In general, a Lorentz transformation is a product of a boost and a spatial rotation. (It is assumed, by analogy with Chapter 6, that we have a *proper* Lorentz matrix with determinant $+1$, which unfolds continuously from the unit matrix.)

It is clear that two boosts along a given direction are equivalent to a third in that direction (see the problem set). But if two successive boosts are in different directions, the equivalent transformation involves rotation. A further generalization of the Lorentz transformation and a discussion of some of its group properties are given in Appendix F.

9.3 THE BREHME DIAGRAM; CAUSALITY

The solution of many complex one-dimensional relativistic problems is facilitated through the introduction of a particular graphic approach by means of the **Brehme** (or **Lorentz**) **diagram**. The rod and clock results of the preceding section are easily visualized in this fashion. Owing to the fact that both time and x-coordinates are needed in each of two frames of reference, the Brehme diagram utilizes four axes placed in a symmetrical manner, an angle α separating both the x-axes and the time axes. As Figure 9.2 illustrates, for either CS we are dealing with a skewed system as discussed in Appendix E. The various components are measured by dropping perpendiculars to the axes (the corresponding diagram in which parallel components are selected instead is known as a **Loedel diagram**, but it has rarely been used).

In such a figure, the world line of a light beam originating at the origin O of either system at time zero must bisect the angle between either primed or

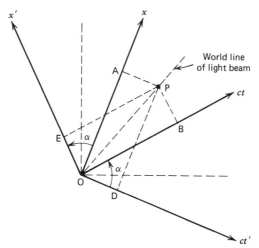

Figure 9.2 Geometry of the Brehme (Lorentz) diagram.

unprimed axes; this is clearly true, since

$$\frac{OA}{OB} = \frac{\Delta x}{c\Delta t} = 1 = \frac{OE}{OD} = \frac{\Delta x'}{c\Delta t'}, \quad \text{or} \quad \frac{\Delta x}{\Delta t} = \frac{\Delta x'}{\Delta t'} = c$$

when the coordinates of an arbitrary point P on the line are analyzed.

To understand the relation of the angle α to the relative speed of the two frames, consider an object at rest at some point in the primed CS. Its world line will naturally be perpendicular to the x'-axis, and its speed in the unprimed frame will be the same as the relative speed v. From the geometry of Figure 9.3, we see that, for points A and B on the world line,

$$\beta = \frac{v}{c} = \frac{\Delta x}{c\Delta t} = \frac{PQ}{RS} = \frac{AB\sin\alpha}{AB} = \sin\alpha. \tag{9.14}$$

Thus, as α approaches zero, the x- and x'-axes (and the time axes also) approach one another and the two frames merge, while as α goes to $\pi/2$, it is the x- and ct-axes which near one another. In view of the fact that $\beta = \sin\alpha$, we also have

$$\gamma = \frac{1}{\sqrt{1-\beta^2}} = \frac{1}{\cos\alpha} = \sec\alpha. \tag{9.15}$$

Consulting the geometry of Figure 9.3 once again, note that for the event at point B (taken as an arbitrary point),

$$x = TO + PT = OD\sec\alpha + PB\tan\alpha = x'\gamma + ct'\beta\gamma$$

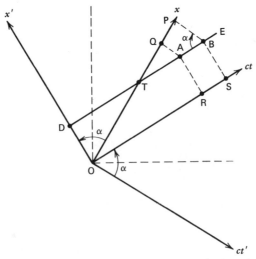

Figure 9.3 World line DE of an object at rest in the primed CS.

and

$$ct = OR + RS = DT + TB = OD \tan \alpha + PB \sec \alpha = x'\beta\gamma + ct'\gamma,$$

which are in agreement with the inverse relations (9.8). Thus, we have proved that the Brehme diagram correctly reproduces the features of the Lorentz transformation for an arbitrary event.

As a simple illustration, consider the rigid rod at rest in the unprimed system, as discussed earlier. In Figure 9.4, we give examples of Brehme diagrams for the ends of the rod in the two cases where the relative velocity of the two systems is small compared to c and where it is almost equal to c. In each case, the straight lines AB and DE are the world lines of the ends of the rod, while the points F and G are the x'-coordinates of the end points as measured simultaneously in the primed CS. This geometrical approach clearly is in accord with (9.9), with $l' = l \cos \alpha$.

Now, let us define a quantity

$$d\tau^2 \equiv -\frac{1}{c^2}\left(\sum_\beta dx_\beta^2\right) = dt^2 - \frac{1}{c^2}(dx^2 + dy^2 + dz^2). \quad (9.16)$$

$d\tau^2$ is a quantity which is an invariant under a Lorentz transformation, because it is the scalar product of the position differential (times i/c) with itself in the

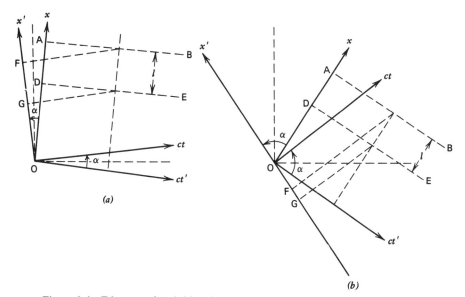

Figure 9.4 Diagrams for rigid rod of length l, with relative velocity: (*a*) small compared to c; (*b*) almost equal to c.

four-space, and we have shown in Chapter 5 that scalar products are invariants of an orthogonal transformation.

Consider a particle at rest in the primed frame of the two Lorentz frames used in the derivation of (9.7). Its squared velocity divided by c^2 will then be *identical* to β^2 from the viewpoint of an unprimed observer, who could write, from (9.16),

$$d\tau^2 = dt^2 \left\{ 1 - \frac{1}{c^2} \left[\left(\frac{dx}{dt}\right)^2 + \left(\frac{dy}{dt}\right)^2 + \left(\frac{dz}{dt}\right)^2 \right] \right\} = dt^2(1 - \beta^2),$$

or

$$d\tau = \sqrt{1 - \beta^2}\, dt. \qquad (9.17)$$

But in the primed frame, the particle is at rest; there, one obtains merely $d\tau = dt'$ either from (9.16) or from (9.10). That is, $d\tau$ is an increment of time as measured on a primed clock, one which moves with the particle. It is therefore called an **interval of proper time** and is a very important invariant.

Another invariant of significance is the square of the position four-vector difference

$$S_{21}^2 \equiv (x_2 - x_1)^2 + (y_2 - y_1)^2 + (z_2 - z_1)^2 - c^2(t_2 - t_1)^2, \qquad (9.18)$$

which connects two events with coordinates (x_1, y_1, z_1, t_1) and (x_2, y_2, z_2, t_2). It is an invariant like $d\tau^2$ because it is also a scalar product, or because S_{21} represents the magnitude of a difference vector which is not affected by rotation in four-space. Unlike a magnitude in real space, however, S_{21}^2 can be *either* positive or negative in Minkowski space (it is zero when a signal propagated at the speed of light connects the two events), depending on the relation between the time and the spatial terms. Other four-vectors have this property in common with S_{21}^2, as will be seen.

When $S_{21}^2 > 0$, the position vector is said to be **space-like**, while it is designated as **time-like** for $S_{21}^2 < 0$. But since it is an *invariant*,

$$S_{21}^2 = (x_2' - x_1')^2 + (y_2' - y_1')^2 + (z_2' - z_1')^2 - c^2(t_2' - t_1')^2$$

for any other CS moving at v relative to the unprimed system. In the case where S_{21}^2 is positive (space-like case), we can always find a CS in which $t_2' = t_1'$, that is, a frame in which the two events are simultaneous, but the events can never occur at the same position in space. On the other hand, if S_{21}^2 is negative (time-like case), a frame can be found in which the two events occur at the same location but never simultaneously. The main point is that the invariance requires the time-like or space-like nature to be an absolute; no transformation to any other system can change this basic property.

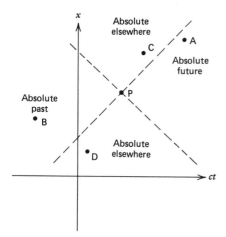

Figure 9.5 Regions of space time for an event at P.

Let us try to understand these concepts better through the diagrammatic approach. For simplicity, in Figure 9.5 we show an event occurring at P in a single system. The dashed lines through P represent light signals propagated along the two directions of the x-axis. Suppose another event takes place to the right of P, say, at A. In this region between the dashed light lines, we have

$$|x_A - x_P| < c|t_A - t_P|,$$

or that the interval between A and P is time-like. Events at A and P can never occur simultaneously in any frame. In fact, it turns out that P always occurs before A to any observer, so the region of A is called the **absolute future** (relative to P). In a similar vein, the region to the left between the dashed lines (where point B is located) is called the **absolute past**. But in the regions above and below P, at points like C and D, we find instead that

$$|x_C - x_P| > c|t_C - t_P|,$$

and so on, so the interval from P to C (or D) is space-like. Thus, in some frame, P and C may occur simultaneously, but never at the same site along the x-axis; these regions are called the **absolute elsewhere** (again relative to P).

Returning now to the usual form of the Brehme diagram, we can draw in the perpendicular dashed lines through P as before, thus separating the space-like and time-like regions (Figure 9.6). An event occurring in the shaded part of the "elsewhere" region above P must occur after P to the unprimed observer but before P to a primed observer, but they agree that it happened at a location other than that of P. In the remainder of the "elsewhere" region, the two observers can agree on the time sequence of events, but other observers in different frames than these may not agree. However, any observer in any frame will agree that the event

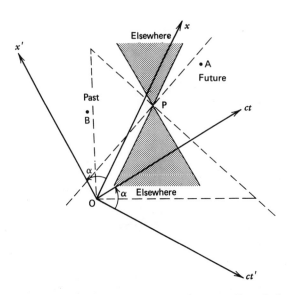

Figure 9.6 Brehme diagram demonstrating causality relations.

at A occurred after P, while that at B happened before P, so these regions to right and left are appropriately named. (To verify these conclusions, consider various diagrams as α is varied from 0 to $\pi/2$.)

In our usual scheme of causality, the cause of an event must always precede it. Thus, the event at B might be the cause of that effect produced at P, or P might be the cause of the event at A. It would certainly violate common sense if a CS could be found in which the effect preceded the cause! But relativity preserves the causality ordering in the regions of absolute future and past, as has been seen above. And these are the very regions in which a particle or signal of some type propagates from cause to effect at less than or up to the speed of light! In other words, if causality is really necessary, relativity tells us that no signal or material body can travel faster than c.

9.4 SOME PARADOXES OF RELATIVITY THEORY

Everyday experience suggests that time flows in an absolute fashion, the same in all frames; relativistic effects are not generally apparent in phenomena with velocities which are small compared to c. As a result, a number of seeming paradoxes have arisen in the literature in an attempt to discredit the theory (or perhaps to strengthen our understanding of it), and we now briefly examine some of the more famous ones.

Example 1 The Twin Paradox

At some point in the future, twin brothers consider a rocket ride to a distant star. Brother A decides to stay at home, while B makes the trip, traveling both ways at constant speed. If B's clock runs slow compared to A's on both the outgoing and return parts of the journey, and physiological processes such as aging are affected accordingly, then B should be somewhat younger than A upon their reunion. (The exact difference in age would depend on the duration of the trip and the rocket's velocity.) But if we examine this journey from B's frame of reference, as relativity apparently allows us to do, it is A who travels outward and finally returns to B. Therefore, it is A's clock which should run slow and A who would be younger upon reuniting. Both of these conclusions cannot be correct, as one brother or the other must actually be the younger if time dilation has taken place. The symmetry assumed in relativity has led to an apparent contradiction.

The key to the resolution of this paradox lies in the fact that the situation is actually asymmetric. Brother A has remained in a single inertial frame throughout the journey, whereas B has undergone accelerations at the beginning and end of the trip, as well as in turning around upon reaching the destination. The acceleration at the start and deceleration at the conclusion are not essential to the discussion; it might as well be assumed that B begins his trip from a point in the opposite direction from the eventual destination, attains full rocket speed, and compares clock readings with A at the instant he passes home on his way outward. But there is no clever way of avoiding the acceleration on turnaround—B has to go from one inertial frame on the outward journey to another on the return trip. The time required for the acceleration at midpoint may be only a small fraction of the total duration of the trip, but in view of our ignorance in dealing with the change of inertial CS, the situation can only be analyzed simply and correctly from A's vantage point. The conclusion that B should be younger upon his return is therefore the valid one. (A little more detailed analysis is really required to make this conclusion rigorous.)

For the outward portion of the journey, the situation is depicted in a Brehme diagram (Figure 9.7). The clock of B is placed at the origin of the B frame, so its world line coincides with the ct_A-axis. The time interval for the outward trip from O to the destination at point D is Δt in the A frame and $\Delta \tau$ in the B frame. These are related through (9.17):

$$\Delta \tau = \Delta t \sqrt{1 - \beta^2} = \Delta t \cos \alpha.$$

For the return trip, we must draw a second Brehme diagram, as shown in the figure, with primed axes for the B frame, which A now sees as moving in the

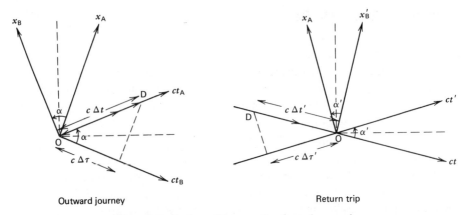

Figure 9.7 Brehme diagrams for the twin paradox.

opposite sense. Again, the time intervals are related as

$$\Delta\tau' = \Delta t' \cos \alpha',$$

allowing for the possibility of a return speed different from the outgoing speed by priming the various quantities. Neglecting the turnaround time, the total trip duration for B is

$$\Delta\tau + \Delta\tau' = \Delta t \cos \alpha + \Delta t' \cos \alpha',$$

which is always less than the corresponding interval ($\Delta t + \Delta t'$) for brother A. At the conclusion of the journey, the twins will agree that B is younger, but B will maintain that A actually aged less than B during the uniform parts of the trip, so therefore must have aged *significantly* during the brief interval required for the turnaround!

Example 2   The Pole in the Barn Paradox

Another famous paradox concerns a pole vaulter who runs rapidly toward the open doors of a barn. (Even the superb athletes of today do not approach the speed of light in sprinting, but assume for the sake of the discussion that this one does.) The pole is carried horizontally and parallel to the barn and has the same rest length l as the barn. An observer in the barn (B) frame sees the runner with pole approaching and views the pole as contracted in length according to (9.9), so believes that he can quickly shut both barn doors, trapping the runner with pole inside. But in the pole vaulter's (P) frame it is the *barn* that is approaching and contracted, so the two ends of the pole can never be inside the barn simultaneously. Which observer is correct?

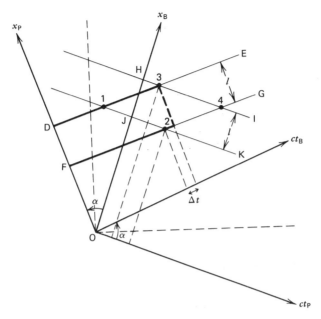

Figure 9.8 Diagram for pole vaulter and barn.

A Brehme diagram (Figure 9.8) resolves the paradox easily. The line DE is the world line of the front end of the pole and FG is that of the back end, while HI is the world line of the second door of the barn and JK that of the first doorway penetrated. Point 1 is the event wherein the pole's forward end has met the first doorway; at point 2, the rear end is at that door. Point 3 marks the meeting of the front end of the pole and second doorway of barn, while at point 4 the rear end reaches that part of the barn. Dropping perpendiculars to the time axes from points 2 and 3, we see that the pole vaulter believes event 3 occurred *before* event 2, hence the pole was at no time entirely contained within the barn. But on the contrary the barn observer thinks that event 3 came *after* event 2, whereupon the pole was entirely within the barn for a short interval Δt. *Both* observers are then correct within their own frames of reference; the apparent paradox arises because of the failure to understand that the relative order of two events in time can depend on the CS used, as discussed in the last section.

There are several variations on the general theme outlined above. In one version, a rod approaches an open manhole at high speed, the rest length of the rod being the same as the hole diameter. Will the rod fall into the hole? It might seem at first to depend upon the frame used for the analysis, as with the pole and barn, but actually the answer is *yes*. Even in the rod frame, the front end of the rod will bend under gravity as it slides over the hole, eventually dragging the entire rod into the hole. (The concept of a perfectly rigid rod cannot be

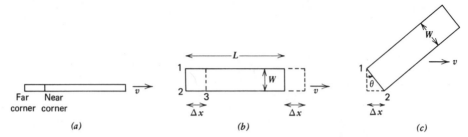

Figure 9.9 Appearance of a rapidly moving rectangular object: (*a*) viewed edge-on; (*b*) viewed from above; (*c*) apparent rotated view.

maintained in relativity, as that would always require an infinite speed of propagation for waves in an elastic medium, which is not possible with finite *c*.)

Example 3 The Appearance of an Object at High Speeds

If you are unfamiliar with the theory of relativity, ask yourself the question: "What is the appearance of a rectangular object (moving almost at *c*) when viewed edge-on in its plane along the long side?" Your answer might be: "It will look the same as if it were moving slowly, except for a contraction along the direction of motion." That answer is quite incorrect, as was pointed out by J. Terrell[1] in 1959, but don't feel badly about it—for more than half a century, any good physicist would have given the same reply!

To understand the true state of affairs, consult Figure 9.9. The object is being viewed edge-on, as in (a) of the illustration. Light leaving point 1 (far corner) and proceeding downward toward the observer in (b) requires a time $\Delta t = W/c$ to reach point 2 (near corner). But in that time, the object has moved to the right a distance

$$\Delta x \equiv v \Delta t = \beta c \Delta t = \beta W,$$

so the near corner is no longer located at point 2 but is now at point 3. The light originally from 1 now mixes with the light coming from 3 at the later instant; and with equal distances left to travel, these waves reach the observer simultaneously. The entire end of the object from 1 to 2 can then be seen, much as if the object were rotated through an angle θ as in (c). (Physically, the reason for this is that at high speeds the object moves out of the way and lets light from the end points spread toward the observer.) Clearly, the angle θ is given by

$$\sin \theta = \frac{\Delta x}{W} = \frac{\beta W}{W} = \beta.$$

The object actually appears rotated as well as Lorentz-contracted. At nonrelativistic speeds, however, the rotation angle is essentially zero and the phenomenon is not observed; near *c*, the rotation is extreme indeed (approaching 90°).

---

[1] J. Terrell, Invisibility of the Lorentz Contraction, *Phys. Rev.* **116**, 1041 (1959).

9.5 FOUR-VECTORS

In this section, we define some four-vectors which are useful in relativistic mechanics and partially explore their properties. First of all, the four quantities

$$\mu_1 \equiv \frac{dx}{d\tau}, \quad \mu_2 \equiv \frac{dy}{d\tau}, \quad \mu_3 \equiv \frac{dz}{d\tau}, \quad \mu_4 \equiv ic\frac{dt}{d\tau},$$

or

$$\mu_\alpha = \frac{dx_\alpha}{d\tau} \qquad (9.19)$$

form the components of what is called the **proper velocity**. From (9.17), we see that we have an invariant given by

$$\sum_{\alpha=1}^{4} \mu_\alpha^2 = -c^2\left(\frac{dt}{d\tau}\right)^2 + \left(\frac{dx}{d\tau}\right)^2 + \left(\frac{dy}{d\tau}\right)^2 + \left(\frac{dz}{d\tau}\right)^2$$

$$= \left(\frac{dt}{d\tau}\right)^2\left[-c^2 + \left(\frac{dx}{dt}\right)^2 + \left(\frac{dy}{dt}\right)^2 + \left(\frac{dz}{dt}\right)^2\right] = \left(\frac{dt}{d\tau}\right)^2[v^2 - c^2]$$

$$= \frac{1}{1-\beta^2}[v^2 - c^2] = -c^2,$$

where

$$v_1 \equiv \frac{dx}{dt}, \quad v_2 \equiv \frac{dy}{dt}, \quad v_3 \equiv \frac{dz}{dt} \quad \left(\text{with } v^2 = v_1^2 + v_2^2 + v_3^2\right)$$

are the usual components of velocity, which will be called the **coordinate velocity**. (A traveler who covers a prescribed distance between two marks on the x-axis would divide this distance by the time interval as measured on a clock *carried along* to get the proper velocity, but the coordinate velocity would be found by noting the time interval as given by clocks which are *fixed in place* along the route.) Again, from (9.17), it follows that

$$\mu_1 = \frac{v_1}{\sqrt{1-\beta^2}} = \gamma v_1, \quad \mu_2 = \gamma v_2, \quad \mu_3 = \gamma v_3, \quad \mu_4 = ic\gamma. \quad (9.20)$$

By applying the Lorentz transformation (9.7) to infinitesimals and then dividing by $d\tau$, we see at once that for motion along the x direction,

$$\frac{dx'}{d\tau} \equiv \mu_1' = \gamma\left(\frac{dx}{d\tau} - v\frac{dt}{d\tau}\right) = \gamma(\mu_1 + i\beta\mu_4),$$

and similarly

$$\mu'_4 = \gamma(\mu_4 - i\beta\mu_1).$$

In other words, the components of the proper velocity transform in exactly the same way as do the four components of the position vector, hence the proper velocity is a legitimate four-vector in every sense.

But suppose we had divided (9.7) by dt rather than $d\tau$:

$$\frac{dx'}{dt} \equiv v'_1 = \gamma\left(\frac{dx}{dt} - v\frac{dt}{dt}\right) = \gamma(v_1 - v).$$

Defining $v_4 \equiv ic$ as the fourth component of the coordinate velocity, we have

$$\frac{dx'}{dt} = \gamma(v_1 + i\beta v_4),$$

which again has the desired form. But in this case one can write

$$\sum_{\alpha=1}^{4} v_\alpha^2 = v^2 + v_4^2 = v^2 - c^2,$$

which is *not* an invariant. (In fact, it is v_4 that is invariant.) The coordinate velocity, in conclusion, does *not* have the transformation properties required of a four-vector; multiplication of each component by γ as in (9.20) gives such a vector.

It seems eminently reasonable now to define the **proper acceleration** as

$$a_\alpha \equiv \frac{d\mu_\alpha}{d\tau}. \qquad (9.21)$$

This quantity behaves as a decent four-vector should under a Lorentz transformation. We have shown that the proper velocity is time-like:

$$\sum_\alpha \mu_\alpha^2 = -c^2.$$

Differentiating both sides with respect to τ, one obtains

$$\sum_\alpha 2\mu_\alpha \frac{d\mu_\alpha}{d\tau} = 0, \quad \text{or} \quad \sum_\alpha \mu_\alpha a_\alpha = 0.$$

Since this relation is valid in any frame of reference, we choose to evaluate it in the rest frame of the particle being discussed. In that CS, we have $\mu_1 = \mu_2 = \mu_3 = 0$, and so $\mu_4 a_4 = 0$. But $\mu_4 \neq 0$, therefore $a_4 \equiv 0$; hence, in this frame

$$\sum_\alpha a_\alpha^2 = (a_1^2 + a_2^2 + a_3^2) \geq 0$$

Four-Vectors

is the result. *All* acceleration components generally will not vanish, even in a rest situation; thus, we conclude that the proper acceleration is a space-like four-vector, if not zero.

The four-component force (often called the **Minkowski force**) is defined as

$$\mathscr{F}_\alpha \equiv m_0 \frac{d\mu_\alpha}{d\tau} = m_0 a_\alpha, \qquad (9.22)$$

where m_0 is the rest mass of the particle in question (the mass m of earlier chapters), a quantity which is a fixed parameter for a point mass with no internal structure. The Minkowski force must be space-like, inasmuch as it is proportional to the proper acceleration. Its definition was selected so that the spatial components reduce to the "ordinary" force components in the limit as $\beta \to 0$. With the aid of eqs (9.20) and (9.17), we see that these components are

$$\mathscr{F}_i = m_0 \gamma \frac{d}{dt}(\gamma v_i) \qquad (i = 1, 2, 3),$$

or

$$\sqrt{1 - \beta^2}\, \mathscr{F}_i = \frac{d}{dt}\left(\frac{m_0 v_i}{\sqrt{1 - \beta^2}}\right). \qquad (9.23)$$

Defining the time-like four-vector **momentum** as

$$p_\alpha \equiv m_0 \mu_\alpha, \qquad (9.24)$$

with spatial components $p_i = m_0 v_i / \sqrt{1 - \beta^2}$, it becomes clear that

$$\sqrt{1 - \beta^2}\, \mathscr{F}_i = \frac{dp_i}{dt} \equiv F_i, \qquad (9.25)$$

where F_i is the ith component of the **coordinate force**, defined as the regular (not the proper) time derivative of the momentum. In the limit as $\beta \to 0$, the usual Newtonian second law is obtained at once.

The momentum, however, is not quite the familiar concept of earlier chapters, for its spatial components contain the factor of γ as well as $m_0 v_i$. Over a long period of time, the traditional practice was to define a "relativistic mass"

$$m \equiv m_0 \gamma = \frac{m_0}{\sqrt{1 - \beta^2}}, \qquad (9.26)$$

so that the four-vector of (9.24) would be just the product of this mass and the coordinate velocity. In this way, the "mass" was said to increase without bound as the particle approached the speed of light. It seems much more reasonable, in accordance with the modern trend, to include the γ factor with v_i to form the proper velocity component μ_i, treating the mass as simply the invariant physical parameter m_0.

Let us now examine the fourth components of the four-vectors for force and momentum. Taking the dot product of the Minkowski force with the proper velocity gives

$$\sum_{\alpha=1}^{4} \mathscr{F}_\alpha \mu_\alpha = m_0 \sum_\alpha \frac{d\mu_\alpha}{d\tau} \mu_\alpha = \frac{1}{2} m_0 \frac{d}{dt}\left(\sum_\alpha \mu_\alpha^2\right)$$

$$= \frac{1}{2} m_0 \frac{d}{d\tau}(-c^2) = 0,$$

hence

$$\mathscr{F}_4 = \frac{-1}{\mu_4}\left(\sum_{i=1}^{3} \mathscr{F}_i \mu_i\right) = -\frac{1}{\mu_4}\left(\sum_i \gamma F_i \gamma v_i\right) = \frac{i\gamma}{c}(\vec{F}\cdot\vec{v}). \qquad (9.27)$$

From (9.22), one has

$$\frac{i\gamma}{c}(\vec{F}\cdot\vec{v}) = m_0 \frac{d}{d\tau}(ic\gamma),$$

$$\vec{F}\cdot\vec{v} = \frac{d}{dt}\frac{m_0 c^2}{\sqrt{1-\beta^2}} = \frac{dE}{dt}, \qquad (9.28)$$

which is the rate of doing work on the particle, $\vec{F}\cdot\vec{v}$ being the time rate of increase of its energy dE/dt. Thus, we have

$$E = \frac{m_0 c^2}{\sqrt{1-\beta^2}} = m_0 c^2\left[1 + \frac{1}{2}\frac{v^2}{c^2} + \cdots\right] \qquad (9.29)$$

within a constant of integration. At low v, higher terms in v^2/c^2 can be ignored; the second term is the usual kinetic energy $\frac{1}{2}m_0 v^2$, and the first term is the "rest energy" $m_0 c^2$. The energy is chosen in this form, with no additional constant, so that

$$p_4 \equiv m_0 \mu_4 = im_0 c\gamma = i\frac{m_0 c}{\sqrt{1-\beta^2}} = \frac{iE}{c} \qquad (9.30)$$

will hold for the time component of momentum. The separate laws for conservation of momentum and energy in classical mechanics become in relativity the

Four-Vectors

single law of conservation of the momentum-energy four-vector. The squared magnitude of this quantity is

$$\sum_\alpha p_\alpha^2 = p^2 - \frac{E^2}{c^2} \equiv \sum_\alpha (m_0 \mu_\alpha)^2 = m_0^2(-c^2),$$

giving

$$E^2 - p^2 c^2 = m_0^2 c^4, \qquad (9.31)$$

where p^2 is the sum of squares of the spatial components. This is the famous **momentum-energy invariant relation**, so called because the right side is always the same constant for a given particle regardless of the frame in which the left side is evaluated. As we shall see below, the equation can even be applied fruitfully to a system of particles.

It is clear that the momentum-energy four-vector has components which must satisfy the Lorentz transformation (9.7); but for convenience, we list these relations:

$$p'_x = \gamma\left(p_x - \frac{vE}{c^2}\right),$$

$$p'_y = p_y, \qquad (9.32)$$

$$p'_z = p_z,$$

$$E' = \gamma(E - vp_x).$$

The main point of this discussion will now be emphasized, lest the details of the four-vectors somewhat obscure it. Any legitimate fundamental equation of physics must be expressed in a **covariant formulation**, by which we mean that each term in the equation must be the same rank of tensor in the four-dimensional representation. Each term of the basic law might be in scalar form, or perhaps appear as a four-vector; but whatever the rank, all terms must agree in their transformation properties. If that is the case, then numerical values of the physical quantities involved may vary widely from one CS to another, but the equation will nevertheless preserve its form under a Lorentz transformation between any two frames in uniform relative motion. The covariance makes it possible for different reference frames to exhibit equivalent physical laws, in accordance with the first postulate of Einstein.

Finally, it seems appropriate to write a little about Lagrangian–Hamiltonian formulations in relativistic mechanics. Consider first the simplest situation, that of a single free particle. If we assume that Hamilton's principle is still valid, then the integral which is to have an extreme value is

$$\int_{t_1}^{t_2} L \, dt.$$

Written in this way, as an integral over time, the quantity in question is clearly dependent on the particular Lorentz frame to be used. For a truly relativistic calculation, the integral should be over the proper time instead. By taking the very simple expression of a constant k times the interval $d\tau$, we have an integral which is certainly invariant under any Lorentz transformation. But with the aid of (9.17), there is then suggested a tentative identification

$$\int k \, d\tau = k \int \sqrt{1 - (v^2/c^2)} \, dt \to \int L \, dt,$$

or

$$L = k\sqrt{1 - (v^2/c^2)}.$$

Expanding for $v \ll c$ to obtain the nonrelativistic limit, one finds

$$L \approx k\left[1 - \frac{1}{2}\frac{v^2}{c^2}\right].$$

If the approach is to have consistency in this limit, we must identify $k = -m_0 c^2$ so that

$$L = -m_0 c^2 + \frac{1}{2} m_0 v^2.$$

The constant rest energy term $-m_0 c^2$ is of no interest, for its variation immediately gives zero; the remaining term is, of course, the usual kinetic energy. From this analysis, we see that the relativistic Lagrangian for a free particle can be taken as

$$L = -m_0 c^2 \sqrt{1 - (v^2/c^2)}.$$

Generalizing to the case of a particle in a conservative force field with potential V independent of velocity, we postulate that

$$L = -m_0 c^2 \sqrt{1 - (v^2/c^2)} - V$$

will suffice. To demonstrate the correctness of this conclusion, note that Lagrange's equation for the x-coordinate becomes

$$\frac{d}{dt}\left(\frac{\partial L}{\partial \dot{x}}\right) - \frac{\partial L}{\partial x} = 0 = \frac{d}{dt}\left(\frac{m_0 \dot{x}}{\sqrt{1-(v^2/c^2)}}\right) + \frac{\partial V}{\partial x},$$

which agrees with (9.25) on setting $F_x = -\partial V/\partial x$. Furthermore, the conjugate momentum $p_x \equiv \partial L/\partial \dot{x}$ is in agreement with (9.24), and the Hamiltonian is

$$H \equiv \sum_i q_i p_i - L = \frac{m_0(\dot{x}^2 + \dot{y}^2 + \dot{z}^2)}{\sqrt{1-(v^2/c^2)}} + m_0 c^2 \sqrt{1-(v^2/c^2)} + V$$

$$= \frac{m_0 c^2}{\sqrt{1-(v^2/c^2)}} + V,$$

which, with the aid of (9.29), is seen to be the total energy of the particle in the field.

The procedure that has been followed is one that has led to the standard form of the relativistic Lagrangian for a particle, yet it is not completely satisfactory. The covariance of the approach was destroyed as soon as the potential V was introduced. Indeed, classical potential functions have always given nightmares to relativity theorists, for in the usual Newtonian sense, they imply instantaneous "action at a distance." But in the Einstein theory, no influence can propagate at a speed greater than that of light, and so the classical potential should be banned from the outset. In fact, there is a "no interaction" theorem which has been proved in recent years.[2] It states, under some general assumptions that are quite reasonable for physical systems, that there can be no direct interaction between two particles in a strictly covariant formulation, except for actual collision contact forces. As yet the true meaning of this theorem and its implications are not fully understood.

Example 4 Relativistic Elastic Collision of Two Particles

For the relativistic collision, we take particle 2 at rest in the unprimed frame before the collision, so the initial momenta along x are p_{1B} and $p_{2B} = 0$ in this CS. In the primed frame (often called the **barycentric**, or **center of momentum** frame), the corresponding momenta are initially p'_{1B} and p'_{2B}, which are equal and

---

[2] E. Sudarshan and N. Mukunda, *Classical Dynamics: A Modern Perspective*, John Wiley, New York, 1974, p. 535.

opposite by the choice of CS (see Figure 9.10):

$$p'_{1B} = -p'_{2B}.$$

After the collision, the various momenta will be designated p_{1A}, p_{2A}, p'_{1A}, p'_{2A}. (The energies will be denoted in a similar fashion.) The interaction is taken to be such that the collision is head-on and reverses the direction of the CM momenta, as shown in the figure. From the inverse of (9.32), we find that

$$E_{1B} + E_{2B} = \gamma(E'_{1B} + E'_{2B})$$

and

$$p_{1B} + p_{2B} = \frac{\gamma v}{c^2}(E'_{1B} + E'_{2B}) = \frac{v}{c^2}(E_{1B} + E_{2B}),$$

where we have used the condition $p'_{1B} + p'_{2B} = 0$. Then,

$$v = \frac{c^2(p_{1B} + p_{2B})}{(E_{1B} + E_{2B})}$$

is the relative velocity of the frames. Initially, particle 2, being at rest in the unprimed frame, has a velocity $-v$ in the primed, so that

$$p'_{2B} = -\frac{m_2 v}{\sqrt{1 - (v^2/c^2)}} = -p'_{2A}, \qquad E'_{2B} = E'_{2A} = \frac{m_2 c^2}{\sqrt{1 - (v^2/c^2)}},$$

$$p'_{1B} = -p'_{2B} = -p'_{1A},$$

where the rest masses are denoted m_1, m_2. It follows from (9.32) that

$$E_{2A} = \gamma(E'_{2A} + vp'_{2A}) = \frac{m_2 c^2 [1 + (v^2/c^2)]}{1 - (v^2/c^2)}$$

and, of course, $E_{2B} = m_2 c^2$, whereupon, from (9.31),

$$p_{2A} = \frac{\sqrt{E_{2A}^2 - m_2^2 c^4}}{c}, \qquad p_{2B} = 0.$$

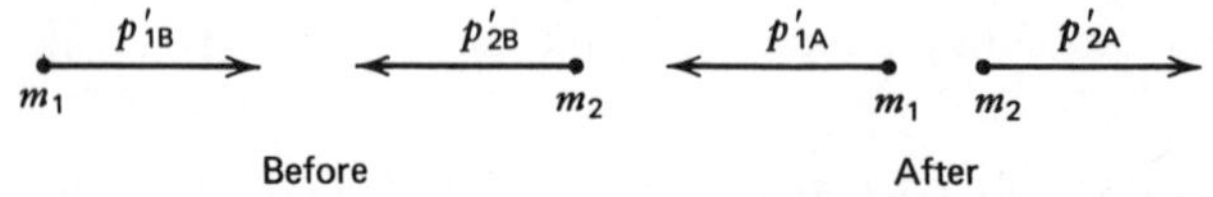

Figure 9.10 Momenta before and after collision in primed frame.

Four-Vectors

All quantities for particle 2 have now been determined as a function of v. But

$$v = \frac{c^2 p_{1B}}{E_{1B} + m_2 c^2} = c\frac{\sqrt{E_{1B}^2 - m_1^2 c^4}}{E_{1B} + m_2 c^2}$$

depends only on the incoming "projectile" energy E_{1B}. Eliminating v from the E_{2A} expression produces

$$E_{2A} = m_2 c^2 + \frac{2m_2(E_{1B}^2 - m_1^2 c^4)}{(m_1^2 + m_2^2)c^2 + 2m_2 E_{1B}}$$

after a little algebra. Energy conservation assures that

$$E_{1B} + m_2 c^2 = E_{1A} + E_{2A},$$

so that

$$E_{1A} = E_{1B} - \frac{2m_2(E_{1B}^2 - m_1^2 c^4)}{(m_1^2 + m_2^2)c^2 + 2m_2 E_{1B}}.$$

The fraction of the kinetic energy transferred to the particle at rest by the incident particle becomes

$$\frac{E_{2A} - m_2 c^2}{E_{1B} - m_1 c^2} = \frac{2m_2(E_{1B} + m_1 c^2)}{(m_1^2 + m_2^2)c^2 + 2m_2 E_{1B}} \equiv f.$$

For $m_2 \gg m_1$,

$$f \simeq \frac{E_{1B} + m_1 c^2}{E_{1B} + (m_2 c^2/2)};$$

f approaches unity when E_{1B} is large, the velocity of m_1 being comparable to c. Classically, of course, such a transfer could not take place from light to heavy particle. Nor could it occur classically for $m_1 \gg m_2$, in which case

$$f \simeq \frac{E_{1B} + m_1 c^2}{E_{1B} + (m_1^2 c^2/2m_2)},$$

but relativistically it *can* occur for sufficiently large E_{1B}.

Example 5 Antiproton Production

A high-energy particle accelerator, the bevatron, was designed some years ago to produce antiprotons (it is no longer in use). The collision of a rapidly moving proton with one at rest creates a proton–antiproton pair along with two normal protons, the reaction being

$$p + p \to p + p + p + \bar{p}.$$

In order for this reaction to occur, the two protons on the left must have at least enough energy to be equivalent to the rest masses of the four particles on the right. Calling the rest energy of a proton or antiproton m_0c^2, the energy of the incident fast particle mc^2, and the minimum required momentum p, we have from eq. (9.31):

$$(mc^2 + m_0c^2)^2 - p^2c^2 = (4m_0c^2)^2,$$

or

$$p^2c^2 = (mc^2)^2 + 2(mc^2)(m_0c^2) - 15(m_0c^2)^2.$$

Here, we have applied (9.31) to the system of particles on each side of the reaction, the right side of the original equation having been evaluated in a frame in which the four particles are produced at rest. But if instead (9.31) is applied only to the incident proton, then

$$(mc^2)^2 - p^2c^2 = (m_0c^2)^2,$$

or

$$p^2c^2 = (mc^2)^2 - (m_0c^2)^2.$$

Equating the expressions for p^2c^2 gives

$$2(mc^2)(m_0c^2) = 14(m_0c^2)^2,$$

or

$$mc^2 = 7m_0c^2.$$

The incoming proton must then have a kinetic energy of at least six times its rest energy for it to be capable of antiproton production. Inasmuch as this rest energy is 0.94 GeV, the accelerator used must produce particles with about 6 GeV kinetic energy.

9.6 THE GENERAL THEORY OF RELATIVITY

In preceding sections, we have learned how to use the Lorentz transformation in the treatment of motions as seen in two reference frames which are moving uniformly relative to one another. A key feature of the special theory of relativity is that $\sum_{\alpha=1}^{4} dx_\alpha^2$ is preserved as an invariant by this linear transformation. The implication is that it is possible to formulate the theory with a set of four orthogonal Cartesian coordinates. One then refers to the spacetime structure as "flat," or as having the **Lorentz metric**.

Einstein found it advantageous to introduce more general coordinate transformations and a "curved" space-time in order to analyze the motion of masses in a gravitational field. After a decade of intense labor, his work culminated in 1915 with the elegant geometrical theory of gravitation known as the **general theory of relativity**.

The physical reasoning behind Einstein's work is fairly straightforward, but its mathematical structure is often tedious and cumbersome. The reader who wishes to master this material thoroughly should first obtain the necessary mathematical background by consulting Appendix E. For those who desire to acquire a familiarity with the physical concepts but do not wish to examine the mathematical details, we have tried to write this chapter in such a way that the concepts can be followed even with considerable omission of mathematical steps.

We have pointed out in Chapter 1 that Newtonian mechanics presupposed the existence of absolute space, which provided a background setting for an inertial CS. We also know that all inertial frames are equivalent; one cannot say that *this* inertial frame is really at rest while *that* one is really moving. If, however, there is relative acceleration between two frames, the moving one is readily distinguished from the other by virtue of the forces manifested in it.

As an example, in Chapter 6 we noted the appearance of "fictitious" centrifugal and Coriolis forces in the equation of motion for a rotating CS. In the rotating framework, these forces are very real indeed, but they are called fictitious because a transformation to another frame (an inertial one) will eliminate them. The laws of physics are simpler in the inertial frame.

It occurred to Einstein that forces caused by a uniform gravitational field could be transformed away as readily as fictitious forces due to rotation. He demonstrated this fact with his famous illustration of an observer in an enclosed elevator located far out in space. A uniform gravitational field directed *downward* in the region of a fixed cage could not be distinguished in its effect on the observer (and other objects inside the elevator) from that produced by a uniform acceleration of the proper magnitude directed *upward*, a consideration known as the **principle of equivalence**. In either case, the observer's feet would press downward against the floor of the elevator, and the floor in turn would react upward on the observer. Alternatively, if the elevator were accelerated downward (put in free fall) along the direction of a uniform field, no pressure would result between floor and feet. Without being able to view external objects from inside, the observer would conclude ideally that neither gravity nor acceleration was present; the vital point is that the transformation to an accelerated frame has *eliminated* the force of gravity.

That the principle of equivalence should hold at all is a puzzle in the Newtonian theory, for it assumes the equality of inertial and gravitational masses. That is, the mass m that appears as the measure of inertia in Newton's second law and the mass that serves as a kind of "gravitational charge" in the law of gravity are one and the same. But why should this be? In the Einstein theory, this problem does not arise, for gravity is clearly related to accelerated motion.

Actually, the above remarks are a bit idealistic. Nature does not produce exactly uniform gravitational fields, but merely ones which approximate that behavior over a small region of space. A realistic field, such as that of the Earth, is as shown in Figure 9.11. The lines of the field are radially inward, toward the Earth's center. As a consequence, objects on either side of the elevator experience a small sidewise component of force toward the elevator center. Further, while the elevator as a whole moves down at some average rate, an object near the floor (where the field is a little stronger) has a slightly greater acceleration than one near the ceiling. Gravity can therefore be replaced by an accelerated reference frame only in a very localized region of space-time.

Einstein's original concept of the principle of equivalence may have been a giant intellectual achievement which eventually enabled him to formulate his general theory, but today, we accept uniform gravitational fields as an idealization not realized in nature. In fact, we shall see that the Riemann curvature tensor is the measure of the curvature of space-time in general relativity, and it is zero for a uniform field (but its nonzero components actually measure the strength of

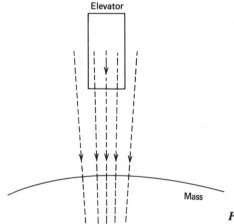

Figure 9.11 Gravitational field on the elevator near a massive sphere.

The General Theory of Relativity 349

a real field)! This means that the uniform field, which is completely equivalent to an accelerated frame, can be treated by the methods of special relativity for a flat space-time.

The modern version of the principle of equivalence might therefore be stated as follows: "At the origin of a locally inertial CS in a gravitational field, the laws of physics (excluding gravity) have the same form as in the inertial reference frames of special relativity." (The **locally inertial CS** is a freely falling and nonrotating frame such as the elevator.) A real field is never entirely transformed away, but this can be done approximately over a limited region. In any case, the principle makes it possible to replace the concept of forces on a body by that of "natural motion" along a curved trajectory.

Perhaps it would be appropriate here to make a few remarks on "natural motions." Consider a freely falling body or a satellite in motion near the Earth. If it is assumed that these objects are undisturbed by other bodies, no contact forces are acting on them. In order to account for their observed motions, Newton had to develop the theory of gravitation. Einstein simply took a different point of view — that the space-time is warped around a large mass like Earth and that an object such as the satellite is simply following its natural trajectory in that vicinity, much like a phonograph needle in the groove of a record. From the Einstein theory, a transformation to a system of coordinates fixed on the satellite can be made, such a system being just as permissible as one attached to Earth. In the accelerated frame, the gravitational field has been eliminated, as has been amply demonstrated by the weightless condition of astronauts in orbit.

These comments are somewhat abstract at this point. Shortly, we shall derive the equations which explicitly show how the matter distribution in the vicinity of a test mass influences its motion.

Suppose we are given a set of independent continuously varying coordinates $x_1, x_2, \ldots x_n$. Each such set can be taken to specify the location of a point, and the totality of points represented by these variables is called a **space**. (The Cartesian coordinates of a Euclidean n-space serve as a good example.) But it is always possible to describe the space in other sets of coordinates as well, provided that the transformation to the new set is one to one and possesses a unique inverse. The complete description of a space with all such transformations is called a **geometry**.

The spaces that are of greatest interest to us are known as **metric spaces**, in which a **metric tensor** can be defined. In such a space, a vector exists as an objective physical reality which can be expressed quantitatively in terms of either its contravariant or covariant components, the relation between which is supplied by the metric tensor as discussed in Appendix E. (More general **affine spaces** which have no metric tensor have been studied extensively but will not concern us here.)

If the events of physical interest are viewed as taking place in a four-dimensional space-time continuum as was the case with special relativity, the metric relations involve four indices rather than three. Suppose that a transformation $q = q(q')$ and its inverse are known for two sets of coordinates q and q'. Then, the basic metric relation (E.7) gives the square of the line element for the space (and defines the metric tensor—see Appendix E):

$$ds^2 = \sum_{i,j=1}^{4} g'_{ij} dq^{i'} dq^{j'} = \sum_{\substack{i,j,\\k,l}} g'_{ij} \frac{\partial q^{i'}}{\partial q^k} \frac{\partial q^{j'}}{\partial q^l} dq^k dq^l$$

$$\equiv \sum_{k,l} g_{kl} dq^k dq^l, \qquad (9.33)$$

with

$$g_{kl} \equiv \sum_{i,j} g'_{ij} \frac{\partial q^{i'}}{\partial q^k} \frac{\partial q^{j'}}{\partial q^l}. \qquad (9.33')$$

If the set q is taken as the usual Cartesian one with x, y, z, and ict as in eq. (9.1), then

$$ds^2 = dx_1^2 + dx_2^2 + dx_3^2 + dx_4^2,$$

and so, from (9.33), we have

$$g_{kl} = \delta_{k,l}.$$

The Cartesian coordinates give what is called the Lorentz metric, which is just the identity in matrix form and corresponds to a *flat* (Euclidean) space. The Lorentz metric is quite adequate for special relativity; but in general relativity, the treatment of curved spaces will require much more sophistication.

The mathematics of curved spaces is known as **Riemannian geometry**. In order to clarify the essential difference between Euclidean and Riemannian spaces, let us ignore the time factor momentarily and consider some simple two-dimensional surfaces as examples. If a circular cylinder of radius R is oriented along the z-axis, such that

$$ds^2 = R^2 d\phi^2 + dz^2$$

holds in cylindrical coordinates, then the metric elements can be identified as

$$g_{11} = R^2, \qquad g_{22} = 1, \qquad g_{12} = g_{21} = 0 \qquad (q^1 \equiv \phi, \qquad q^2 \equiv z).$$

The General Theory of Relativity 351

A scale change of variable, namely, $x = R\phi$, produces the Euclidean form

$$ds^2 = dx^2 + dz^2,$$

which has the identity metric. On the other hand, on the surface of a sphere of radius R, ds^2 is given by

$$ds^2 = R^2 d\theta^2 + R^2 \sin^2\theta \, d\phi^2$$

in spherical coordinates, and so using these as the primed quantities

$$g'_{11} = R^2, \qquad g'_{22} = R^2 \sin^2\theta, \qquad g'_{12} = g'_{21} = 0 \qquad (q^{1'} \equiv \theta, \qquad q^{2'} \equiv \phi).$$

It is certainly not obvious how this metric can be transformed into the flat space form, so let us be a little more general in approach. Introducing $q^1 \equiv x(\theta, \phi)$, $q^2 \equiv y(\theta, \phi)$ as new functions of θ and ϕ so chosen as to satisfy the flat metric form

$$ds^2 = dx^2 + dy^2 \qquad (g_{11} = g_{22} = 1, \qquad g_{12} = g_{21} = 0),$$

we can write from (9.33'),

$$1 = R^2 \left(\frac{\partial \theta}{\partial x}\right)^2 + R^2 \sin^2\theta \left(\frac{\partial \phi}{\partial x}\right)^2, \qquad 1 = R^2 \left(\frac{\partial \theta}{\partial y}\right)^2 + R^2 \sin^2\theta \left(\frac{\partial \phi}{\partial y}\right)^2,$$

$$0 = R^2 \frac{\partial \theta}{\partial x}\frac{\partial \theta}{\partial y} + R^2 \sin^2\theta \frac{\partial \phi}{\partial x}\frac{\partial \phi}{\partial y}.$$

Inasmuch as these equations must hold for any value of θ, setting $\theta = 0$ gives

$$\left(\frac{\partial \theta}{\partial x}\right)^2 = \frac{1}{R^2} = \left(\frac{\partial \theta}{\partial y}\right)^2,$$

while

$$R^2 \left(\frac{\partial \theta}{\partial x}\right)\left(\frac{\partial \theta}{\partial y}\right) = 0.$$

These relations are clearly inconsistent with one another, so we conclude that the metric for a spherical surface is different in a basic sense from that for a flat space, and no transformation to the latter form is possible. (A cylinder can simply be "unrolled" into a plane, but this is not so for a sphere, without some kind of distortion or stretching.)

In his development of a theory of gravitation, Einstein recognized that a uniform field could be transformed away through a transition to an accelerated coordinate frame. But for a general field the situation is much more complicated; however, in the limit as such a field is reduced to zero, the flat space Lorentz metric must be obtained once more. When the field is present, the square of the line element given in eq. (9.33) must have metric g_{ij} which are not simply unit constants but are *in some way* related to the gravitational field.

The goal of general relativity is then to express the laws of physics as covariant tensor relations in a "curved space," that is, one with a complex metric that generally cannot be cast into a flat space Lorentz form. In such a space, the motion of a body can be viewed in a "natural sense" where no gravitational field is needed, while a description in terms of the usual Cartesian coordinates requires the introduction of gravitational field parameters. In effect, general relativity modifies the metric of the space so as to eliminate such parameters.

In a region free of matter, classical theory tells us that

$$\vec{F} = m\frac{d^2\vec{r}}{dt^2} = -m\nabla\Phi, \qquad \nabla^2\Phi = 0,$$

according to eqs. (1.24) and (1.29). Just as second derivatives of the potential enter the Laplace equation, so it is that the field equations of general relativity involve second derivatives of the metric tensor, and these enter in linear fashion so as to guarantee uniqueness of the solution. The metric elements g_{ij} must be the components of a covariant tensor and must reduce to $\delta_{i,j}$ (the Lorentz metric) in spaces that contain no matter.

9.7 THE RIEMANN CURVATURE TENSOR

In a flat space having the Lorentz metric, the elements g_{ij} are all constants, so the derivatives of g_{ij} will vanish. In the language of Appendix E, the Christoffel symbols of both kinds must also vanish. From (E.46), this means that covariant derivatives and ordinary derivatives are equal, and this will be true to all orders of differentiation. Thus, for any covariant component V_η of a vector **V**,

$$V_{\eta;\beta} \equiv \frac{\partial V_\eta}{\partial q^\beta}, \qquad V_{\eta;\beta;\gamma} \equiv \frac{\partial^2 V_\eta}{\partial q^\beta \partial q^\gamma},$$

and so forth, the symbol $V_{\eta;\beta;\gamma}$ indicating a second covariant derivative. From ordinary derivative properties, the order of differentiation is irrelevant:

$$V_{\eta;\beta;\gamma} = V_{\eta;\gamma;\beta}. \qquad (9.34)$$

But (9.34) is a tensor relation which holds in *any* coordinate system in the space; it is then a necessary condition for the flatness of the space. The two second

The Riemann Curvature Tensor

derivatives in this relation are certainly not equal in general, however. In fact, a few cumbersome steps of tensor algebra lead in the general case to

$$V_{\alpha;\beta;\gamma} - V_{\alpha;\gamma;\beta}$$

$$= -\sum_\eta \left[\frac{d\begin{Bmatrix}\eta\\ \beta\ \alpha\end{Bmatrix}}{dq^\gamma} - \frac{d\begin{Bmatrix}\eta\\ \gamma\ \alpha\end{Bmatrix}}{dq^\beta} + \sum_\tau \left(\begin{Bmatrix}\eta\\ \tau\ \gamma\end{Bmatrix}\begin{Bmatrix}\tau\\ \beta\ \alpha\end{Bmatrix} - \begin{Bmatrix}\eta\\ \tau\ \beta\end{Bmatrix}\begin{Bmatrix}\tau\\ \gamma\ \alpha\end{Bmatrix} \right) \right] V_\eta$$

$$\equiv -\sum_\eta R^\eta_{\alpha\beta\gamma} V_\eta, \qquad (9.35)$$

where the quantity $R^\eta_{\alpha\beta\gamma}$ is known as the **Riemann curvature tensor**, and is defined by the expression in square brackets. Since **V** is arbitrary, the flat space condition (9.34) becomes simply

$$R^\eta_{\alpha\beta\gamma} = 0. \qquad (9.36)$$

The curvature tensor must vanish in a (flat) space with a Lorentz metric. The tensor has 4^4, or 256, components (although only 20 are independent), and each must vanish if the space contains no gravitational field. This statement will be true in any curvilinear coordinate system, so that the vanishing of the Riemann tensor is a necessary condition if a transformation that takes the metric into Lorentz form is to be found.

The curvature tensor has a number of interesting symmetry properties which we shall not prove[3], although some of them are readily established by examining its definition:

$$R_{\eta\alpha\beta\gamma} = -R_{\eta\alpha\gamma\beta},$$

$$R_{\eta\alpha\beta\gamma} = -R_{\alpha\eta\beta\gamma}, \qquad (9.37)$$

$$R_{\eta\alpha\beta\gamma} = +R_{\beta\gamma\eta\alpha},$$

$$R_{1023} + R_{2031} + R_{3012} = 0.$$

Here, we have used the metric to lower one index (see the discussion on raising and lowering indices given in Appendix E), giving the fully covariant form often employed. It is because of the properties (9.37) that only 20 components are independent.

---
[3] R. Adler, M. Bazin, and M. Schiffer, *Introduction to General Relativity*, 2nd edition, McGraw-Hill, New York, 1975, p. 151.

By contraction of the Riemann tensor, we obtain the **Ricci tensor**

$$R_{\alpha\beta} \equiv \sum_{\eta} R^{\eta}{}_{\alpha\beta\eta} = R_{\beta\alpha}, \qquad (9.38)$$

the symmetry being a consequence of (9.37). A further contraction yields the invariant

$$R \equiv \sum_{\alpha,\beta} g^{\alpha\beta} R_{\alpha\beta}. \qquad (9.39)$$

The Ricci tensor is an important quantity which is linear in the second derivatives of the metric, and vanishes not only when the Riemann tensor does but also in some cases when the curvature tensor does *not* (allowing for the existence of a gravity field in a curved space). As a simple criterion for the determination of the metric components, Einstein chose to set

$$R_{\alpha\beta} \equiv 0. \qquad (9.40)$$

Since the physical consequences of this assumption agree with experiment, and in particular do not differ greatly from Newtonian mechanics, the simplicity and elegance of Einstein's treatment cannot fail to be appreciated.

Now that we understand how the metric of space-time is determined, we shall obtain the equation of motion for a particle brought into the region. Einstein's basic assumption was that such a mass would move along a geodesic in the four-dimensional continuum, thus satisfying the stationary (often minimizing) principle

$$\delta \int ds = 0. \qquad (9.41)$$

Since ds is an invariant length, this criterion is obviously covariant. Through the relation (9.33), however, the way in which ds is expressed in a particular CS depends on the elements of the metric. The variation of ds in (9.41) can be performed along the lines of earlier such calculations in this text, beginning from eq. (9.33) and proceeding through several algebraic steps much like those which are found in Appendix E. The variational principle finally leads to the set of four equations

$$\frac{d^2 q^{\alpha}}{ds^2} + \sum_{\beta,\gamma} \left\{ \begin{matrix} \alpha \\ \beta\ \gamma \end{matrix} \right\} \frac{dq^{\beta}}{ds} \frac{dq^{\gamma}}{ds} = 0. \qquad (9.42)$$

In order to set up and solve these equations, we must first know what the elements g_{ij} are. The latter are such as to cause the Ricci tensor $R_{\alpha\beta}$ to vanish; and (from its symmetry) $R_{\alpha\beta} = 0$ gives a set of 10 partial differential equations to be solved for the g_{ij}. This is generally an impossible task; so, in practice, one uses physical insight to *guess* at the form of the g_{ij} which will satisfy (9.40) and then give reasonable solutions for (9.42). The end results have the desired properties, justifying this somewhat awkward procedure.

9.8 THE SCHWARZSCHILD SOLUTION OF THE FIELD EQUATIONS

One of the most useful applications of general relativity came in 1916 with the exact solution by Schwarzschild of the field equations in the space around a central mass at the origin. We shall now describe the essential features of that solution.

First, we note that in general relativity it is customary to use the real coordinate differential $dx_0 \equiv c\, dt$ instead of dx_4, since the coordinate transformations are usually much more complicated than an orthogonal transformation in a complex space. In addition, the expression for the path element squared conventionally writes the dx_0 term as positive, then subtracts the space differential terms. If the line element for this problem is chosen in a form with unknown coefficients which are assumed to be spherically symmetric and time independent, we have, in spherical coordinates;

$$ds^2 = f(r)c^2\, dt^2 - g(r)\, dr^2 - r^2\, d\theta^2 - r^2 \sin^2\theta\, d\phi^2. \tag{9.43}$$

The functions $f(r)$ and $g(r)$ are yet to be determined; they must approach unity at large r so that the Lorentz metric can be recovered there. It is reasonable to expect no cross terms of the form $dt\, d\theta$, $d\theta\, d\phi$, and so on, because a replacement of dt, $d\theta$, or $d\phi$ by their negatives should not affect ds^2 under the above assumptions on symmetry and time independence. (Other unknown radial functions could be included as factors in the two angular terms, but it can be shown that they are unnecessary.)

Setting $f(r) \equiv e^v$ and $g(r) \equiv e^\lambda$ for algebraic convenience, we find that the diagonal elements of the metric tensor are

$$g_{00} = e^v, \qquad g_{11} = -e^\lambda, \qquad g_{22} = -r^2, \qquad g_{33} = -r^2 \sin^2\theta.$$

The nonvanishing Christoffel symbols are then given by (E.42) as

$$\Gamma^1_{11} = \frac{1}{2}\frac{d\lambda}{dr}, \quad \Gamma^1_{22} = -e^{-\lambda}r, \quad \Gamma^1_{33} = -e^{-\lambda}r \sin^2\theta, \quad \Gamma^1_{00} = \frac{1}{2}e^{v-\lambda}\frac{dv}{dr},$$

$$\Gamma^2_{21} = \Gamma^2_{12} = \frac{1}{r}, \quad \Gamma^2_{33} = -\sin\theta\cos\theta, \quad \Gamma^3_{13} = \Gamma^3_{31} = \frac{1}{r},$$

$$\Gamma^3_{23} = \Gamma^3_{32} = \cot\theta, \quad \Gamma^0_{10} = \Gamma^0_{01} = \frac{1}{2}\frac{dv}{dr}.$$

It turns out that these surviving symbols can be evaluated rather quickly in a simple fashion. From eqs. (9.41) and (9.43),

$$\delta \int ds = 0 = \delta \int \left[e^v c^2 \left(\frac{dt}{ds}\right)^2 - e^\lambda \left(\frac{dr}{ds}\right)^2 - r^2 \left(\frac{d\theta}{ds}\right)^2 - r^2 \sin^2\theta \left(\frac{d\phi}{ds}\right)^2 \right]^{1/2} ds$$

$$= \delta \int \left[e^v (\dot{q}^0)^2 - e^\lambda (\dot{q}^1)^2 - (q^1)^2 (\dot{q}^2)^2 - (q^1)^2 \sin^2 q^2 \cdot (\dot{q}^3)^2 \right]^{1/2} ds,$$

where $q^1 \equiv r$, $q^0 \equiv ct$, $\dot{q}^1 \equiv dr/ds$, and so on. The problem of variation is somewhat easier to handle if the square root is removed from the integrand, leaving

$$-\delta \int \left[-e^v (\dot{q}^0)^2 + e^\lambda (\dot{q}^1)^2 + (q^1)^2 (\dot{q}^2)^2 + (q^1)^2 \sin^2 q^2 (\dot{q}^3)^2 \right] ds = 0,$$

which can be shown to be an equivalent expression.[4] Denoting the integrand by F, the Euler–Lagrange equations for the variation take on the form

$$\frac{d}{ds}\left(\frac{\partial F}{\partial \dot{q}^\alpha}\right) - \frac{\partial F}{\partial q^\alpha} = 0$$

for any of the four coordinates, by analogy with Hamilton's principle. As an example, for q^0 one can write

$$\frac{d}{ds}\left(\frac{\partial F}{\partial \dot{q}^0}\right) - \frac{\partial F}{\partial q^0} = 0 = \frac{d}{ds}[-2e^v \dot{q}^0] - 0 = e^v \ddot{q}^0 + e^v \frac{dv}{dr}\frac{dr}{ds}\dot{q}^0,$$

or

$$\ddot{q}^0 + \frac{dv}{dr}\dot{q}^1 \dot{q}^0 = 0.$$

Comparing coefficients in the last relation with (9.42), we find that

$$\left\{ \begin{matrix} 0 \\ 1 \ 0 \end{matrix} \right\} = \left\{ \begin{matrix} 0 \\ 0 \ 1 \end{matrix} \right\} = \frac{1}{2}\frac{dv}{dr}, \qquad \left\{ \begin{matrix} 0 \\ \beta \ \gamma \end{matrix} \right\} = 0$$

for other values of β, γ. The other Euler–Lagrange equations give the remaining coefficients similarly.

The next step is to substitute the Christoffel symbols into the Ricci tensor and equate its components to zero in accordance with (9.40). Let us choose R_{00} alone

---

[4] R. Adler, M. Bazin, and M. Schiffer, *Introduction to General Relativity*, Second Edition, McGraw-Hill, New York, 1975, p. 188.

to illustrate the procedure:

$$R_{00} = \sum_\eta R^\eta_{00\eta} = \sum_\eta \left[\frac{d\left\{{\eta \atop 0\ 0}\right\}}{dq^\eta} - \frac{d\left\{{\eta \atop 0\ 0}\right\}}{dq^0} + \sum_\tau \left(\left\{{\eta \atop \tau\ \eta}\right\}\left\{{\tau \atop 0\ 0}\right\} - \left\{{\eta \atop \tau\ 0}\right\}\left\{{\tau \atop \eta\ 0}\right\} \right) \right]$$

$$= \frac{d\Gamma^1_{00}}{dr} + \left[\left\{{1 \atop 1\ 1}\right\} + \left\{{2 \atop 1\ 2}\right\} + \left\{{3 \atop 1\ 3}\right\} + \left\{{0 \atop 1\ 0}\right\} \right]\left\{{1 \atop 0\ 0}\right\}$$

$$- 2\left\{{1 \atop 0\ 0}\right\}\left\{{0 \atop 1\ 0}\right\}$$

$$= -\frac{1}{2}e^{v-\lambda}\left(\frac{d^2v}{dr^2} + \left(\frac{dv}{dr}\right)^2 - \frac{d\lambda}{dr}\frac{dv}{dr} \right)$$

$$+ \left[\frac{1}{2}\frac{d\lambda}{dr} + \frac{2}{r} - \frac{1}{2}\frac{dv}{dr} \right]\left[-\frac{1}{2}e^{v-\lambda}\frac{dv}{dr} \right] = 0,$$

or

$$\frac{d^2v}{dr^2} + \frac{1}{2}\left(\frac{dv}{dr}\right)^2 - \frac{1}{2}\frac{d\lambda}{dr}\frac{dv}{dr} + \frac{2}{r}\frac{dv}{dr} = 0.$$

The end result is a differential equation involving v and λ, the unknown functions of r. A similar equation arises upon evaluation of R_{11}, and combining of the two yields

$$\frac{dv}{dr} = -\frac{d\lambda}{dr},$$

or

$$v = -\lambda \quad \text{(constant taken as zero)},$$

whereupon on eliminating dv/dr from the R_{00} relation we have

$$\frac{d^2\lambda}{dr^2} - \left(\frac{d\lambda}{dr}\right)^2 + \frac{2}{r}\frac{d\lambda}{dr} = 0,$$

or

$$\frac{d^2}{dr^2}(re^{-\lambda}) = 0.$$

On integration, this equation produces

$$\frac{d}{dr}(re^{-\lambda}) = \text{constant} = 1,$$

the value unity of the constant being selected from evaluation of R_{22}. A further integration gives

$$re^{-\lambda} = r - 2m,$$

or

$$e^{-\lambda} = 1 - \frac{2m}{r} = e^{v} \quad (2m = \text{constant}).$$

The functions λ and v, or $f(r)$ and $g(r)$, are now fully determined, and the solutions are consistent with the vanishing of all other components of the Ricci tensor. The line element (9.43) is now

$$ds^2 = \left(1 - \frac{2m}{r}\right)c^2 dt^2 - \frac{dr^2}{1 - (2m/r)} - r^2 d\theta^2 - r^2 \sin^2\theta\, d\phi^2.$$

Only one detail remains—the evaluation of the constant $2m$.

From (9.42) recall that the equation of motion from general relativity is

$$\frac{d^2 q^\alpha}{ds^2} + \sum_{\beta,\gamma}\left\{\begin{matrix}\alpha\\ \beta\ \gamma\end{matrix}\right\}\frac{dq^\beta}{ds}\frac{dq^\gamma}{ds} = 0,$$

which bears a close resemblance to the classical radial equation of motion in a gravitational potential Φ (setting $\alpha = 1$),

$$\frac{d^2 q^1}{dt^2} + \frac{\partial \Phi}{\partial q^1} = 0,$$

and must reduce to the latter in the low-velocity nonrelativistic limit. Taking velocity components such as

$$dq^1/dt \ll c,$$

we find from eq. (9.43) that

$$|ds| \simeq \sqrt{g_{00}}\, c\, dt \equiv \sqrt{g_{00}}\, dq^0 \quad (dq^0 \equiv c\, dt),$$

and so the $\alpha = 1$ term of eq. (9.42) is

$$\frac{d^2 q^\alpha}{ds^2} + \sum_{\beta,\gamma}\left\{\begin{matrix}\alpha\\ \beta\ \gamma\end{matrix}\right\}\frac{dq^\beta}{ds}\frac{dq^\gamma}{ds} \simeq \frac{d^2 q^\alpha}{dt^2}\frac{1}{g_{00}c^2} + \left\{\begin{matrix}\alpha\\ 0\ 0\end{matrix}\right\}\left(\frac{dq^0}{\sqrt{g_{00}}\, dq^0}\right)^2 = 0,$$

The Schwarzschild Solution of the Field Equations

or

$$\frac{d^2 q^1}{dt^2} + \begin{Bmatrix} 1 \\ 0\ 0 \end{Bmatrix} c^2 = 0.$$

Since

$$\begin{Bmatrix} 1 \\ 0\ 0 \end{Bmatrix} \simeq \frac{1}{2} \frac{\partial g_{00}}{\partial q^1},$$

$$\frac{d^2 q^1}{dt^2} + \frac{1}{2} c^2 \frac{\partial g_{00}}{\partial q^1} = 0$$

results (we have assumed large $r \equiv q^1$); only $\alpha = 1$ is really of interest. On identifying terms with the Newtonian equation,

$$\frac{\partial \Phi}{\partial q^1} = \frac{1}{2} c^2 \frac{\partial g_{00}}{\partial q^1}, \qquad \Phi = \frac{1}{2} c^2 g_{00} + \text{constant}.$$

In flat space with no gravity, $g_{00} = 1$ and $\Phi = 0$, so the constant of integration is $-\frac{1}{2}c^2$ and

$$\Phi = \frac{1}{2} c^2 (g_{00} - 1), \qquad \text{or} \qquad g_{00} = 1 + \frac{2\Phi}{c^2}. \tag{9.44}$$

We see directly the relation of the metric to the gravity field. In the spherically symmetric field around a mass M at the origin,

$$\Phi = \frac{-GM}{r}$$

in the Newtonian theory, giving

$$g_{00} = 1 - \frac{2GM}{c^2 r}.$$

At last, we can evaluate the constant of integration in the line element:

$$m = \frac{GM}{c^2},$$

whereupon

$$ds^2 = \left(1 - \frac{2GM}{c^2 r}\right) c^2 dt^2 - \frac{dr^2}{\left(1 - \frac{2GM}{c^2 r}\right)} - r^2 d\theta^2 - r^2 \sin^2 \theta \, d\phi^2.$$

$$\tag{9.45}$$

(Birkhoff was able to show in 1923 that this same line element is obtained for any spherically symmetric distribution of matter, even if in radial oscillation; time independence is not necessary.)

Let us summarize what has been accomplished in this section. We have used both physical insight and recourse to the classical approximation to obtain the elements of the metric which cause the vanishing of the Ricci tensor, and finally have been led to the expression (9.45) for ds^2.

To further investigate this general relativistic analogue of the Kepler problem, we must see that the geodesic principle (9.41) (with the line element as given above) is satisfied:

$$\delta \int ds = 0.$$

It is simpler to proceed as in the evaluation of the Christoffel symbols, squaring the integrand once more, with dots indicating derivatives with respect to s:

$$\delta \int \left[\left(1 - \frac{2GM}{c^2 r}\right) c^2 \dot{t}^2 - \frac{\dot{r}^2}{\left(1 - \frac{2GM}{c^2 r}\right)} - r^2 \dot{\theta}^2 - r^2 \dot{\phi}^2 \sin^2 \theta \right] ds = 0.$$

The Euler–Lagrange equation for θ is

$$\frac{d}{ds}(-2r^2 \dot{\theta}) + 2r^2 \dot{\phi}^2 \sin \theta \cos \theta = 0.$$

Choosing the axes so that $\theta = \pi/2$ and letting $\dot{\theta} = 0$ at the initial s value, we see that $\theta = \pi/2$ then holds for all s, so the orbit lies in a plane as expected. The ϕ-equation is now

$$\frac{d}{ds}(-2r^2 \dot{\phi} \sin^2 \theta) = 0, \quad \text{or} \quad r^2 \dot{\phi} \equiv h = \text{constant},$$

upon insertion of the θ-condition. The Euler-Lagrange t-equation becomes

$$\frac{d}{ds}\left[2c^2\left(1 - \frac{2GM}{c^2 r}\right)\dot{t}\right] = 0, \quad \text{or} \quad \left(1 - \frac{2GM}{c^2 r}\right)\dot{t} \equiv l = \text{constant}.$$

Since the r-equation is more complicated, let us instead divide (9.45) by ds^2 to obtain

$$1 = \left(1 - \frac{2GM}{c^2 r}\right) c^2 \dot{t}^2 - \frac{\dot{r}^2}{1 - (2GM/c^2 r)} - r^2 \dot{\theta}^2 - r^2 \dot{\phi}^2 \sin^2 \theta$$

$$= \frac{c^2 l^2 - \dot{r}^2}{1 - (2GM/c^2 r)} - \frac{h^2}{r^2},$$

The Schwarzschild Solution of the Field Equations

a differential equation which gives r as a function of s. Using the insight of the classical Keplerian analysis, let us change variables in this last relation from s to ϕ and from r to $\mu \equiv 1/r$, noting that

$$\frac{dr}{d\phi} = \frac{\dot{r}}{\dot{\phi}} = \frac{r^2}{h}\dot{r}, \quad \frac{d\mu}{d\phi} = -\frac{1}{r^2}\frac{dr}{d\phi} = -\frac{\dot{r}}{h}.$$

We therefore obtain

$$1 = \frac{c^2 l^2 - h^2 \left(\frac{d\mu}{d\phi}\right)^2}{1 - (2GM\mu/c^2)} - h^2\mu^2,$$

or

$$\left(\frac{d\mu}{d\phi}\right)^2 = \left(\frac{c^2 l^2 - 1}{h^2}\right) + \frac{2GM}{h^2 c^2}\mu - \mu^2 + \frac{2GM}{c^2}\mu^3. \tag{9.46}$$

As with the Kepler problem, this equation can now be integrated to yield ϕ as an integral over a function of μ. It is more enlightening, however, to differentiate again with respect to ϕ:

$$2\frac{d\mu}{d\phi}\frac{d^2\mu}{d\phi^2} = \frac{2GM}{h^2 c^2}\frac{d\mu}{d\phi} - 2\mu\frac{d\mu}{d\phi} + \frac{6GM}{c^2}\mu^2\frac{d\mu}{d\phi},$$

or

$$\frac{d^2\mu}{d\phi^2} + \mu = \frac{GM}{h^2 c^2} + \frac{3GM}{c^2}\mu^2, \tag{9.47}$$

on dividing out $2(d\mu/d\phi)$ and disregarding the circular solution with $d\mu/d\phi = 0$.

This last relation is identical with (4.10) applied to the inverse square case, except for the term $(3GM/c^2)\mu^2$. A rough numerical estimate readily shows this term to be very small in comparison with the term GM/h^2c^2, so that the classical elliptical motion of the planets will remain valid to a high degree of approximation. The effect of the small perturbing term is to cause the ellipse to rotate through a small angle in the plane of the orbit for each revolution, a phenomenon known as the **advance of the perihelion**. Only for the planet Mercury is this effect of significance, the unexplained part of the advance amounting to about 43 seconds of arc per century, a small quantity but one which has been measured and found to be in excellent agreement with general relativity. (Most of the observed advance can be accounted for with Newtonian theory by allowing for perturbations of other planets, etc.) In recent years, attempts have been made to explain discrepancies in the advance of Mercury's perihelion through deviations in the shape of the sun from a perfect sphere; both experimental and theoretical difficulties have prevented the idea from gaining general acceptance.

The Schwarzschild solution we have obtained can also be applied to the case of a light ray, or photon, passing close to a source of gravity M, another traditional experimental test of general relativity. A comparison of eqs. (9.47) and (4.10) shows that the GM/h^2c^2 term should drop out for a particle of zero rest mass like a photon; in other words, h is then infinite. (Alternatively, since $r^2(d\phi/ds) = h$ and $ds = 0$ for the photon, the result also follows.) The photon trajectory is described by the differential equation

$$\frac{d^2\mu}{d\phi^2} + \mu = \frac{3GM}{c^2}\mu^2. \tag{9.48}$$

A solution of this equation, treating the term on the right as a small perturbation, shows that the gravity field of a massive body M with radius R deflects the light ray through an angle $4GM/c^2R$. The deflection of starlight on grazing the sun should then be 1.75 seconds of arc, a prediction again in excellent accord with the best recent experimental values.

Finally, consider an atom at rest in a gravitational field and emitting a spectral line of frequency ν. With $dr = d\theta = d\phi = 0$, we have, from (9.45),

$$ds^2 = \left(1 - \frac{2GM}{c^2r}\right)c^2\,dt^2.$$

The generalization of the proper time concept from special relativity requires that

$$d\tau \equiv \frac{ds}{c} = \sqrt{1 - \frac{2GM}{c^2r}}\,dt;$$

this refers to an increment of time at the local field position. The frequency of the atomic oscillator is

$$\nu \sim \frac{1}{d\tau} = \frac{1}{\sqrt{1 - (2GM/c^2r)}\,dt}.$$

The frequency ratio for two such atoms at different radial positions (denoted by r_1 and r_2, where $r_1 < r_2$) is then upon expansion

$$\frac{\nu_1}{\nu_2} = \frac{[1 - (2GM/c^2r_2)]^{1/2}}{[1 - (2GM/c^2r_1)]^{1/2}} \approx 1 + \frac{GM}{c^2}\left(\frac{1}{r_1} - \frac{1}{r_2}\right) = 1 + \frac{\Phi_2 - \Phi_1}{c^2},$$

$$\tag{9.49}$$

assuming that the coordinate-time intervals dt are equal and therefore cancel and that the expanded radical is essentially unity. Since the gravitational potential Φ is negative and increases in magnitude as one approaches the surface of the source M, the frequency ν_1 is greater at the surface than ν_2 at larger radial distances. In viewing the light from some distance away, we would observe a reduction of frequency or a **red shift** of the spectral lines, which can be interpreted as a photon loss of energy in escaping the gravitational pull of the source M. The Pound–Rebka experiment[5] of 1960 used the Mössbauer effect to measure this loss of energy between the two ends of a vertical tower, verifying the red shift formula to a considerable accuracy.

It is to be noted from the first member of eq. (9.49) that $\nu_2 \to 0$ as $r_1 \to 2GM/c^2$. This particular radius (called the **Schwarzschild radius** or **event horizon**), dependent only on the mass of the source, is such that an atom radiating there suffers an infinite red shift and is not visible to observers farther out. Indeed, inspection of the line element (9.45) would already have warned us of the uniqueness of this radial position. Contained within the Schwarzschild radius we have the famous "black hole," about which modern astronomers have written so much. It is theorized that a massive star, aged and undergoing its death throes through gravitational collapse, may form such a black hole. The name is appropriate, for neither matter nor even light rays can escape the boundary of the hole once inside, so it appears black. As no form of communication with its interior is possible to us "outsiders," it is literally like a hole in our space.

The Schwarzschild solution has been known since 1916, and it has been extensively investigated. A solution of still greater interest to astronomers, however, was developed in 1963 by R. P. Kerr.[6] Most collapsing stars are in rotation, and Kerr was able to solve the exterior field equations exactly for a rotating black hole. The added complications introduced and insights gained in the Kerr solution are beyond the scope of the present treatment.

PROBLEMS

9.1 Show that the scalar wave equation

$$\nabla^2 f - \frac{1}{c^2} \frac{\partial^2 f}{\partial t^2} = 0$$

is invariant under a Lorentz transformation but not under a Galilean transformation. Here, f represents any scalar function of position and time, such as a particular component of a field.

---

[5] R. Pound and G. Rebka, "Apparent Weight of Photons," *Phys. Rev. Lett.* **4**, 337 (1960).

[6] R. P. Kerr, "Gravitational Field of a Spinning Body as an Example of Algebraically Special Metrics," *Phys. Rev. Lett.* **11**, 237 (1963).

9.2 (a) If a particle moves with a constant velocity in one frame of reference, show from eqs. (9.7) that it will also move with constant velocity in another frame moving with speed v along the x-axis of the first. You may restrict the motion of the particle to the x-axis. (This property of the Lorentz transformation has its counterpart in the Galilean transformation, where it provides a basis for the application of Newton's first law in any inertial frame.)
(b) Show that the result of part (a) follows from the linearity of (9.3), without using the specific form of (9.7).
(c) How would the acceleration transform?

9.3 Prove the Einstein addition formula through the successive application of two Lorentz transformation matrices. This formula states that
$$v_3 = \frac{v_1 + v_2}{1 + (v_1 v_2 / c^2)},$$
where v_1 is the speed of a second Lorentz frame relative to the first, v_2 is the speed of a third CS relative to the second, and v_3 is the speed of the third CS relative to first, all such relative motions being along the $+x$-axes. Then prove that $v_3 < c$ even if $v_1 \simeq c$ and $v_2 \simeq c$.

9.4 Use rotation matrices (of the type discussed in Chapter 5) with the Lorentz transformation to show that
$$\vec{r}' = \vec{r} - \gamma t \vec{v} + (\gamma - 1)(\hat{n} \cdot \vec{r})\hat{n},$$
$$t' = \gamma \left(t - \frac{\vec{r} \cdot \vec{v}}{c^2} \right),$$
where the relative velocity between the two frames is given by $\vec{v} = v\hat{n}$, $\hat{n}$ being a unit vector in an arbitrary direction.

9.5 An astronaut in the space shuttle remains in orbit around the Earth for five days. If the orbit has a radius of 10^7 meters, how much younger will the astronaut be upon landing than a twin who stayed on Earth? (Neglect effects of general relativity in this problem.)

9.6 Consider a relativistic "photon rocket" with initial rest mass m_0 and payload mass some fraction of m_0. (This rocket ejects photons at speed c in order to provide thrust.) If the rocket accelerates from rest on departure, decelerates to a stop at its destination, and then returns similarly, show that the final payload is $m_0/16\gamma^4$, where $\gamma \gg 1$ is the appropriate γ-factor at maximum velocity of the rocket.

9.7 An elastic collision occurs in which a particle has its four-momentum altered such that q is the difference between final and initial four-momenta. Prove that q is either a space-like four-vector or is zero.

9.8 (a) Consider a pair annihilation process in which a positron and electron interact with one another. Calculate the total four-momentum of the pair and show from your result that the pair cannot annihilate and thereby produce a *single* photon.

(b) The decay of a Λ-particle produces a proton and a π-meson according to the reaction

$$\Lambda \to p + \pi^-.$$

If the four-momenta of the particles created are $(\vec{p}_p, iE_p/c)$ and $(\vec{p}_\pi, iE_\pi/c)$ and the rest masses m_Λ, m_p, and m_π, show that the angle θ between $\vec{p}_p$ and $\vec{p}_\pi$ is given by the expression

$$\cos\theta = \frac{(m_\Lambda^2 - m_p^2 - m_\pi^2)c^4 - 2E_p E_\pi}{2c^2 |\vec{p}_p||\vec{p}_\pi|}.$$

9.9 Two identical particles of rest mass m_0 suffer a completely *inelastic* collision, sticking together afterward to form a single composite particle. If both particles move with speed v in the center of momentum frame before the collision (one being at rest in the lab frame), find the relativistic mass of the composite particle in both the *CM* and laboratory frames. Note that the result is *not* $2m_0$ in either frame!

9.10 A particle starts from rest at the origin and moves along the x-axis with a constant acceleration g. Calculate the components of the coordinate force and the Minkowski force. (Assume the speed remains less than c.)

9.11 Consider the line element in a transformation from rectangular to polar coordinates on a Euclidean plane. Work out all the Christoffel symbols of both kinds and show that the curvature tensor vanishes. Then do the same for the line element on the surface of a sphere, showing that the curvature tensor will not vanish there.

9.12 In the three-dimensional transformation from rectangular to spherical coordinates, use the metric tensor to work out all the Christoffel symbols of both kinds.

9.13 Using the symbols found in the previous problem, evaluate all components of the Riemann curvature tensor and show that each of them vanishes. How do you interpret this result?

9.14 In the text, the Euler–Lagrange equations are used to derive the basic eq. (9.47). Show that it can also be obtained by direct application of (9.42).

9.15 (a) Use straightforward considerations of energy to derive the red-shift formula (9.49).

(b) A mass rotates with constant angular velocity ω. Calculate the centrifugal potential field and use it in connection with proper time considerations from special relativity to derive eq. (9.44).

Appendix A

VECTOR OPERATIONS AND IDENTITIES

A.1 DEFINITIONS AND BASIC RELATIONSHIPS

At the level of this text, it is assumed that the reader is quite familiar with vector operations. Nevertheless, we have included this appendix for the benefit of those who need a little brushing up or who want a handy reference source for vector relationships.

A **vector** $\vec{V}$ is a quantity with both magnitude and direction. It is specified completely by giving its components V_x, V_y, and V_z along rectangular axes. Unit vectors along the xyz-axes are labelled $\hat{i}, \hat{j}$, and $\hat{k}$, respectively, so that an arbitrary $\vec{V}$ can be written as

$$\vec{V} = V_x\hat{i} + V_y\hat{j} + V_z\hat{k}.$$

The **scalar product**, or **dot product**, of two vectors is defined as the product of the magnitudes of each and the cosine of the angle θ between them. Thus, for two vectors $\vec{A}$ and $\vec{B}$,

$$\vec{A} \cdot \vec{B} \equiv |\vec{A}| \cdot |\vec{B}| \cdot \cos\theta. \tag{A.1}$$

The dot product is a scalar; it therefore has no direction associated with it. The two vectors commute under this type of multiplication. Applying the definition to the unit vectors along the axes, we find that

$$\hat{i} \cdot \hat{i} = \hat{j} \cdot \hat{j} = \hat{k} \cdot \hat{k} = 1,$$

while

$$\hat{i} \cdot \hat{j} = \hat{i} \cdot \hat{k} = \hat{j} \cdot \hat{k} = 0,$$

Definitions and Basic Relationships

and, in general,

$$\vec{A} \cdot \vec{B} = \left(A_x \hat{i} + A_y \hat{j} + A_z \hat{k}\right) \cdot \left(B_x \hat{i} + B_y \hat{j} + B_z \hat{k}\right)$$
$$= A_x B_x + A_y B_y + A_z B_z. \tag{A.2}$$

Physically, the dot product projects one vector onto the direction of the other.

The **vector product**, or **cross product**, of two vectors is another vector whose magnitude is given by

$$\vec{A} \times \vec{B} \equiv |\vec{A}| \cdot |\vec{B}| \cdot \sin\theta \cdot \hat{\varepsilon}, \tag{A.3}$$

where $\hat{\varepsilon}$ is a unit vector perpendicular to the plane of $\vec{A}$ and $\vec{B}$, its sense determined by a rotation of $\vec{A}$ into $\vec{B}$ by a right-handed screw rule. Clearly, two vectors anticommute under this definition: $\vec{A} \times \vec{B} = -\vec{B} \times \vec{A}$. The physical interpretation of the cross product is that of the area of a parallelogram with adjacent sides $\vec{A}$ and $\vec{B}$. Here, the unit vectors along the axes satisfy

$$\hat{i} \times \hat{i} = \hat{j} \times \hat{j} = \hat{k} \times \hat{k} = 0$$

and

$$\hat{i} \times \hat{j} = \hat{k}, \hat{j} \times \hat{k} = \hat{i}, \hat{k} \times \hat{i} = \hat{j},$$

so that, in general,

$$\vec{A} \times \vec{B} = \left(A_x \hat{i} + A_y \hat{j} + A_z \hat{k}\right) \times \left(B_x \hat{i} + B_y \hat{j} + B_z \hat{k}\right)$$
$$= \left(A_y B_z - A_z B_y\right)\hat{i} + \left(A_z B_x - A_x B_z\right)\hat{j} + \left(A_x B_y - A_y B_x\right)\hat{k}$$
$$= \begin{vmatrix} \hat{i} & \hat{j} & \hat{k} \\ A_x & A_y & A_z \\ B_x & B_y & B_z \end{vmatrix}. \tag{A.4}$$

This last determinantal form is convenient as an aid to memory.

Suppose one has a scalar function $f(x, y, z)$. The **gradient** of this scalar is defined as

$$\text{grad } f \equiv \nabla f \equiv \frac{\partial f}{\partial x}\hat{i} + \frac{\partial f}{\partial y}\hat{j} + \frac{\partial f}{\partial z}\hat{k}. \tag{A.5}$$

Note that

$$\nabla f \cdot d\vec{r} = \frac{\partial f}{\partial x}dx + \frac{\partial f}{\partial y}dy + \frac{\partial f}{\partial z}dz = df.$$

Now $d\vec{r} = dx\hat{i} + dy\hat{j} + dz\hat{k}$ may be chosen as a small displacement on a surface where f is a constant. Since $df = 0$ on such a surface, it follows from eq. (A.1) that ∇f is perpendicular to the surface. Therefore, ∇f gives the (greatest) rate of change of a function f, perpendicular at each point to that surface on which f is constant.

When each component of a vector $\vec{V}$ is a function of the coordinates x, y, z, one can define the **divergence** of $\vec{V}$ by taking the dot product of the vector operator ∇ with $\vec{V}$:

$$\text{div } \vec{V} \equiv \nabla \cdot \vec{V} = \frac{\partial V_x}{\partial x} + \frac{\partial V_y}{\partial y} + \frac{\partial V_z}{\partial z}. \tag{A.6}$$

The divergence itself is a scalar, but the div operation is performed only on a vector. If $\vec{v}$ is taken to represent the flow velocity at a point in a fluid, the integral of the divergence over a certain volume gives the excess volume of fluid flowing out of the region over that flowing into the volume, per unit of time.

The **curl** is an operator which acts on a vector and results in another vector. It is obtained by taking the cross product of the gradient operator and $\vec{V}$:

$$\text{curl } \vec{V} \equiv \nabla \times \vec{V} \equiv \begin{vmatrix} \hat{i} & \hat{j} & \hat{k} \\ \frac{\partial}{\partial x} & \frac{\partial}{\partial y} & \frac{\partial}{\partial z} \\ V_x & V_y & V_z \end{vmatrix}. \tag{A.7}$$

Consider $\vec{V}$ as a displacement vector tangent to a circle in the xy-plane (Figure A.1). Then we have

$$\vec{V} = -V_0(\sin\theta)\hat{i} + V_0(\cos\theta)\hat{j} = \frac{V_0}{r}[-y\hat{i} + x\hat{j}] \quad (r = \text{constant}),$$

$$\text{curl } \vec{V} = \frac{V_0}{r} \begin{vmatrix} \hat{i} & \hat{j} & \hat{k} \\ \frac{\partial}{\partial x} & \frac{\partial}{\partial y} & \frac{\partial}{\partial z} \\ -y & x & 0 \end{vmatrix} = \frac{2V_0}{r}\hat{k}.$$

If the fingers encircle the curve in the direction of $\vec{V}$, the thumb will point in the curl direction, to use the famous "right-hand rule."

Two important vector theorems, proved in many undergraduate texts, are known as **Gauss' and Stokes' theorems**. The first of these is

$$\int_V \nabla \cdot \vec{A} \, dV = \int_S \vec{A} \cdot d\vec{S}, \tag{A.8}$$

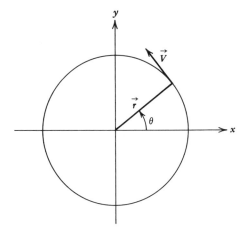

Figure A.1 Displacement vector.

where the surface S encloses the volume V; this theorem allows transformation of a volume integral into a surface integral, or vice versa. The physical significance is clear if we again view $\vec{A}$ as representing the flow velocity of a fluid. The second theorem, that of Stokes, is given by

$$\int_S \operatorname{curl} \vec{B} \cdot \vec{dS} = \int_L \vec{B} \cdot \vec{dL}, \tag{A.9}$$

where the line integral is taken along the curve L which forms the boundary of the surface S. Both theorems are vital to potential theory.

By applying the various concepts of this section, numerous useful vector identities can be derived. The more frequently needed identities are summarized in a table at the end of this appendix. As an example of how these relations can be established, consider

$$\operatorname{div} \operatorname{curl} \vec{V} = \frac{\partial}{\partial x}\left(\frac{\partial V_z}{\partial y} - \frac{\partial V_y}{\partial z}\right) + \frac{\partial}{\partial y}\left(\frac{\partial V_x}{\partial z} - \frac{\partial V_z}{\partial x}\right) + \frac{\partial}{\partial z}\left(\frac{\partial V_y}{\partial x} - \frac{\partial V_x}{\partial y}\right)$$

$$= \left(\frac{\partial^2 V_z}{\partial x \partial y} - \frac{\partial^2 V_z}{\partial y \partial x}\right) + \left(\frac{\partial^2 V_y}{\partial z \partial x} - \frac{\partial^2 V_y}{\partial x \partial z}\right) + \left(\frac{\partial^2 V_x}{\partial y \partial z} - \frac{\partial^2 V_x}{\partial z \partial y}\right) = 0,$$

the order of differentiation in the partial derivatives being irrelevant.

A.2 VECTOR RELATIONSHIPS IN ORTHOGONAL CURVILINEAR COORDINATES

In the previous section, we demonstrated the basic vector operations purely in terms of a rectangular coordinate system. It is clear, however, that the symmetry of a physical problem often dictates the use of other kinds of coordinates. We shall now show how the vector operations are transformed into other systems, restricting our considerations to spherical and cylindrical coordinates as examples. The method adopted here is not the usual one, but serves to bring forth the key ideas in a concise fashion.

A three-dimensional coordinate system is determined by three sets of surfaces; the three coordinates of a given point select one member of each set, the intersection of these three surfaces (on each of which one coordinate is constant) being the location of the point. Thus, for rectangular systems, the surfaces are orthogonal planes. In spherical systems, the surfaces are concentric spheres for r, circular cones for θ, and planes for ϕ, and so on.

If the coordinates are denoted by q_1, q_2, and q_3, the orthogonality of the surfaces makes it possible to construct an orthogonal triplet of unit vectors at any point. These will be called $\hat{\epsilon}_1$, $\hat{\epsilon}_2$, and $\hat{\epsilon}_3$, and they will be in the direction of increase of coordinates, perpendicular to the aforementioned surfaces; therefore, in general, they will vary in direction from point to point. Since the coordinates do not always have units of length, a small displacement ds along one of the unit vectors must be written as

$$ds_i = h_i dq_i \qquad (i = 1, 2, \text{ or, } 3),$$

implying that the gradient operator must have the form

$$\nabla = \frac{\hat{\epsilon}_1}{h_1}\frac{\partial}{\partial q_1} + \frac{\hat{\epsilon}_2}{h_2}\frac{\partial}{\partial q_2} + \frac{\hat{\epsilon}_3}{h_3}\frac{\partial}{\partial q_3}. \qquad (A.10)$$

To perform div or curl operations on a vector expressed in the new system, this expression for ∇ must be used with allowance for the variation in direction of the unit vectors. To clarify this point, consider a spherical system (Figure A.2). The appropriate unit vectors may be written in terms of the constant rectangular basis

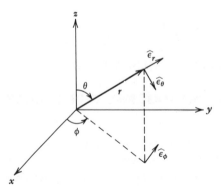

Figure A.2 Unit vectors of spherical coordinates.

Vector Relationships in Orthogonal Curvilinear Coordinates

set $\hat{i}, \hat{j}, \hat{k}$ as

$$\hat{\varepsilon}_r = \sin\theta\cos\phi\,\hat{i} + \sin\theta\sin\phi\,\hat{j} + \cos\theta\,\hat{k},$$

$$\hat{\varepsilon}_\theta = \cos\theta\cos\phi\,\hat{i} + \cos\theta\sin\phi\,\hat{j} - \sin\theta\,\hat{k},$$

$$\hat{\varepsilon}_\phi = -\sin\phi\,\hat{i} + \cos\phi\,\hat{j}.$$

Therefore, differentiation yields

$$\frac{\partial \hat{\varepsilon}_i}{\partial r} = 0 \quad \text{for all } i;$$

$$\frac{\partial \hat{\varepsilon}_r}{\partial \theta} = \hat{\varepsilon}_\theta, \quad \frac{\partial \hat{\varepsilon}_\theta}{\partial \theta} = -\hat{\varepsilon}_r, \quad \frac{\partial \hat{\varepsilon}_\phi}{\partial \theta} = 0;$$

$$\frac{\partial \hat{\varepsilon}_r}{\partial \phi} = \sin\theta\,\hat{\varepsilon}_\phi, \quad \frac{\partial \hat{\varepsilon}_\theta}{\partial \phi} = \cos\theta\,\hat{\varepsilon}_\phi, \quad \frac{\partial \hat{\varepsilon}_\phi}{\partial \phi} = -\sin\theta\,\hat{\varepsilon}_r - \cos\theta\,\hat{\varepsilon}_\theta.$$

The calculation of the divergence, for example, then proceeds as follows:

$$\text{div}\,\vec{A} = \nabla \cdot \vec{A} = \left[\hat{\varepsilon}_r \frac{\partial}{\partial r} + \frac{\hat{\varepsilon}_\theta}{r}\frac{\partial}{\partial \theta} + \frac{\hat{\varepsilon}_\phi}{r\sin\theta}\frac{\partial}{\partial \phi}\right] \cdot \left[A_r\hat{\varepsilon}_r + A_\theta\hat{\varepsilon}_\theta + A_\phi\hat{\varepsilon}_\phi\right]$$

$$= \frac{\partial A_r}{\partial r} + \frac{\hat{\varepsilon}_\theta}{r} \cdot \left[A_r\hat{\varepsilon}_\theta + \frac{\partial A_r}{\partial \theta}\hat{\varepsilon}_r - A_\theta\hat{\varepsilon}_r + \frac{\partial A_\theta}{\partial \theta}\hat{\varepsilon}_\theta + \frac{\partial A_\phi}{\partial \theta}\hat{\varepsilon}_\phi\right]$$

$$+ \frac{\hat{\varepsilon}_\phi}{r\sin\theta} \cdot \left[A_r\sin\theta\,\hat{\varepsilon}_\phi + \frac{\partial A_r}{\partial \phi}\hat{\varepsilon}_r + A_\theta\cos\theta\,\hat{\varepsilon}_\phi + \frac{\partial A_\theta}{\partial \phi}\hat{\varepsilon}_\theta + \frac{\partial A_\phi}{\partial \phi}\hat{\varepsilon}_\phi\right.$$

$$\left. - A_\phi\sin\theta\,\hat{\varepsilon}_r - A_\phi\cos\theta\,\hat{\varepsilon}_\theta\right]$$

$$= \frac{\partial A_r}{\partial r} + \frac{A_r}{r} + \frac{1}{r}\frac{\partial A_\theta}{\partial \theta} + \frac{A_r}{r} + \frac{A_\theta\cos\theta}{r\sin\theta} + \frac{1}{r\sin\theta}\frac{\partial A_\phi}{\partial \phi}$$

$$= \frac{1}{r^2}\frac{\partial}{\partial r}(r^2 A_r) + \frac{1}{r\sin\theta}\frac{\partial}{\partial \theta}(A_\theta\sin\theta) + \frac{1}{r\sin\theta}\frac{\partial A_\phi}{\partial \phi}.$$

Other such calculations would be similar.

A more common but less illuminating method makes use of Gauss' and Stokes' theorems to yield the general results for an orthogonal coordinate system:

$$\operatorname{div} \vec{A} = \frac{1}{h_1 h_2 h_3} \left[\frac{\partial}{\partial q_1}(A_1 h_2 h_3) + \frac{\partial}{\partial q_2}(A_2 h_1 h_3) + \frac{\partial}{\partial q_3}(A_3 h_1 h_2) \right],$$

(A.11)

$$\operatorname{curl} \vec{A} = \frac{\hat{\varepsilon}_1}{h_2 h_3} \left[\frac{\partial}{\partial q_2}(h_3 A_3) - \frac{\partial}{\partial q_3}(h_2 A_2) \right] + \text{other components,} \quad \text{(A.12)}$$

$$\nabla^2 f = \frac{1}{h_1 h_2 h_3} \left[\frac{\partial}{\partial q_1}\left(\frac{h_2 h_3}{h_1}\frac{\partial f}{\partial q_1}\right) + \frac{\partial}{\partial q_2}\left(\frac{h_1 h_3}{h_2}\frac{\partial f}{\partial q_2}\right) + \frac{\partial}{\partial q_3}\left(\frac{h_1 h_2}{h_3}\frac{\partial f}{\partial q_3}\right) \right].$$

(A.13)

A still more general technique utilizing the metric tensor and valid even in nonorthogonal systems is discussed in the last section of Appendix E.

TABLE A.1 *Useful Vector Relationships*

1. $(\vec{A} \times \vec{B}) \cdot \vec{C} = \vec{A} \cdot (\vec{B} \times \vec{C}) = -(\vec{A} \times \vec{C}) \cdot \vec{B}$ (triple scalar product).
2. $\vec{A} \times (\vec{B} \times \vec{C}) = \vec{B}(\vec{A} \cdot \vec{C}) - \vec{C}(\vec{A} \cdot \vec{B})$ (triple vector product-"back cab" rule).
3. $\nabla(\vec{A} \cdot \vec{B}) = (\vec{A} \cdot \nabla)\vec{B} + (\vec{B} \cdot \nabla)\vec{A} + \vec{A} \times \operatorname{curl} \vec{B} + \vec{B} \times \operatorname{curl} \vec{A}$.
4. $\nabla(\operatorname{div} \vec{A}) = \operatorname{curl} \operatorname{curl} \vec{A} + \nabla^2 \vec{A}$.
5. $\operatorname{div}(f\vec{A}) = f \operatorname{div} \vec{A} + \vec{A} \cdot \nabla f$ (f is a scalar function of the coordinates).
6. $\operatorname{div}(\vec{A} \times \vec{B}) = \vec{B} \cdot \operatorname{curl} \vec{A} - \vec{A} \cdot \operatorname{curl} \vec{B}$.
7. $\operatorname{div}(\operatorname{curl} \vec{A}) = 0$.
8. $\operatorname{curl}(f\vec{A}) = f \operatorname{curl} \vec{A} + \nabla f \times \vec{A}$.
9. $\operatorname{curl}(\vec{A} \times \vec{B}) = \vec{A} \operatorname{div} \vec{B} - \vec{B} \operatorname{div} \vec{A} + (\vec{B} \cdot \nabla)\vec{A} - (\vec{A} \cdot \nabla)\vec{B}$.
10. $\operatorname{curl}(\nabla f) = 0$.
11. $\int_V \nabla \cdot \vec{A} \, dV = \int_S \vec{A} \cdot \vec{dS}$ (Gauss' theorem).
12. $\int_S \operatorname{curl} \vec{B} \cdot \vec{dS} = \int_L \vec{B} \cdot \vec{dL}$ (Stokes' theorem).

Vector Operators in Spherical Coordinates

1. $\nabla f = \dfrac{\partial f}{\partial r}\hat{\varepsilon}_r + \dfrac{1}{r}\dfrac{\partial f}{\partial \theta}\hat{\varepsilon}_\theta + \dfrac{1}{r \sin \theta}\dfrac{\partial f}{\partial \phi}\hat{\varepsilon}_\phi$.

2. $\nabla \vec{A} = \dfrac{1}{r^2}\dfrac{\partial}{\partial r}(r^2 A_r) + \dfrac{1}{r \sin \theta}\dfrac{\partial}{\partial \theta}(A_\theta \sin \theta) + \dfrac{1}{r \sin \theta}\dfrac{\partial A_\phi}{\partial \phi}$.

3. $\nabla \times \vec{A} = \dfrac{\hat{\varepsilon}_r}{r \sin \theta}\left[\dfrac{\partial}{\partial \theta}(A_\phi \sin \theta) - \dfrac{\partial A_\theta}{\partial \phi}\right] + \dfrac{\hat{\varepsilon}_\theta}{r \sin \theta}\left[\dfrac{\partial A_r}{\partial \phi} - \dfrac{\partial}{\partial r}(rA_\phi \sin \theta)\right]$
$\quad + \dfrac{\hat{\varepsilon}_\phi}{r}\left[\dfrac{\partial}{\partial r}(rA_\theta) - \dfrac{\partial A_r}{\partial \theta}\right]$.

4. $\nabla^2 f = \dfrac{1}{r}\dfrac{\partial^2}{\partial r^2}(rf) + \dfrac{1}{r^2 \sin \theta}\dfrac{\partial}{\partial \theta}\left(\sin \theta \dfrac{\partial f}{\partial \theta}\right) + \dfrac{1}{r^2 \sin^2 \theta}\dfrac{\partial^2 f}{\partial \phi^2}$.

Vector Operators in Cylindrical Coordinates

1. $\nabla f = \dfrac{\partial f}{\partial \rho}\hat{\varepsilon}_\rho + \dfrac{1}{\rho}\dfrac{\partial f}{\partial \phi}\hat{\varepsilon}_\phi + \dfrac{\partial f}{\partial z}\hat{\varepsilon}_z$.

2. $\nabla \cdot \vec{A} = \dfrac{1}{\rho}\left[\dfrac{\partial}{\partial \rho}(\rho A_\rho) + \dfrac{\partial A_\phi}{\partial \phi} + \rho \dfrac{\partial A_z}{\partial z}\right]$.

3. $\nabla \times \vec{A} = \hat{\varepsilon}_\rho\left[\dfrac{1}{\rho}\dfrac{\partial A_z}{\partial \phi} - \dfrac{\partial A_\phi}{\partial z}\right] + \hat{\varepsilon}_\phi\left[\dfrac{\partial A_\rho}{\partial z} - \dfrac{\partial A_z}{\partial \rho}\right] + \dfrac{\hat{\varepsilon}_z}{\rho}\left[\dfrac{\partial}{\partial \rho}(\rho A_\phi) - \dfrac{\partial A_\rho}{\partial \phi}\right]$.

4. $\nabla^2 f = \dfrac{\partial^2 f}{\partial \rho^2} + \dfrac{1}{\rho}\dfrac{\partial f}{\partial \rho} + \dfrac{1}{\rho^2}\dfrac{\partial^2 f}{\partial \phi^2} + \dfrac{\partial^2 f}{\partial z^2}$.

Appendix B
DIRAC DELTA FUNCTIONS AND THEIR ROLE IN PHYSICS

B.1 THE CONCEPT OF THE DELTA FUNCTION

In reality, physical processes do not occur at a point in space or in an instant of time. Nevertheless, it is convenient when working idealized problems to treat natural phenomena as localized in space or instantaneous in time. In this way, we may acquire much insight into significant aspects of the situation (as in the discussion of point masses and gravitational fields early in the book).

It is therefore advantageous to use the Dirac delta function, defined in one dimension as follows:

$$\delta(x) \equiv 0 \quad (x \neq 0),$$
$$\delta(x) \equiv \infty \quad (x = 0),$$

such that

$$\int_{-\infty}^{\infty} \delta(x) f(x)\, dx = f(0) \tag{B.1}$$

for any function $f(x)$. The expression (B.1) is called the **sifting** property, since it selects, from all the possible values of $f(x)$ at different points along the x-axis, that value $f(0)$ which corresponds to the only point where $\delta(x)$ does not vanish. From the sifting property, many useful relationships for the δ-function will shortly be derived.

Of course, $\delta(x)$ is not the well-behaved continuous function that a mathematician would like to see. There is, in fact, no function that has the property (B.1) for all $f(x)$. Just as our applications are physical idealizations, so is the δ-function a mathematical idealization. It turns out, however, that if we view the δ-function as the *limit* of a function such as

$$\lim_{n \to \infty} \frac{n}{\sqrt{\pi}} e^{-n^2 x^2}$$

or

$$\lim_{n \to \infty} \frac{n}{\pi} \frac{1}{1 + n^2 x^2}$$

(the factors of π simply normalize these expressions to unity), then it is possible to satisfy the sifting property for all $f(x)$ in a rigorous fashion. The reader who is interested in pursuing these mathematical details should consult a modern text on the theory of distributions.[1] We shall assume here that the δ-function has meaning within an integrand as in the sifting property and work with it much like any normal function, understanding that our results can be justified in a more rigorous yet time-consuming approach.

B.2 DELTA FUNCTION CALCULUS

Let us derive a few basic identities which involve the δ-function. Setting

$$\delta'(x) \equiv \frac{d\delta(x)}{dx}$$

and integrating by parts, we find that

$$\int_{-\infty}^{\infty} \delta'(x) f(x)\, dx = [\delta(x) f(x)]_{-\infty}^{\infty} - \int_{-\infty}^{\infty} \delta(x) f'(x)\, dx = -f'(0),$$

since $\delta(x)$ vanishes at the infinite limits of the first term, and the second term obeys the sifting property. The result is easily generalized.

Next, consider

$$\int_{-\infty}^{\infty} \delta(x - a) f(x)\, dx.$$

Letting $x - a \equiv y$ gives

$$\int_{-\infty}^{\infty} \delta(y) f(y + a)\, dy = f(a),$$

where a is a constant. Thus, we conclude that the function $f(x)$ is to be evaluated at the point where the δ-function has a zero argument, in all cases.

Often, one sees an identity like $x\delta(x) = 0$. This statement means that

$$\int_{-\infty}^{\infty} x\delta(x) f(x)\, dx = \int_{-\infty}^{\infty} \delta(x)[xf(x)]\, dx = 0 \cdot f(0) = 0,$$

provided that $f(x)$ does not "blow up" as $x \to 0$ (at least not too rapidly).

---

[1] E. Butkov, *Mathematical Physics*, Addison-Wesley, Reading, Mass., 1968, Chap. 6.

Another such identity is

$$\delta(ax) = \frac{1}{a}\delta(x),$$

where $a > 0$; it is derived by substitution of $y \equiv ax$, obtaining

$$\int_{-\infty}^{\infty} \delta(ax)f(x)\,dx = \int_{-\infty}^{\infty} \delta(y)f\left(\frac{y}{a}\right)\left(\frac{dy}{a}\right) = \frac{1}{a}f(0),$$

establishing the result. For $a < 0$, we find similarly that

$$\delta(ax) = -\frac{1}{a}\delta(x),$$

so, in general,

$$\delta(ax) = \frac{1}{|a|}\delta(x),$$

the special case $\delta(-x) = \delta(x)$ being needed quite frequently.

As a further example,

$$\delta(f(x)) = \frac{1}{|df/dx|_{x=x_0}}\delta(x - x_0),$$

where x_0 is a root of the function $f(x)$; that is, $f(x_0) = 0$. To prove this identity, expand $f(x)$ in a Taylor series about $x = x_0$:

$$f(x) = f(x_0) + \left(\frac{df}{dx}\right)_{x=x_0}(x - x_0) + \cdots \doteq \left(\frac{df}{dx}\right)_{x=x_0}(x - x_0),$$

keeping only the linear term. Then,

$$\delta(f(x)) = \delta\left[\left(\frac{df}{dx}\right)_{x=x_0}(x - x_0)\right] = \frac{1}{|df/dx|_{x=x_0}}\delta(x - x_0),$$

which follows upon realizing that the derivative evaluated at x_0 is a constant and applying previous identities. The result is exact, for the δ-functions contribute only at $x = x_0$; hence, the higher terms of the Taylor series are unimportant. If the function $f(x)$ has other roots than x_0 on the real axis, another term of this same form will have to be added for each of them.

Sometimes, one may define a unit step function $u(x)$ as

$$u(x) \equiv 0 \quad (x < 0),$$

$$u(x) \equiv 1 \quad (x > 0).$$

Delta Function Calculus

Note that the *derivative* of $u(x)$ is zero everywhere except at the origin, where it is infinite. Since

$$\int_{-\infty}^{\infty} \frac{du(x)}{dx} dx = [u(x)]_{-\infty}^{\infty} = 1 \quad \text{and} \quad \int_{-\infty}^{\infty} \delta(x) dx = 1,$$

this suggests that $du(x)/dx$ be identified with $\delta(x)$.

TABLE B.1 Useful Relationships Involving Dirac Delta Functions

Identities

1. $\int_{-\infty}^{\infty} \delta(x) f(x) dx = f(0)$ (sifting property).

2. $\int_{-\infty}^{\infty} \delta(x - a) f(x) dx = f(a)$ (shifted sifting).

3. $\int_{-\infty}^{\infty} \delta'(x) f(x) dx = -f'(0)$ (derivative sifting).

4. $\int_{-\infty}^{\infty} \frac{d^m \delta(x)}{dx^m} f(x) dx = (-1)^m \left[\frac{d^m f}{dx^m} \right]_{x=0}$.

5. $\delta(x) = \frac{du(x)}{dx}$ (relation to step function).

6. $x\delta(x) = 0$.

7. $\delta(x) = \delta(-x)$ (even function).

8. $\delta(ax) = \frac{1}{|a|} \delta(x)$.

9. $\delta(x^2 - a^2) = \frac{1}{2a} [\delta(x + a) + \delta(x - a)]$, $a > 0$.

10. $\delta[f(x)] = \frac{1}{|df/dx|_{x=x_0}} \delta(x - x_0)$, x_0 a root of $f(x)$ (sum over roots).

Representations

1. $\delta(x) = \lim_{n \to \infty} \frac{n}{\sqrt{\pi}} e^{-n^2 x^2}$.

2. $\delta(x) = \lim_{n \to \infty} \frac{n}{\pi} \frac{1}{1 + n^2 x^2}$.

3. $\delta(x) = \lim_{n \to \infty} \frac{n}{\pi} \frac{\sin^2 nx}{n^2 x^2}$.

4. $\delta(x) = \frac{1}{2L} + \frac{1}{L} \sum_{n=1}^{\infty} \cos\left(\frac{n\pi x}{L}\right)$ (Fourier series for region $+L$ to $-L$).

5. $\delta(x) = \frac{1}{2\pi} \int_{-\infty}^{\infty} e^{ikx} dk$ (integral representation).

We summarize this section with a table of useful expressions, most of which have already been proved. Included for the sake of thoroughness are integral and series representations for the δ-function, as these are often needed in practical manipulations.

B.3 DELTA FUNCTIONS IN MORE THAN ONE DIMENSION

In rectangular coordinates, the generalization of the delta function concept is straightforward. A unit point mass at the origin, for example, would be represented in three dimensions by

$$\delta(\vec{r}) = \delta(x)\delta(y)\delta(z)$$

and the integration would now be performed over all three coordinates independently.

But suppose we want to make a transformation to another coordinate system, as discussed in the text. Calling the new coordinates u, v, and w, we have

Direct Relations	Inverse Relations
$u = u(x, y, z)$	$x = x(u, v, w)$
$v = v(x, y, z)$	$y = y(u, v, w)$
$w = w(x, y, z)$	$z = z(u, v, w)$

These are the equations of the transformation.

It is known from calculus[2] that the volume elements in the two systems are related by

$$dV = dx\,dy\,dz = |J|\,du\,dv\,dw, \qquad (B.2)$$

where J is the **Jacobian** of the transformation, defined by

$$J(u, v, w) \equiv \begin{vmatrix} \dfrac{\partial x}{\partial u} & \dfrac{\partial y}{\partial u} & \dfrac{\partial z}{\partial u} \\ \dfrac{\partial x}{\partial v} & \dfrac{\partial y}{\partial v} & \dfrac{\partial z}{\partial v} \\ \dfrac{\partial x}{\partial w} & \dfrac{\partial y}{\partial w} & \dfrac{\partial z}{\partial w} \end{vmatrix}. \qquad (B.3)$$

The Jacobian will have physical dimensions whenever the new coordinates are not

---

[2] I. Sokolnikoff, *Advanced Calculus*, McGraw-Hill, New York, 1939, p. 150.

all in length units; it is related closely to the metric tensor. The u-coordinate might be an angle, for instance; to maintain the dimensional consistency, there will then be a length factor needed in J. Any $\delta(u)$ factor to be integrated must then be divided by such a factor (so as to cancel that which is present in J) and such other factors as may be required to make $\int \delta(\vec{r}) \, dx \, dy \, dz = 1$.

To illustrate this point, let us consider spherical coordinates. The inverse relations are $x = r \sin\theta \cos\phi$, $y = r \sin\theta \sin\phi$, $z = r \cos\theta$, and $r \equiv |\vec{r}|$. Taking $r = u$, $\theta = v$, $\phi = w$, and differentiating, we find that the Jacobian is

$$\begin{vmatrix} \sin\theta \cos\phi & \sin\theta \sin\phi & \cos\theta \\ r\cos\theta \cos\phi & r\cos\theta \sin\phi & -r\sin\theta \\ -r\sin\theta \sin\phi & r\sin\theta \cos\phi & 0 \end{vmatrix}$$

which equals $r^2 \sin\theta$, as would be expected. The volume element is then $r^2 \sin\theta \, dr \, d\theta \, d\phi$. For the point mass above,

$$\delta(\vec{r}) = \frac{\delta(r)}{4\pi r^2},$$

giving the result

$$\int \delta(\vec{r}) \, dx \, dy \, dz = \int \frac{\delta(r)}{4\pi r^2} r^2 \sin\theta \, dr \, d\theta \, d\phi = 1,$$

the angular integrals contributing the 4π solid angle factor. Only a factor of $4\pi r^2$ is needed in the denominator here, because the volume element is effectively $4\pi r^2 \, dr$ in this spherically symmetrical situation. Had there also been delta functions of θ and ϕ present, then division by the Jacobian $r^2 \sin\theta$ would have been required. In general, we mean that one can always write

$$\delta(x - x_0)\delta(y - y_0)\delta(z - z_0) \, dx \, dy \, dz$$
$$= \frac{\delta(u - u_0)\delta(v - v_0)\delta(w - w_0)}{|J|} (|J| \, du \, dv \, dw)$$

(B.4)

for the three-dimensional delta function centered at some arbitrary point specified by the vector $\vec{r}_0 \equiv x_0 \hat{i} + y_0 \hat{j} + z_0 \hat{k}$.

Appendix C
DETERMINANTS AND MATRICES

C.1 DEFINITION AND EVALUATION OF DETERMINANTS

A determinant of nth order is a set of n^2 numbers (or symbols representing numbers), which are called the **elements**, arranged between two vertical bars in the form of a square array of n rows and n columns. The numerical value of the determinant is obtained by taking the algebraic sum of all possible products of n factors formed by taking one and only one element from each row and column, then affixing the proper sign to each such product.

In order to understand the choice of sign, we must know what an **inversion** is. If a set of numbers or symbols has a particular arrangement which might be designated as the "natural order," then any other arrangement of this set has a number of inversions equal to the number of ways in which the natural order has been reversed. Thus, taking 1234 as the natural order of positive integers, the arrangement 4231 has five inversions: 4 before 2, 4 before 3, 4 before 1, 2 before 1, and 3 before 1. The set 67532 has nine inversions, and so on. It can be seen that the number of inversions is simply the number of interchanges of adjacent digits that will be required to bring the set back to the natural order.

Now suppose we write the determinant as

$$D \equiv \begin{vmatrix} a_1 & b_1 & \ldots & x_1 \\ a_2 & b_2 & \ldots & x_2 \\ \vdots & & & \\ a_n & b_n & \ldots & x_n \end{vmatrix}$$

Each of the product terms mentioned above will be assigned a plus or minus sign, depending on whether the number of inversions in the *subscripts* is even or odd, respectively, *after* the letters in the various factors have been put into the natural alphabetical order in each term.

Definition and Evaluation of Determinants

The element a_1 is said to be in the **leading position**. The term having factors $(a_1 b_2 \ldots x_n)$, going from upper left to lower right, is called the **principal diagonal** term.

When the letters of a particular product term are in natural order, the first subscript can be any number from 1 to n, but the next factor must be from a different row and hence must have a different subscript, which can be any of the remaining $n-1$ possibilities. If this argument is continued with the factors that follow in the product term, we see that the expansion of the determinant must contain $n!$ terms. (It is also clear from symmetry that a transposition, or complete interchange of corresponding rows with columns, will not change the value of the determinant.)

If in a determinant we delete the row and column in which a particular element such as f_i lies, the determinant of order $n-1$ formed from the remaining elements in their same relative positions is called the **minor** of f_i; we shall designate it as F_i. Thus, in the example where

$$D = \begin{vmatrix} a_1 & b_1 & c_1 & d_1 \\ a_2 & b_2 & c_2 & d_2 \\ a_3 & b_3 & c_3 & d_3 \\ a_4 & b_4 & c_4 & d_4 \end{vmatrix}$$

the minor of the element b_2 is

$$B_2 \equiv \begin{vmatrix} a_1 & c_1 & d_1 \\ a_3 & c_3 & d_3 \\ a_4 & c_4 & d_4 \end{vmatrix}.$$

Sometimes, the minor of a given element is multiplied by the sign factor $(-1)^{i+j}$ and the product is referred to as the **cofactor** of the element. Here, i is the number of the row and j is the number of the column in which the element is found. This leads us to the following important theorem:

Theorem C.1 A determinant of order n may be expressed as the sum of the n products obtained by multiplying each element of any row or column by its cofactor.

The general proof of this basic proposition is indicated in the next section. Here, we merely illustrate its application to the case of a 3 × 3 determinant:

$$\begin{vmatrix} a_1 & b_1 & c_1 \\ a_2 & b_2 & c_2 \\ a_3 & b_3 & c_3 \end{vmatrix} = a_1 b_2 c_3 + a_2 b_3 c_1 + a_3 b_1 c_2 - a_1 b_3 c_2 - a_2 b_1 c_3 - a_3 b_2 c_1.$$

Factoring the a-elements, we have

$$a_1(b_2c_3 - b_3c_2) + a_2(b_3c_1 - b_1c_3) + a_3(b_1c_2 - b_2c_1) = a_1A_1 - a_2A_2 + a_3A_3$$

in terms of minors of the elements from the first column. The choice of sign for each term need only be determined once at the beginning of row or column and will alternate in the successive terms following. If one starts with the leading position, the first term will be positive.

The usefulness of this theorem is apparent in reducing the order of a determinant so as to make its evaluation simpler in practice.

C.2 PROPERTIES OF DETERMINANTS

In this section, we state some properties that are vital both in formal manipulations and in explicit evaluation:

Property 1 For every theorem on the rows of a determinant, there is a corresponding theorem on the columns, and conversely.

The validity of this statement clearly follows from the symmetry between letters and subscripts inherent in the determinantal definition.

Property 2 If any two rows (or columns) are interchanged, the sign of the determinant is changed.

An interchange of two *adjacent* rows is such as to interchange an adjacent pair of subscripts in each term of the expansion, which in turn changes the number of inversions by one, or changes the sign of each term, hence the determinant.

But if the two rows in question are separated by m intermediate rows, there must be m interchanges of adjacent rows to bring the upper row down to the position just above the lower. The two rows can then be interchanged with one another, and, finally, the row initially lower (now above the other) must be moved to the initial upper row position, requiring m further interchanges. In all, $2m + 1$ interchanges of adjacent rows have been carried out, each of which causes a sign change. Since $2m + 1$ is odd, the sign of the determinant will always be changed.

Property 3 If two rows (columns) of a determinant are identical, the value of the determinant is zero.

The determinant is certainly invariant under an interchange of identical rows, yet by Property 2, such an action must change its sign. Only if the determinant vanishes can this conflict be reconciled. (The student of quantum mechanics will be familiar with this property in connection with the Pauli exclusion principle.)

Property 4 If each element of a given row (column) is multiplied by the same number k, the value of the determinant is multiplied by k.

Properties of Determinants

This property follows at once from the determinant definition, there being one and only one element from a particular row (containing the factor k) in each term of the expansion. Thus, as an example,

$$\begin{vmatrix} ka_1 & b_1 \\ ka_2 & b_2 \end{vmatrix} = \begin{vmatrix} ka_1 & kb_1 \\ a_2 & b_2 \end{vmatrix} = k \begin{vmatrix} a_1 & b_1 \\ a_2 & b_2 \end{vmatrix}.$$

It is apparent that a determinant vanishes if all elements of a given row (column) are zero. Further, by Properties 3 and 4, we see that it will also vanish if corresponding elements of two rows (columns) are proportional to one another.

Property 5 If each element of a given row (column) is written as the sum of two terms, the determinant may be written as the sum of two determinants.

To illustrate, note that

$$\begin{vmatrix} (a_1 + c_1) & b_1 \\ (a_2 + c_2) & b_2 \end{vmatrix} = \begin{vmatrix} a_1 & b_1 \\ a_2 & b_2 \end{vmatrix} + \begin{vmatrix} c_1 & b_1 \\ c_2 & b_2 \end{vmatrix}.$$

The validity of the statement is readily established by expanding both sides of the equation term by term and observing the equality of corresponding terms.

Property 6 The value of a determinant is not affected if the elements of a given row (column) are increased by k times the corresponding elements of another row (column).

Using a 3×3 determinant as an example and applying Properties 5, 4, and 3 in that order, the proof proceeds in this manner:

$$\begin{vmatrix} a_1 & b_1 & (c_1 + kb_1) \\ a_2 & b_2 & (c_2 + kb_2) \\ a_3 & b_3 & (c_3 + kb_3) \end{vmatrix} = \begin{vmatrix} a_1 & b_1 & c_1 \\ a_2 & b_2 & c_2 \\ a_3 & b_3 & c_3 \end{vmatrix} + \begin{vmatrix} a_1 & b_1 & kb_1 \\ a_2 & b_2 & kb_2 \\ a_3 & b_3 & kb_3 \end{vmatrix}$$

$$= \begin{vmatrix} a_1 & b_1 & c_1 \\ a_2 & b_2 & c_2 \\ a_3 & b_3 & c_3 \end{vmatrix} + k \begin{vmatrix} a_1 & b_1 & b_1 \\ a_2 & b_2 & b_2 \\ a_3 & b_3 & b_3 \end{vmatrix}$$

$$= \begin{vmatrix} a_1 & b_1 & c_1 \\ a_2 & b_2 & c_2 \\ a_3 & b_3 & c_3 \end{vmatrix}.$$

These properties are very useful in numerical evaluation. The usual procedure is to apply them in such a way as to reduce to zero all elements of a particular row (column) except one, expansion by minors from this row then being easily accomplished. The order of the determinant is thereby effectively reduced, and the procedure may be repeated.

Example
Evaluate

$$\begin{vmatrix} 5 & -5 & 4 \\ -1 & 2 & 3 \\ -3 & 1 & -2 \end{vmatrix}.$$

Solution
By Property 6, doubling the first column and adding it to the second gives

$$\begin{vmatrix} 5 & 5 & 4 \\ -1 & 0 & 3 \\ -3 & -5 & -2 \end{vmatrix}.$$

Now triple the first column and add to the third:

$$\begin{vmatrix} 5 & 5 & 19 \\ -1 & 0 & 0 \\ -3 & -5 & -11 \end{vmatrix}.$$

Expanding from the second row yields

$$(-1)(-1)\begin{vmatrix} 5 & 19 \\ -5 & -11 \end{vmatrix}.$$

Direct evaluation now produces $(-55 + 95) = 40$ for the value. Alternatively, one can add second row to first to get

$$\begin{vmatrix} 0 & 8 \\ -5 & -11 \end{vmatrix} = (-1)(8)(-5) = 40,$$

expanding by minors of the first row.

C.3 SYSTEMS OF n LINEAR EQUATIONS IN n UNKNOWNS

Consider as an example a system of three equations in three unknowns x, y, z:

$$a_1 x + b_1 y + c_1 z = d_1,$$

$$a_2 x + b_2 y + c_2 z = d_2,$$

$$a_3 x + b_3 y + c_3 z = d_3.$$

The symbols with subscripts represent numerical constants. If $d_1 = d_2 = d_3 = 0$, the system is said to be **homogeneous**; we shall assume nonhomogeneity. Define the determinant

$$\Delta \equiv \begin{vmatrix} a_1 & b_1 & c_1 \\ a_2 & b_2 & c_2 \\ a_3 & b_3 & c_3 \end{vmatrix}.$$

Using the minors A_i of the elements a_i in Δ, multiply the first equation by A_1, the second by $-A_2$, and the third by A_3, then add the three equations:

$$(a_1 A_1 - a_2 A_2 + a_3 A_3)x + (b_1 A_1 - b_2 A_2 + b_3 A_3)y + (c_1 A_1 - c_2 A_2 + c_3 A_3)z$$
$$= d_1 A_1 - d_2 A_2 + d_3 A_3.$$

By Theorem C.1, the coefficient of x is Δ, while the coefficients of y and z must vanish; the y-coefficient, for instance, is equivalent to the expansion of the determinant

$$\begin{vmatrix} b_1 & b_1 & c_1 \\ b_2 & b_2 & c_2 \\ b_3 & b_3 & c_3 \end{vmatrix},$$

which is zero by Property 3. The resulting solution for x is then

$$x = \frac{d_1 A_1 - d_2 A_2 + d_3 A_3}{\Delta} = \frac{\begin{vmatrix} d_1 & b_1 & c_1 \\ d_2 & b_2 & c_2 \\ d_3 & b_3 & c_3 \end{vmatrix}}{\Delta}, \qquad \Delta \neq 0.$$

Similar solutions are found for y and z, the numerator determinant in each case being the same as Δ, except that the column containing the coefficients of the variable is replaced by the column of d-constants. Thus, we have what is known as Cramer's rule:

Theorem C.2 A system of n linear equations in n unknowns has a unique solution provided the determinant Δ of the coefficients is not zero. If Δ is in fact zero, there will be *no* finite solution unless the system of equations is homogeneous, in which case the numerator determinants also vanish and there are an unlimited number of solutions.

C.4 PROPERTIES OF MATRICES AND INVERSES

In Chapter 5, we saw that linear transformations could be treated conveniently by means of matrices. Here we should like to review and extend the matrix concepts a little further.

Suppose that there are matrices **A**, **B**, and **C**, with elements a_{ij}, b_{ij}, and c_{ij}, respectively. The addition of **A** and **B** to form the sum **C** is defined in terms of normal addition of corresponding elements: $\mathbf{C} = \mathbf{A} + \mathbf{B}$ means that $c_{ij} \equiv a_{ij} + b_{ij}$ for all i and j.

Subtraction is defined similarly by a difference of elements, and multiplication by the rule $\mathbf{C} = \mathbf{AB}$ when $c_{ij} \equiv \sum_k a_{ik} b_{kj}$.

The inverse $\mathbf{A}^{-1}$ of any matrix **A** has the property that multiplication by **A** itself from either side yields the identity matrix **1**: $\mathbf{A}^{-1}\mathbf{A} = \mathbf{1} = \mathbf{A}\mathbf{A}^{-1}$.

We now wish to examine methods for constructing the inverse of a given matrix and how these methods are related to determinants. Recalling the determinant Δ from the last section, let us form a matrix **D** with the elements of Δ, so that

$$d_{i1} \equiv a_i, \qquad d_{i2} \equiv b_i, \qquad d_{i3} \equiv c_i,$$

and

$$\mathbf{D} \equiv \begin{pmatrix} a_1 & b_1 & c_1 \\ a_2 & b_2 & c_2 \\ a_3 & b_3 & c_3 \end{pmatrix}, \qquad \Delta = |\mathbf{D}|.$$

If we denote the minor of an element d_{ij} by M_{ij} and the cofactor by $(-1)^{i+j}M_{ij}$, then

$$\sum_i d_{ik}(-1)^{i+j}M_{ij} = \Delta \delta_{j,k}$$

from the results of the previous section. Now, define a matrix **T** with elements

$$t_{ji} \equiv (-1)^{i+j}M_{ij}$$

(**T** is simply the transpose of the matrix of cofactors), and note that

$$\sum_i d_{ik}(-1)^{i+j}M_{ij} = \sum_i d_{ik}t_{ji} = \sum_i t_{ji}d_{ik}$$

$$= (\mathbf{TD})_{jk} = \Delta \delta_{j,k} = \Delta (\mathbf{1})_{jk},$$

Properties of Matrices and Inverses

or

$$TD = \Delta 1,$$

whereupon

$$D^{-1} = T/\Delta.$$

If $\Delta \neq 0$, then **D** is called **nonsingular**, and its inverse can always be found in this way.

Theorem C.3 The inverse of any nonsingular matrix **D** is obtained by constructing a matrix of cofactors for each element in **D**; the inverse is the transpose of the cofactor matrix divided by the determinant of **D**.

As an example, let us again consider

$$D = \begin{pmatrix} 5 & -5 & 4 \\ -1 & 2 & 3 \\ -3 & 1 & -2 \end{pmatrix}, \quad \text{with } |D| = 40.$$

The matrix of cofactors is

$$C \equiv \begin{pmatrix} -7 & -11 & 5 \\ -6 & 2 & 10 \\ -23 & -19 & 5 \end{pmatrix},$$

whereupon

$$D^{-1} = \frac{\tilde{C}}{\Delta} = \frac{1}{40} \begin{pmatrix} -7 & -6 & -23 \\ -11 & 2 & -19 \\ 5 & 10 & 5 \end{pmatrix},$$

as is readily verified.

There is another way of finding inverse matrices which proceeds by analogy with the determinantal properties discussed earlier. We define three types of **elementary row operations** on matrices as follows:

1. Interchange of any two rows.
2. Multiplication of some row by any nonzero constant k.
3. Addition of any multiple of one row to any other row.

An **elementary matrix** is one that is obtained from the identity matrix through one such elementary row operation. In the 2 × 2 case, the only possible elementary matrices are

$$\begin{pmatrix} 0 & 1 \\ 1 & 0 \end{pmatrix} \text{ (Type 1 operation)},$$

$$\begin{pmatrix} k & 0 \\ 0 & 1 \end{pmatrix} \quad \text{or} \quad \begin{pmatrix} 1 & 0 \\ 0 & k \end{pmatrix} \text{ (Type 2 operation)},$$

$$\begin{pmatrix} 1 & 0 \\ k & 1 \end{pmatrix} \quad \text{or} \quad \begin{pmatrix} 1 & k \\ 0 & 1 \end{pmatrix} \text{ (Type 3 operation)}.$$

The application of such a matrix to some arbitrary nonsingular **A** performs the corresponding row operation on the elements of **A**. A standard procedure can be developed for reducing **A** to the identity matrix through a product of n elementary matrices $\mathbf{E}_i$:

$$\mathbf{E}_n \mathbf{E}_{n-1} \cdots \mathbf{E}_2 \mathbf{E}_1 \mathbf{A} = \mathbf{1}.$$

Once such a product is determined (which is the hard part to accomplish), clearly,

$$\mathbf{A}^{-1} = \mathbf{E}_n \mathbf{E}_{n-1} \cdots \mathbf{E}_2 \mathbf{E}_1.$$

Furthermore, inverting the last equation gives ($\mathbf{F}_1 \equiv \mathbf{E}_1^{-1}$, etc.)

$$\mathbf{A} = \mathbf{E}_1^{-1} \mathbf{E}_2^{-1} \cdots \mathbf{E}_{n-1}^{-1} \mathbf{E}_n^{-1} \equiv \mathbf{F}_1 \mathbf{F}_2 \cdots \mathbf{F}_n.$$

A is composed of a product of inverses $\mathbf{F}_i$, which are themselves elementary. (It can be shown that every elementary matrix is nonsingular; the inverse must be elementary so as to perform a row operation which reverses the original one.)

As an example, consider the matrix

$$\mathbf{A} = \begin{pmatrix} 2 & -1 \\ -1 & 1 \end{pmatrix}.$$

Successive application of Type 3 elementary matrices produces

$$\begin{pmatrix} 1 & 0 \\ 1 & 1 \end{pmatrix} \begin{pmatrix} 1 & 1 \\ 0 & 1 \end{pmatrix} \begin{pmatrix} 2 & -1 \\ -1 & 1 \end{pmatrix} = \begin{pmatrix} 1 & 0 \\ 1 & 1 \end{pmatrix} \begin{pmatrix} 1 & 0 \\ -1 & 1 \end{pmatrix} = \begin{pmatrix} 1 & 0 \\ 0 & 1 \end{pmatrix}.$$

Then,

$$\mathbf{A}^{-1} = \begin{pmatrix} 1 & 0 \\ 1 & 1 \end{pmatrix} \begin{pmatrix} 1 & 1 \\ 0 & 1 \end{pmatrix} = \begin{pmatrix} 1 & 1 \\ 1 & 2 \end{pmatrix}.$$

The two prefactors were found by reducing the off-diagonal elements of **A** to zeroes, much like the earlier determinantal procedure.

In addition, the determinantal theorems immediately give the following results for the determinants of a product of a particular type of elementary matrix with an arbitrary **A**:

Type 1 $|\mathbf{E}| = -1$, hence $|\mathbf{EA}| = -|\mathbf{A}|$.
Type 2 $|\mathbf{E}| = k|\mathbf{1}| = k$, hence $|\mathbf{EA}| = k|\mathbf{A}|$.
Type 3 $|\mathbf{E}| = 1$, hence $|\mathbf{EA}| = |\mathbf{A}|$.

In all cases, we can then write that

$$|\mathbf{EA}| = |\mathbf{E}||\mathbf{A}|.$$

Continuing this process, we also have from above that

$$|\mathbf{A}| = |\mathbf{F}_1\mathbf{F}_2 \cdots \mathbf{F}_n| = |\mathbf{F}_1||\mathbf{F}_2\mathbf{F}_3 \ldots \mathbf{F}_n|$$
$$= \cdots = |\mathbf{F}_1||\mathbf{F}_2| \cdots |\mathbf{F}_n|.$$

Another matrix **B** has a determinant which is the product of those for some other set of elementary matrices $\mathbf{G}_i$:

$$|\mathbf{B}| = |\mathbf{G}_1||\mathbf{G}_2| \cdots |\mathbf{G}_m|, \quad \text{where } \mathbf{B} = \mathbf{G}_1\mathbf{G}_2 \cdots \mathbf{G}_m.$$

Then,

$$|\mathbf{A}||\mathbf{B}| = |\mathbf{F}_1| \cdots |\mathbf{G}_m| = |\mathbf{F}_1 \cdots \mathbf{G}_m| = |\mathbf{AB}|.$$

The determinant of a matrix product is the product of the determinants, a conclusion true in general and often used in the text; it follows at once from the decomposition in terms of elementary matrices.

C.5 THE CHARACTERISTIC EQUATION OF A MATRIX

The determinantal product theorem just established implies that

$$|\mathbf{AA}^{-1}| = |\mathbf{1}| = 1 = |\mathbf{A}||\mathbf{A}^{-1}|;$$

any nonsingular matrix has a determinant which is reciprocal to that of its inverse. Suppose now that two matrices **A** and **B** are related by a similarity transformation with the matrix **T**, so that

$$\mathbf{B} = \mathbf{T}^{-1}\mathbf{AT}, \quad \mathbf{B} - \lambda\mathbf{1} = \mathbf{T}^{-1}(\mathbf{A} - \lambda\mathbf{1})\mathbf{T},$$

and

$$|\mathbf{B} - \lambda\mathbf{1}| = |\mathbf{T}^{-1}| \cdot |\mathbf{A} - \lambda\mathbf{1}| \cdot |\mathbf{T}| = |\mathbf{A} - \lambda\mathbf{1}|,$$

using the reciprocal rule above. Matrices such as **A** and **B** are said to be **similar**;

what we have shown is that the characteristic (eigenvalue) equation for the allowed λ values is the same whether it is expressed in terms of the elements of **A** or of **B**. This means that the coefficients in the characteristic equation (which is a polynomial in λ equated to zero) are invariants under a similarity transformation.

To clarify this point, note for a 2×2 matrix that

$$|\mathbf{A} - \lambda \mathbf{1}| = 0 = \lambda^2 - (a_{11} + a_{22})\lambda + (a_{11}a_{22} - a_{12}a_{21}).$$

The last term (in parentheses) is just $|\mathbf{A}|$, which is clearly the same as $|\mathbf{B}|$ by the product rule. The coefficient of the λ term is the trace of **A** (denoted Tr(**A**)), well known as an invariant of a similarity transformation.

For a 3×3 matrix **R**, the characteristic equation is a cubic polynomial which can always be written as

$$\lambda^3 - \text{Tr}(\mathbf{R}) \cdot \lambda^2 + \text{Tr}(\mathbf{C}) \cdot \lambda - |\mathbf{R}| = 0,$$

where **C** is the matrix of the cofactors of **R** which was discussed in the previous section. While the algebraic expressions are more complicated in the 3×3 case, the coefficients will still be invariants under any similarity transformation applied to **R**.

Among relationships involving λ, one of the most famous and useful theorems was developed about 1858, and is known today as the **Cayley–Hamilton theorem**[1]:

Theorem C.4 A matrix satisfies its own characteristic equation.

We shall not give a proof of this theorem, but merely indicate the manner in which it is utilized. As a simple example, consider again the characteristic equation for the 2×2 matrix **A**:

$$\lambda^2 - \text{Tr}(\mathbf{A}) \cdot \lambda + |\mathbf{A}| = 0.$$

In order to apply the theorem, **A** itself is substituted for λ, giving

$$\mathbf{A}^2 - \text{Tr}(\mathbf{A}) \cdot \mathbf{A} + |\mathbf{A}| \cdot \mathbf{1} = \mathbf{0}, \quad \text{or} \quad \mathbf{A}^2 = \text{Tr}(\mathbf{A}) \cdot \mathbf{A} - |\mathbf{A}| \cdot \mathbf{1}.$$

If we now wanted an expression for $\mathbf{A}^3$, we would just multiply the last equation by **A** on both sides, then substitute for $\mathbf{A}^2$ on the right with the original relation. If the order of a matrix is n, the highest power of λ in the characteristic equation is λ^n. The Cayley–Hamilton theorem then gives a matrix equation for $\mathbf{A}^n$, which allows us to write any positive integral power $\geq n$ as a linear combination of powers $< n$ (including the unit matrix).

---

[1] T. Wade, *The Algebra of Vectors and Matrices*, Addison-Wesley, 1951, p. 109.

Negative integral powers of **A** can be treated similarly, which leads to another convenient method for finding inverses. Thus, multiplying the $\mathbf{A}^2$ relation above by $\mathbf{A}^{-1}$ and solving it, one obtains

$$|\mathbf{A}| \cdot \mathbf{A}^{-1} = \text{Tr}(\mathbf{A}) \cdot \mathbf{1} - \mathbf{A}, \quad \text{or} \quad \mathbf{A}^{-1} = \frac{1}{|\mathbf{A}|} \begin{pmatrix} a_{22} & -a_{12} \\ -a_{21} & a_{11} \end{pmatrix}$$

for the general 2×2 case. The method is easily generalized to more complicated situations.

Appendix D
COMPLEX VARIABLES AND CONTOUR INTEGRATION

D.1 BASIC CONCEPTS

A complex number z is composed of a pair of real numbers x and y in the linear combination

$$z \equiv x + iy, \qquad (D.1)$$

where $i \equiv \sqrt{-1}$ is the imaginary unit, which performs the function of a 90° rotation operator. Thus, z may be conveniently represented as a vector from the origin to a point in the xy-plane with coordinates (x, y), the vertical component of any complex number always being indicated by the factor of i which precedes it. The **real part** (horizontal component) is given by the expression

$$\text{Re}\, z \equiv x = r \cos \theta, \qquad (D.2)$$

while the **imaginary part** (vertical component) is

$$\text{Im}\, z \equiv y = r \sin \theta, \qquad (D.3)$$

where r and θ are the usual polar coordinates in the plane. Then,

$$z = x + iy = r(\cos \theta + i \sin \theta) = re^{i\theta} \qquad (D.4)$$

is the **polar form** of z, the exponential form often facilitating computations. The quantity

$$|z| \equiv \sqrt{x^2 + y^2} = r \qquad (D.5)$$

is the **magnitude** of z, while

$$\theta = \arctan \frac{y}{x} \qquad (D.6)$$

Basic Concepts

is its **phase angle**, or **argument**. Closely related to z is the complex number

$$z^* \equiv x - iy = r(\cos\theta - i\sin\theta) = re^{-i\theta}, \tag{D.7}$$

called the **complex conjugate** of z.

Many elementary functions of complex argument can be treated straightforwardly once the student has become sufficiently familiar with these concepts. Thus, we have

$$\cos\theta = \frac{e^{i\theta} + e^{-i\theta}}{2} = \cosh i\theta, \tag{D.8}$$

and

$$\sin\theta = \frac{e^{i\theta} - e^{-i\theta}}{2i} = -i\sinh i\theta, \tag{D.9}$$

whereupon it follows that

$$\cos z = \cos(x + iy) = \cos x \cos iy - \sin x \sin iy$$
$$= \cos x \cosh y - i\sin x \sinh y. \tag{D.10}$$

The cosine (or sine) of a complex number can therefore have a magnitude *greater* than unity, whereas the cosine of a real angle never does.

Some functions such as

$$w(z) \equiv \ln z = \ln(re^{i\theta}) = \ln r + i\theta$$

are **multivalued**. That is, for integral n, z can be written

$$z = re^{i(\theta + 2\pi n)}$$

hence,

$$w(z) = \ln z = \ln r + i(\theta + 2\pi n). \tag{D.11}$$

There are, then, an infinite number of values for $\ln z$, all of which correspond to the *same* point (x, y) in the complex z-plane; the value for $n = 0$ is known as the **principal value**.

Another example of this type is the square root of z. As the vector z sweeps completely around the origin, θ increases from zero to 2π, but the function

$$w(z) = \sqrt{z} = \sqrt{r}\, e^{i\theta/2}$$

has an argument which goes only from zero to π. Writing z as $re^{i(\theta + 2\pi)}$, we see there is another branch of

$$w(z) = \sqrt{z} = \sqrt{r} \cdot e^{i(\theta + 2\pi)/2}$$

which is required to have the phase angle between π and 2π. (One branch, of course, is just the negative of the other.)

If an arbitrary function is written as $w(x, y)$, then by inversion of eqs. (D.4) and (D.7), giving

$$x = \frac{z + z^*}{2}, \qquad y = \frac{z - z^*}{2i}, \tag{D.12}$$

we can express the function as $w(z, z^*)$ in general. In order to have no explicit dependence on z^* in such a function (as with $\sin z$ or $\sqrt{z}$, for example), certain necessary conditions must be satisfied. These can be determined by setting

$$\frac{\partial w}{\partial z^*} = 0 = \frac{\partial w}{\partial x}\frac{\partial x}{\partial z^*} + \frac{\partial w}{\partial y}\frac{\partial y}{\partial z^*} = \frac{\partial w}{\partial x}\left(\frac{1}{2}\right) + \frac{\partial w}{\partial y}\left(\frac{1}{2}i\right).$$

Letting $\mu \equiv \text{Re}(w)$ and $v \equiv \text{Im}(w)$, so that

$$w = \mu + iv, \tag{D.13}$$

we have

$$\frac{1}{2}\frac{\partial w}{\partial x} + \frac{1}{2}i\frac{\partial w}{\partial y} = 0 = \frac{1}{2}\left(\frac{\partial \mu}{\partial x} + i\frac{\partial v}{\partial x}\right) + \frac{1}{2}i\left(\frac{\partial \mu}{\partial y} + i\frac{\partial v}{\partial y}\right),$$

or, on equating both real and imaginary parts to zero,

$$\frac{\partial \mu}{\partial x} = \frac{\partial v}{\partial y}, \qquad \frac{\partial \mu}{\partial y} = -\frac{\partial v}{\partial x}. \tag{D.14}$$

These necessary relations are the **Cauchy–Riemann equations**; a function $w(z)$ which satisfies them is said to be **analytic**. As such, it can be shown to have a unique z-derivative found by ordinary rules of differentiation.

Suppose we next examine a line integral around a closed contour such as a circle of radius r about the point z_0. The integral chosen is

$$\oint (z - z_0)^n \, dz \qquad (n = \text{integer}).$$

Putting $z - z_0 \equiv re^{i\theta}$, $dz = ire^{i\theta}\, d\theta$, the integral becomes

$$\oint (re^{i\theta})^n (ire^{i\theta}\, d\theta) = ir^{n+1}\oint e^{i(n+1)\theta}\, d\theta = 0$$

Basic Concepts

so long as $n \neq -1$. But, if $n = -1$, one obtains

$$\oint \frac{dz}{z - z_0} = i\oint d\theta = 2\pi i.$$

The contour can be deformed into a noncircular shape around z_0, but the integral will still be given by

$$\oint (z - z_0)^n \, dz = 2\pi i \, \delta_{n,-1}. \tag{D.15}$$

Contour integrals, in fact, always vanish (the Cauchy–Goursat theorem) if the function $f(z)$ to be integrated is analytic everywhere within and on the boundary curve:

$$\oint f(z) \, dz = 0 \quad (f(z) \text{ analytic}). \tag{D.16}$$

Now let us evaluate another integral by means of a Taylor expression for $f(z)$ about z_0, assuming that the expansion still works for complex argument z and that $f(z)$ is again analytic everywhere within and on the boundary:

$$\oint \frac{f(z) \, dz}{z - z_0} = \oint \frac{dz}{z - z_0} [f(z_0) + f'(z_0)(z - z_0) + \cdots]$$

$$= 2\pi i \, f(z_0), \tag{D.17}$$

where we have used (D.15) to work out each term. This is **Cauchy's integral formula**. We speak of a **simple pole** at z_0 due to the presence of the factor $(z - z_0)$ in the denominator. If we wished instead to evaluate

$$\oint \frac{f(z) \, dz}{(z - z_0)^m} = \oint \frac{dz}{(z - z_0)^m} [f(z_0) + f'(z_0)(z - z_0) + \cdots]$$

$$= 2\pi i \frac{f^{(m-1)}(z_0)}{(m-1)!}, \tag{D.18}$$

we can do this as easily by (D.15), since only the $(m-1)$ derivative term can contribute. One refers then to a **higher-order pole** (order m if $f(z_0) \neq 0$), defining the **residue** of any pole as

$$\lim_{z \to z_0} \left[\frac{d^{m-1}}{dz^{m-1}} f(z) \right] \frac{1}{(m-1)!} \equiv \text{Residue}. \tag{D.19}$$

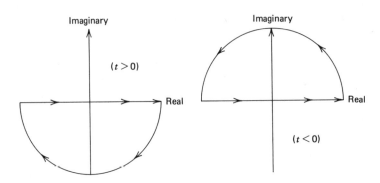

Figure D.1 Contours in the z-plane.

(For $m = 1$ the formula merely gives $f(z_0)$, the simple pole case.) If there are several poles within a contour, each contributes an amount $2\pi i$ times its residue to the value of the integral—this is the important **residue theorem**. To have such "poles within a contour" means that the integrand has factors $(z - z_0)^p$, $(z - z_1)^q$, ... appearing in the denominator, where $p, q, \ldots$, are positive integers.

Many integrals can be performed with the aid of the residue theorem alone. A typical integral of the type that often occurs in oscillation theory or in quantum mechanics is

$$\int_{-\infty}^{\infty} f(z)e^{-izt}\,dz,$$

where z and t are real variables, the integral running along the real axis (Figure D.1). We can loop around the upper or lower half-plane with a large semicircle from $+\infty$ to $-\infty$, so as to form a closed contour and thereby allowing us to apply the residue theorem. If the semicircle has a radius R, $z = Re^{i\theta}$, and $e^{-izt} = e^{-itR\cos\theta}\,e^{Rt\sin\theta}$. The contour integral consists of the path along the real axis plus that of the semicircle; the latter will add nothing in the limit as $R \to \infty$ if $e^{Rt\sin\theta} \to 0$, which only requires that $t\sin\theta$ be negative in sign (the factor $e^{-itR\cos\theta}$ has unit magnitude and therefore is of no interest). Thus, if $t > 0$, we close the contour in the lower half-plane, but if $t < 0$ it must be closed in the upper region. Integrals with exponentials are often treated in this fashion.

Example

Evaluate

$$\int_{-\infty}^{\infty} \frac{e^{-itz}\,dz}{z^2 + 1}.$$

Solution

$$\int_{-\infty}^{\infty} \frac{e^{-itz}\,dz}{z^2 + 1} = \oint \frac{e^{-itz}\,dz}{(z+i)(z-i)}.$$

If $t > 0$, the contour is closed in the lower half-plane, so only the pole at $z = -i$ contributes, giving

$$-2\pi i \cdot \frac{i}{2} e^{-t} = \pi e^{-t}.$$

The minus sign in front is due to the clockwise route of the contour, the positive direction being counterclockwise. For $t < 0$, the contour is closed above the real axis, leading to a contribution from the pole at $z = +i$, and producing a result $2\pi i e^t / 2i = \pi e^t$.

Many other examples could be given. We hope this brief review has somewhat renewed the reader's familiarity with the subject.

Appendix E

GENERAL TENSORS AND THE METRIC TENSOR

E.1 CONTRAVARIANCE AND COVARIANCE

In physics, it is often convenient to use coordinate systems other than rectangular ones. While it is adequate for many purposes, the tensor transformation law (5.20) which relates Cartesian axes through rotation is not the most general definition of a tensor; we now proceed to enlarge that concept somewhat. An important example is the tensor formulation of general relativity theory (see Chapter 9), where the laws of physics are developed in a manner that is not dependent on the specifics of a particular kind of coordinate system.

The coordinate systems most likely to be encountered in addition to the Cartesian are of the orthogonal curvilinear type, such as spherical or cylindrical; occasionally even a nonorthogonal set is employed (in crystallography, for example). Regardless of the coordinates selected, we shall simply represent them by the symbols q^i, supposing that there is an invertible set of equations

$$q^i = q^i(x^1, x^2, x^3) \quad \text{and} \quad x^i = x^i(q^1, q^2, q^3), \tag{E.1}$$

which relate the q's to the rectangular coordinates, and vice versa. More generally, it is possible to transform from an arbitrary set q^i to another set $(q^i)'$, but it will be convenient to keep the familiar Cartesian coordinates on one side of the transformation. In addition, the space will be taken to be three-dimensional although the generalization to spaces of higher dimensionality is not difficult.

In tensor analysis, it is a common practice to label the coordinates with superscripts rather than subscripts, for reasons soon to be discussed. Then, we write

$$x^1 \equiv x, \quad x^2 \equiv y, \quad x^3 \equiv z.$$

The important point is that these superscripts are labels and not exponents.

Contravariance and Covariance

Let us proceed to a more general definition of a tensor. For the zero-rank tensor (scalar), the situation is just as in Chapter 5. In the case of a first-rank tensor, consider the three quantities V^j ($j = 1, 2, 3$) and three "transformed" quantities $(V')^i$ related to the V^j by the equations

$$(V')^i = \sum_{j=1}^{3} \frac{\partial q^i}{\partial x^j} V^j. \tag{E.2}$$

When this relationship is satisfied, the elements V^j, or the $(V')^i$, are said to be the components of a **contravariant vector**, which is one kind of first-rank tensor. There is another type, called a **covariant vector**, the components of which transform according to

$$V'_i = \sum_{j=1}^{3} \frac{\partial x^j}{\partial q^i} V_j. \tag{E.3}$$

In either case, the coefficients in the transformation are partial derivatives which can be found from eq. (E.1). As with Cartesian tensors, it is customary to think of the V^j or V_j on the right of (E.2) and (E.3) as functions of the rectangular coordinates x^i and the primed quantities on the left as corresponding functions of the q^i.

There are classic examples of contravariant and covariant vectors which can be given immediately. Note that the coordinate differential is

$$dq^i = \sum_j \frac{\partial q^i}{\partial x^j} dx^j \tag{E.4}$$

by the chain rule. Upon comparison with (E.2), it is obvious that the differential vector $d\vec{r}$ transforms contravariantly. The notation chosen assigns superscripts to such quantities and subscripts to those which transform covariantly. It turns out that the vector $\vec{r}$ itself (to be examined shortly) does not transform to an arbitrary coordinate system in either of these ways, but because the *differential $d\vec{r}$* transforms contravariantly we put superscript labels on the x^i and q^i.

For a covariant example, consider a scalar function Φ and observe that

$$\frac{\partial \Phi}{\partial q^i} = \sum_j \frac{\partial \Phi}{\partial x^j} \frac{\partial x^j}{\partial q^i}. \tag{E.5}$$

From eq. (E.3), we see that the gradient of Φ transforms covariantly. The position differential and the gradient are the prototypes for contravariant and covariant vectors, respectively.

In theory, the idea is that any given physical vector will have a set of contravariant and a set of covariant components, which generally will not be the same; these can be thought of as two different ways of expressing the same vector in an arbitrary coordinate system. A little later, we shall see how these components are found. In the previous work with Cartesian tensors, no such distinction had to be made, for in rectangular systems the contravariant and covariant components are always identical. To see this, note from eqs. (5.4) and (5.15) that

$$\frac{\partial q^i}{\partial x^j} = \frac{\partial x^j}{\partial q^i} = a_{ij},$$

taking the q^i as the primed coordinates. The coefficients in (E.2) and (E.3) are then equal, showing that the two transformations become the same.

Next, we proceed to justify the assertion that an arbitrary coordinate q^i does not transform in either contravariant or covariant fashion. In order to better understand what is involved, consider as an example a transformation from rectangular to polar coordinates in the xy-plane:

$$q^1 \equiv \rho = \sqrt{x^2 + y^2} = \sqrt{(x^1)^2 + (x^2)^2},$$

$$q^2 \equiv \phi = \arctan\left(\frac{y}{x}\right) = \arctan\left(\frac{x^2}{x^1}\right),$$

with inverse

$$x^1 \equiv x = \rho \cos \phi \equiv q^1 \cos q^2, \qquad x^2 \equiv y = \rho \sin \phi \equiv q^1 \sin q^2.$$

Upon calculation of the various partial derivatives, one finds from (E.2) that the components of a contravariant vector must transform as

$$(V')^1 = \frac{\partial q^1}{\partial x^1} V^1 + \frac{\partial q^1}{\partial x^2} V^2 = \frac{x^1}{q^1} V^1 + \frac{x^2}{q^1} V^2 = V^1 \cos q^2 + V^2 \sin q^2,$$

$$(V')^2 = \frac{\partial q^2}{\partial x^1} V^1 + \frac{\partial q^2}{\partial x^2} V^2 = -\frac{x^2}{(q^1)^2} V^1 + \frac{x^1}{(q^1)^2} V^2$$

$$= -\frac{\sin q^2}{q^1} V^1 + \frac{\cos q^2}{q^1} V^2.$$

Choosing $V^1 \equiv x^1$, $V^2 \equiv x^2$, we obtain with the use of the transformation that

$$(V')^1 = x^1 \cos q^2 + x^2 \sin q^2 = q^1(\cos^2 q^2 + \sin^2 q^2) = q^1,$$

$$(V')^2 = -\frac{\sin q^2}{q^1} x^1 + \frac{\cos q^2}{q^1} x^2 = 0.$$

Contravariance and Covariance

The first component gives q^1 (or ρ) as desired, but the second gives zero instead of q^2 (or ϕ). The position vector then does not transform contravariantly (nor covariantly either, by similar reasoning).

Its differential with $V^1 \equiv dx^1$, $V^2 \equiv dx^2$, yields, on the other hand,

$$(V')^1 = \cos q^2 dx^1 + \sin q^2 dx^2 = \cos q^2 (dq^1 \cos q^2 - q^1 \sin q^2 \, dq^2)$$

$$+ \sin q^2 (dq^1 \sin q^2 + q^1 \cos q^2 \, dq^2) = dq^1,$$

$$(V')^2 = -\frac{\sin q^2}{q^1}(dx^1) + \frac{\cos q^2}{q^1}(dx^2) = -\frac{\sin q^2}{q^1}(dq^1 \cos q^2 - q^1 \sin q^2 \, dq^2)$$

$$+ \frac{\cos q^2}{q^1}(dq^1 \sin q^2 + q^1 \cos q^2 \, dq^2) = dq^2.$$

The differential *does* transform contravariantly, as stated earlier. (Note though that the contravariant components are just $d\rho$ and $d\phi$, not the so-called **physical components** $d\rho$ and $\rho \, d\phi$ which have dimensions of length. Generally, the contravariant and covariant components, to be discussed later, both differ from the normal physical components as well as from each other.)

The explanation for the conclusions just reached lies in the fact that the transformation from xyz to $\rho\phi z$ is not a linear one with constant coefficients as for a rotation; the transformation coefficients of this section are instead partial derivatives which vary from point to point. But these coefficients do behave approximately as constants in a *local* sense, and therefore differential relations between the coordinates can be established in the form of a linear transformation.

It is easy to generalize these ideas to more complex tensors. For the second rank, tensors come in three varieties:

$$(T')^{kl} \equiv \sum_{i,j} \frac{\partial q^k}{\partial x^i} \frac{\partial q^l}{\partial x^j} T^{ij},$$

$$(T')_{kl} \equiv \sum_{i,j} \frac{\partial x^i}{\partial q^k} \frac{\partial x^j}{\partial q^l} T_{ij}, \quad (E.6)$$

and

$$(T')^l_k \equiv \sum_{i,j} \frac{\partial q^l}{\partial x^i} \frac{\partial x^j}{\partial q^k} T^i_j,$$

called **contravariant**, **covariant**, and **mixed** tensors, respectively. Again, for Cartesian coordinates these expressions are in full accord with (5.20). (Note that the Kronecker delta, $\delta_{i,j}$, is a mixed tensor and should really be written δ^i_j.)

E.2 THE METRIC TENSOR

Returning to the differential concept, note that the square of the line element in space is

$$ds^2 = dx^2 + dy^2 + dz^2 = \sum_i (dx^i)^2.$$

Writing the chain rule for dx^i by analogy with (E.4),

$$ds^2 = \sum_{i,j,k} \left(\frac{\partial x^i}{\partial q^j} \frac{\partial x^i}{\partial q^k} \right) dq^j\, dq^k = \sum_{j,k} g_{jk}\, dq^j\, dq^k, \qquad (E.7)$$

where

$$g_{jk} \equiv \sum_i \frac{\partial x^i}{\partial q^j} \frac{\partial x^i}{\partial q^k} \qquad (E.8)$$

defines the covariant elements of the quantity known as the **metric tensor**. Written in matrix form for rectangular coordinates, the metric tensor **G** is simply the unit matrix **1**. In cylindrical coordinates, it is

$$\mathbf{G} = \begin{pmatrix} 1 & 0 & 0 \\ 0 & \rho^2 & 0 \\ 0 & 0 & 1 \end{pmatrix}, \qquad (E.9)$$

and in spherical coordinates

$$\mathbf{G} = \begin{pmatrix} 1 & 0 & 0 \\ 0 & r^2 & 0 \\ 0 & 0 & r^2 \sin^2\theta \end{pmatrix}, \qquad (E.10)$$

both of which follow from either (E.7) or (E.8). It is also clear from the latter that the metric tensor is symmetric. It is of fundamental importance in understanding the geometry of a particular coordinate system.

In addition, the metric tensor is vital to the "raising or lowering" of indices (the conversion of covariant to contravariant components, or vice versa). To see how this feat is accomplished, we first need to learn a little about **base vectors** $\vec{b}_i$ (Figure E.1). Consider that at a given point in space, there must be an intersection

The Metric Tensor

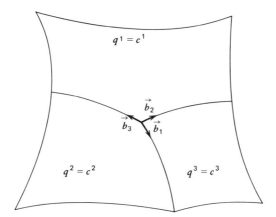

Figure E.1 Coordinate surfaces and base vectors.

of the three surfaces $q^i = c^i$, where the c^i are constants which determine the location of the point by their particular values. (In Cartesian coordinates, these surfaces would simply be planes perpendicular to the axes.) Any two of these surfaces will intersect along some curve; suppose that the tangent vector $\vec{b}$ to this curve is drawn at the given point. The vector $\vec{b}$ and the other two tangents formed in the same manner (from the remaining pairs of surfaces) will form a linearly independent set of basis vectors which spans the space. That is, an arbitrary vector in the space can be expanded as a linear combination of the three tangents $\vec{b}_i$ even though, in general, the $\vec{b}_i$ are not orthogonal to one another.

Thus, along the intersection of surfaces with q^2 and q^3 constant, only q^1 is permitted to vary, and the displacement vector is simply

$$d\vec{r}^1 = \frac{\partial \vec{r}}{\partial q^1} dq^1 \equiv \vec{b}_1 \, dq^1 \qquad (dq^2 = 0 = dq^3),$$

$$\vec{b}_1 \equiv \frac{\partial \vec{r}}{\partial q^1}.$$

Then, by eq. (E.7), we have

$$ds^2 = d\vec{r}^1 \cdot d\vec{r}^1 = g_{11}(dq^1)^2 = (b_1)^2(dq^1)^2,$$

or

$$|\vec{b}_1| = \sqrt{g_{11}},$$

and so on. In other words, one can write

$$\vec{b}_i = \sqrt{g_{ii}}\,\hat{\varepsilon}_i \qquad (E.11)$$

in terms of *unit* tangent vectors $\hat{\varepsilon}_i$, since in a more general situation an arbitrary displacement $d\vec{r}$ can be expanded as

$$d\vec{r} = \sum_i \frac{\partial \vec{r}}{\partial q^i} dq^i \equiv \sum_i \vec{b}_i\, dq^i,$$

$$\vec{b}_i \equiv \frac{\partial \vec{r}}{\partial q^i}. \qquad (E.12)$$

Noting that

$$\vec{b}_i \cdot \vec{b}_j = \frac{\partial \vec{r}}{\partial q^i} \cdot \frac{\partial \vec{r}}{\partial q^j} = \sum_k \frac{\partial x^k}{\partial q^i} \frac{\partial x^k}{\partial q^j} = g_{ij} \qquad (E.13)$$

from (E.8), we see that the angle θ_{ij} between two of the base vectors is simply found with the aid of

$$\cos\theta_{ij} = \frac{\vec{b}_i \cdot \vec{b}_j}{|\vec{b}_i||\vec{b}_j|} = \frac{g_{ij}}{\sqrt{g_{ii}\cdot g_{jj}}}. \qquad (E.14)$$

In an orthogonal system, the metric tensor must therefore be diagonal with $g_{ij} \equiv 0$.

Next, define **reciprocal base vectors** $\vec{b}^i$ (the notation implies that the $\vec{b}^i$ transform contravariantly and the $\vec{b}_i$ covariantly, though these assertions will not be proved) by the the line above relations

$$\vec{b}^1 \equiv \frac{\vec{b}_4 \times \vec{b}_3}{\vec{b}_1 \cdot (\vec{b}_2 \times \vec{b}_3)}, \qquad \vec{b}^2 \equiv \frac{\vec{b}_3 \times \vec{b}_1}{\vec{b}_2 \cdot (\vec{b}_3 \times \vec{b}_1)}, \qquad \vec{b}^3 \equiv \frac{\vec{b}_1 \times \vec{b}_2}{\vec{b}_3 \cdot (\vec{b}_1 \times \vec{b}_2)}.$$

$$(E.15)$$

This set of vectors gets its name from the readily observed property that

$$\vec{b}^i \cdot \vec{b}_j = \delta_{i,j}. \qquad (E.16)$$

In addition, (E.15) will still be valid if subscripts are replaced by superscripts, and vice versa. The reciprocal set provides an alternative scheme for expansion of an arbitrary vector.

The Metric Tensor

The rules for the triple scalar product ensure that the denominator is the same in all three members of (E.15); let us call it

$$D \equiv \vec{b}_1 \cdot (\vec{b}_2 \times \vec{b}_3).$$

A fair amount of algebraic effort will be required to evaluate it; for the moment let us postpone that chore. Instead, consider an arbitrary vector $\vec{V}$. By analogy with the line above (E.12), we can expand $\vec{V}$ as

$$\vec{V} = \sum_i V^i \vec{b}_i \tag{E.17}$$

in terms of the base vectors and the contravariant components V^i. Alternatively, one could expand in terms of the reciprocal set and the covariant components V_i as

$$\vec{V} = \sum_i V_i \vec{b}^i. \tag{E.18}$$

Inasmuch as the base vector sets were specified earlier, the relations (E.17) and (E.18) serve to define the components V^i and V_i, respectively. (Again, the indicated transformation properties will not be shown.) Bear in mind from (E.11) that the $\vec{b}$'s are usually not unit vectors, hence the components just defined are not generally the same as those suggested by physical intuition.

Since any vector may be expanded in terms of either the $\vec{b}_i$ or the $\vec{b}^i$, let us expand $\vec{b}_i$ itself in the $\vec{b}^j$ set, with coefficients C_{ij}:

$$\vec{b}_i = \sum_j C_{ij} \vec{b}^j.$$

Taking the dot product with $\vec{b}_k$, we obtain

$$\vec{b}_i \cdot \vec{b}_k = g_{ik} = \sum_j C_{ij} \vec{b}^j \cdot \vec{b}_k = \sum_j C_{ij} \delta^j_k = C_{ik},$$

so

$$\vec{b}_i = \sum_j g_{ij} \vec{b}^j. \tag{E.19}$$

The dot product of (E.19) with $\vec{b}^k$ gives

$$\vec{b}_i \cdot \vec{b}^k = \delta_i^k = \sum_j g_{ij} \vec{b}^j \cdot \vec{b}^k = \sum_j g_{ij} g^{jk}, \tag{E.20}$$

where
$$g^{jk} \equiv \vec{b}^j \cdot \vec{b}^k. \tag{E.21}$$

The relation (E.20) shows that the contravariant (g^{jk}) could be written in matrix form as the inverse of the covariant tensor (g_{ij}). Further, one can proceed as above to get

$$\vec{b}^i = \sum_j g^{ij} \vec{b}_j. \tag{E.22}$$

The g^{jk} are the contravariant components of the metric tensor. Finally, upon insertion of (E.22) into (E.18) one obtains the results

$$\vec{V} = \sum_{i,j} V_i g^{ij} \vec{b}_j = \sum_j V^j \vec{b}_j, \quad \text{or} \quad V^j = \sum_i V_i g^{ij}, \tag{E.23}$$

and from (E.17) and (E.19),

$$\vec{V} = \sum_{j,k} V^j g_{jk} \vec{b}^k = \sum_k V_k \vec{b}^k, \quad \text{or} \quad V_k = \sum_j V^j g_{jk}. \tag{E.24}$$

Thus, the metric tensor allows us to go from contravariant to covariant components of any vector, or vice versa. For convenience, one can also write

$$\vec{V} = \sum_j (\vec{V} \cdot \vec{b}^j) \vec{b}_j = \sum_j (\vec{V} \cdot \vec{b}_j) \vec{b}^j, \tag{E.25}$$

as is apparent at once from (E.16).

The dot product of two vectors $\vec{u}$ and $\vec{v}$ is given by the expression

$$\vec{u} \cdot \vec{v} = \left(\sum_j u^j \vec{b}_j \right) \cdot \left(\sum_k v^k \vec{b}_k \right) = \sum_{j,k} u^j v^k \vec{b}_j \cdot \vec{b}_k = \sum_{j,k} u^j v^k g_{jk} = \sum_k u_k v^k \tag{E.26}$$

using (E.24). If contravariant components of one vector and covariant components of the other are inserted in this formula, we do not need the metric tensor and have the familiar kind of dot product relation.

One loose end still dangles. We have not as yet evaluated the triple scalar product $D \equiv \vec{b}_1 \cdot (\vec{b}_2 \times \vec{b}_3)$ which occurred in the definition of the $\vec{b}^i$. Defining $\vec{V} \equiv \vec{b}_2 \times \vec{b}_3$ and utilizing (E.25), we find

$$\vec{b}_2 \times \vec{b}_3 = [\vec{b}^1 \cdot (\vec{b}_2 \times \vec{b}_3)] \vec{b}_1 + [\vec{b}^2 \cdot (\vec{b}_2 \times \vec{b}_3)] \vec{b}_2 + [\vec{b}^3 \cdot (\vec{b}_2 \times \vec{b}_3)] \vec{b}_3$$

$$= \frac{1}{D} \{ [(\vec{b}_2 \times \vec{b}_3) \cdot (\vec{b}_2 \times \vec{b}_3)] \vec{b}_1 + [(\vec{b}_3 \times \vec{b}_1)$$

$$\cdot (\vec{b}_2 \times \vec{b}_3)] \vec{b}_2 + [(\vec{b}_1 \times \vec{b}_2) \cdot (\vec{b}_2 \times \vec{b}_3)] \vec{b}_3 \},$$

The Metric Tensor

upon application of (E.15). Expanding the quadruple products by means of the vector identity $(\vec{A} \times \vec{B}) \cdot (\vec{C} \times \vec{D}) = (\vec{A} \cdot \vec{C})(\vec{B} \cdot \vec{D}) - (\vec{A} \cdot \vec{D})(\vec{B} \cdot \vec{C})$ produces

$$\vec{b}_2 \times \vec{b}_3 = \frac{1}{D}\{[(\vec{b}_2 \cdot \vec{b}_2)(\vec{b}_3 \cdot \vec{b}_3) - (\vec{b}_2 \cdot \vec{b}_3)(\vec{b}_3 \cdot \vec{b}_2)]\vec{b}_1$$
$$+ [(\vec{b}_2 \cdot \vec{b}_3)(\vec{b}_3 \cdot \vec{b}_1) - (\vec{b}_2 \cdot \vec{b}_1)(\vec{b}_3 \cdot \vec{b}_3)]\vec{b}_2$$
$$+ [(\vec{b}_2 \cdot \vec{b}_1)(\vec{b}_3 \cdot \vec{b}_2) - (\vec{b}_2 \cdot \vec{b}_2)(\vec{b}_3 \cdot \vec{b}_1)]\vec{b}_3\}.$$

(This lengthy expression arises because $\vec{b}_1$ is not necessarily perpendicular to $\vec{b}_2$ or $\vec{b}_3$.) Taking the dot product of both sides with $\vec{b}_1$ and noting that $\vec{b}_i \cdot \vec{b}_j = g_{ij}$, we get

$$D^2 = [g_{22}g_{33} - g_{23}g_{32}]g_{11} + [g_{23}g_{31} - g_{21}g_{33}]g_{12} + [g_{21}g_{32} - g_{22}g_{31}]g_{13}$$
$$\equiv g,$$

where g is the determinant of the metric tensor (in matrix form). Thus, one has

$$D = \sqrt{g},$$

so that

$$\vec{b}^1 = \frac{\vec{b}_2 \times \vec{b}_3}{\sqrt{g}},$$

and so on. Furthermore, and perhaps of greater significance, the volume element $d\tau$ in the q-coordinates is

$$d\tau = (\vec{b}_1 \, dq^1) \cdot [(\vec{b}_2 \, dq^2) \times (\vec{b}_3 \, dq^3)] = \sqrt{g} \, dq^1 \, dq^2 \, dq^3;$$

for example, $d\tau = \rho \, d\rho \, d\phi \, dz$ in cylindrical coordinates.

Example Base Vectors in Cylindrical Coordinates

Consider the usual transformation

$$x = \rho \cos \phi, \qquad y = \rho \sin \phi, \qquad z = z.$$

From (E.12), one can write

$$\vec{b}_1 = \frac{\partial \vec{r}}{\partial \rho} = \frac{\partial x}{\partial \rho}\hat{i} + \frac{\partial y}{\partial \rho}\hat{j} = \cos\phi\,\hat{i} + \sin\phi\,\hat{j} = \hat{\varepsilon}_\rho,$$

$$\vec{b}_2 = \frac{\partial \vec{r}}{\partial \phi} = -\rho\sin\phi\,\hat{i} + \rho\cos\phi\,\hat{j} = \rho\hat{\varepsilon}_\phi,$$

$$\vec{b}_3 = \frac{\partial \vec{r}}{\partial z} = \hat{k}.$$

In this orthogonal curvilinear system, the base vectors are mutually perpendicular and simply related to the cylindrical unit vectors. The reciprocal base vectors are just

$$\vec{b}^1 = \frac{\vec{b}_2 \times \vec{b}_3}{\sqrt{g}} = \hat{\varepsilon}_\rho, \qquad \vec{b}^2 = \frac{1}{\rho}\hat{\varepsilon}_\phi, \qquad \vec{b}^3 = \hat{k}.$$

They are in the same directions as the corresponding base vectors, again because of the orthogonality of these coordinates. Furthermore, these vectors clearly satisfy all the relations involving the metric tensor.

Consider next the displacement $d\vec{r}$, the contravariant components of which were found earlier. From (E.23),

$$d\vec{r} = d\rho\,\vec{b}_1 + d\phi\,\vec{b}_2 + dz\,\vec{b}_3 = d\rho\,\hat{\varepsilon}_\rho + \rho\,d\phi\,\hat{\varepsilon}_\phi + dz\,\hat{k},$$

as expected. But using (E.24), we may also write

$$d\vec{r} = (d\vec{r})_1 \vec{b}^1 + (d\vec{r})_2 \vec{b}^2 + (d\vec{r})_3 \vec{b}^3;$$

on equating to the previous expression,

$$(d\vec{r})_1 = d\rho, \qquad (d\vec{r})_2 = \rho^2\,d\phi, \qquad (d\vec{r})_3 = dz,$$

which are the **covariant** components of $d\vec{r}$. Because $d\vec{r}$ transforms in a straightforward **contravariant** fashion, these components are rarely seen. They only add confusion to a situation which is bad enough already!

We shall see later that the metric can be quite useful in the differentiation of tensors. Earlier, we have shown how the components of one vector are related to those of another through tensor relations, the emphasis having been on the transformation properties. Let us stress again at this point that the main theoretical motivation for any tensor formulation is that physical equations will thereby remain invariant in form as we transform from one coordinate system to another. The laws of physics are therefore independent of the particular system utilized, as indeed they should be.

Example NonOrthogonal (Skewed or Oblique) Coordinate Systems

The most familiar sets of coordinates are orthogonal, resulting in a diagonal metric tensor with many corresponding algebraic simplifications of analysis. But the tensor formulation can be applied as well to the case of Cartesian axes which make other than right angles with one another. Such skewed axes are capable of

The Metric Tensor

providing some insight into the general situation of nonorthogonality and have numerous technical applications in areas such as crystallography, yet are rarely discussed in the literature in a straightforward fashion.

We shall consider a simple change of basis in the xy-plane. With a set of skewed axes, there are two simple ways in which the coordinates of an arbitrary point P might be measured. (Both of these definitions are equivalent for the usual rectangular axes.) One way is by means of perpendiculars drawn to the xy-axes, the components being the distances OA and OB shown in Figure E.2. For reasons that are a little obscure, these distances are sometimes designated as the "covariant components" of $\vec{r}$, hence we shall denote them for the moment as r_x and r_y. From the geometry, it is clear that

$$r_x = r\cos\phi = x,$$

$$r_y = r\cos(\alpha - \phi) = r[\cos\alpha\cos\phi + \sin\alpha\sin\phi] \quad \text{(E.27)}$$

$$= x\cos\alpha + y\sin\alpha,$$

where $x = r\cos\phi$ and $y = r\sin\phi$ are the usual components in rectangular axes.

An alternative way of measuring the coordinates of P is by means of parallels to the axes, indicated as OC and OD. These distances are called the "contravariant components," and are given by

$$r^x = OA - CA = r\cos\phi - r\sin\phi\cot\alpha = x - y\cot\alpha,$$

$$r^y = r\sin\phi\csc\alpha = y\csc\alpha, \quad \text{(E.28)}$$

so that in either case we have a linear transformation from the old xy-coordinates to the new $r^x r^y$- or $r_x r_y$-coordinates.

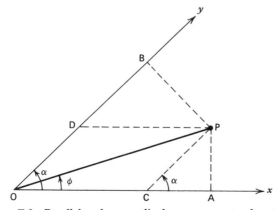

Figure E.2 Parallel and perpendicular components of vector $\vec{r}$.

For the parallel components, we could write a matrix equation as

$$\begin{pmatrix} r^x \\ r^y \end{pmatrix} = \mathbf{T} \begin{pmatrix} x \\ y \end{pmatrix}, \qquad \mathbf{T} = \begin{pmatrix} 1 & -\cot \alpha \\ 0 & \csc \alpha \end{pmatrix} \tag{E.29}$$

to specify the transformation. Then note that we have

$$ds^2 = (dx \ dy) \begin{pmatrix} dx \\ dy \end{pmatrix}$$

in matrix form. Since

$$\begin{pmatrix} x \\ y \end{pmatrix} = \mathbf{T}^{-1} \begin{pmatrix} r^x \\ r^y \end{pmatrix}$$

is the inverse transformation, we need to find $\mathbf{T}^{-1}$, which is readily seen to be

$$\mathbf{T}^{-1} = \begin{pmatrix} 1 & \cos \alpha \\ 0 & \sin \alpha \end{pmatrix}. \tag{E.30}$$

This means that

$$ds^2 = (dr^x \ dr^y)(\widetilde{\mathbf{T}^{-1}})(\mathbf{T}^{-1}) \begin{pmatrix} dr^x \\ dr^y \end{pmatrix},$$

where we have used (5.13). By (E.7), the metric tensor for the transformation to parallel components is:

$$\mathbf{G} = (\widetilde{\mathbf{T}^{-1}})(\mathbf{T}^{-1}) = \begin{pmatrix} 1 & 0 \\ \cos \alpha & \sin \alpha \end{pmatrix} \begin{pmatrix} 1 & \cos \alpha \\ 0 & \sin \alpha \end{pmatrix} = \begin{pmatrix} 1 & \cos \alpha \\ \cos \alpha & 1 \end{pmatrix}. \tag{E.31}$$

Note that $\mathbf{G} \to \mathbf{1}$ when $\alpha \to \pi/2$. Alternatively, the expression for $\mathbf{G}$ also follows immediately from (E.13), as soon as one has worked out the base vectors $\vec{b}_1 = \hat{i}$ and $\vec{b}_2 = \cos \alpha \hat{i} + \sin \alpha \hat{j}$.

Now let us look at the transformation properties of the parallel components. In view of the fact that (E.28) is a linear transformation with constant coefficients, it clearly satisfies (E.2) with r^x and r^y as the components of a contravariant vector. Then lowering the indices by (E.24) with the aid of the metric tensor, the covariant components are given by

$$r_x = r^x g_{11} + r^y g_{21} = (x - y \cot \alpha) + y \csc \alpha \cos \alpha = x,$$

$$r_y = r^x g_{12} + r^y g_{22} = x \cos \alpha + y \sin \alpha,$$

in perfect agreement with (E.27).

The Metric Tensor

If the reciprocal base vectors are calculated from (E.15), giving

$$\vec{b}^1 = \hat{i} - \cot\alpha\,\hat{j}, \qquad \vec{b}^2 = \csc\alpha\,\hat{j},$$

one finds in accordance with (E.25) that

$$\vec{r} = r^x\vec{b}_1 + r^y\vec{b}_2 = r_x\vec{b}^1 + r_y\vec{b}^2.$$

Suppose, however, that we had chosen (E.27) rather than (E.28) for use as the transformation assumed in (E.2). The matrices **T** and **G** would have a different form from those above, and it would appear at first that the transformation properties are thereby *reversed*, with r_x, r_y transforming contravariantly and r^x, r^y covariantly. (To reach this conclusion, simply retrace the above calculations starting from (E.27).) The resolution of this difficulty lies in the fact that r_x and r_y are the components along the reciprocal vectors; it is the parallel components r^x and r^y which are the contravariant components along the actual skewed axes.

As a further illustration of tensor properties in a skewed system, let us next rotate the axes through an angle θ in the xy-plane. The rotation matrix is, by (5.1),

$$\mathbf{R} = \begin{pmatrix} \cos\theta & \sin\theta \\ -\sin\theta & \cos\theta \end{pmatrix}, \quad \text{with} \quad \begin{pmatrix} x' \\ y' \end{pmatrix} = \mathbf{R}\begin{pmatrix} x \\ y \end{pmatrix}. \tag{E.32}$$

In terms of the parallel components of (E.28), this becomes

$$\mathbf{T}^{-1}\begin{pmatrix} r^{x'} \\ r^{y'} \end{pmatrix} = \mathbf{R}\mathbf{T}^{-1}\begin{pmatrix} r^x \\ r^y \end{pmatrix},$$

or

$$\begin{pmatrix} r^{x'} \\ r^{y'} \end{pmatrix} = [\mathbf{T}\mathbf{R}\mathbf{T}^{-1}]\begin{pmatrix} r^x \\ r^y \end{pmatrix}. \tag{E.33}$$

The quantities $r^{x'}$, $r^{y'}$ are the parallel components relative to $x'y'$-axes rotated by angle θ relative to the old xy-axes. In other words, the similarity transformation $\mathbf{T}\mathbf{R}\mathbf{T}^{-1}$ plays its usual role of accomplishing a change of basis: the matrix **R** performs the rotation on the xy-coordinates, but the **T**-matrices are required for conversion to the parallel representation.

Calculating the matrix product by means of (E.29), (E.30), and (E.32), one obtains

$$\mathbf{T}\mathbf{R}\mathbf{T}^{-1} = \csc\alpha\begin{pmatrix} \sin(\alpha+\theta) & \sin\theta \\ -\sin\theta & \sin(\alpha-\theta) \end{pmatrix}, \tag{E.34}$$

whereupon (E.33) can be written out as

$$r^{x'} \sin \alpha = r^x \sin(\alpha + \theta) + r^y \sin \theta,$$

$$r^{y'} \sin \alpha = -r^x \sin \theta + r^y \sin(\alpha - \theta). \tag{E.35}$$

These equations are the generalizations of (5.1) to the skewed system. Their correctness can be established easily through a little geometrical analysis. Consulting Figure E.3,

$$r^{x'} = OA + AB = OA(AC - BC)$$
$$= r^y \cos(\alpha - \theta) + r^x \cos \theta - r^{y'} \cos \alpha,$$
$$r^{y'} \sin \alpha = CP = AD - r^x DE = r^y \sin(\alpha - \theta) - r^x \sin \theta.$$

Solving simultaneously, the set (E.35) is obtained once again.

As a final comment, consider the set (E.35) with the direct substitution of (E.28):

$$r^{x'} = r^x \csc \alpha \sin(\alpha + \theta) + r^y \csc \alpha \sin \theta$$
$$= (x - y \cot \alpha) \csc \alpha \sin(\alpha + \theta) + (y \csc \alpha) \csc \alpha \sin \theta$$
$$= (r \cos \phi - r \sin \phi \cot \alpha) \csc \alpha \sin(\alpha + \theta)$$
$$+ (r \sin \phi \csc \alpha) \csc \alpha \sin \theta$$
$$= r \cos(\phi - \theta) - r \sin(\psi - \theta) \cot \alpha,$$

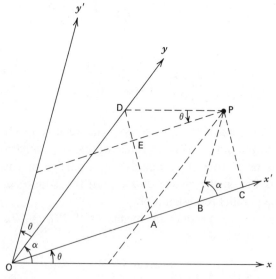

Figure E.3 Geometry for rotation of skewed axes.

Differentiation of Tensors

and similarly,

$$r^{y'} = r\sin(\phi - \theta)\csc\alpha.$$

These results are exactly what we expect from (E.28), ϕ merely being replaced by $\phi - \theta$ in those equations. The contravariant vector of the parallel components has been transformed to a rotated frame where it has exactly the same form as it did originally in (E.28) in terms of unprimed coordinates. This is the basic tensor characteristic.

E.3 DIFFERENTIATION OF TENSORS

From (E.12), the velocity is

$$\vec{v} = \dot{\vec{r}} = \sum_i \dot{q}^i \vec{b}_i, \tag{E.36}$$

and so the acceleration becomes

$$\vec{a} = \dot{\vec{v}} = \sum_i \ddot{q}^i \vec{b}_i + \sum_i \dot{q}^i \dot{\vec{b}}_i.$$

We need to express the derivative $\dot{\vec{b}}_i$ in terms of the generalized coordinates q^i, $\dot{q}^i$, and $\vec{b}_i$. To do so, the chain rule can be employed to yield

$$\dot{\vec{b}}_i = \sum_j \frac{\partial \vec{b}_i}{\partial q^j} \dot{q}^j,$$

and the $\vec{b}_i$ derivatives can be written as linear combinations of the base vectors as

$$\frac{\partial \vec{b}_i}{\partial q^j} \equiv \sum_k \Gamma^k_{ij} \vec{b}_k \equiv \sum_k \left\{ \begin{array}{c} k \\ i \ j \end{array} \right\} \vec{b}_k, \tag{E.37}$$

where the symbols Γ^k_{ij} and $\left\{ \begin{array}{c} k \\ i \ j \end{array} \right\}$ are both used for the expansion coefficients, called **Christoffel symbols of the second kind**. The acceleration now becomes

$$\vec{a} = \sum_k A^k \vec{b}_k, \quad A^k \equiv \ddot{q}^k + \sum_{i,j} \dot{q}^i \dot{q}^j \Gamma^k_{ij}. \tag{E.38}$$

The Christoffel symbols, which abound in relativity theory, have several interesting properties that are easy to develop. First, they are symmetrical in the lower indices, established by noting that

$$\vec{b}_i = \sum_k \frac{\partial x^k}{\partial q^i} \hat{e}_k,$$

whereupon

$$\frac{\partial \vec{b}_i}{\partial q^j} = \sum_k \frac{\partial^2 x^k}{\partial q^j \partial q^i} \hat{\varepsilon}_k,$$

the $\hat{\varepsilon}_k$ being the rectangular unit vectors. Since the order of partial differentiation on the right side can be reversed, we have

$$\frac{\partial \vec{b}_i}{\partial q^j} = \frac{\partial \vec{b}_j}{\partial q^i},$$

or

$$\Gamma^k_{ij} = \Gamma^k_{ji}. \tag{E.39}$$

Secondly, for the evaluation of the Christoffel symbols, we differentiate (E.13):

$$\vec{b}_j \cdot \frac{\partial \vec{b}_k}{\partial q^i} + \frac{\partial \vec{b}_j}{\partial q^i} \cdot \vec{b}_k = \frac{\partial g_{jk}}{\partial q^i}.$$

To this relation add

$$\vec{b}_i \cdot \frac{\partial \vec{b}_k}{\partial q^j} + \frac{\partial \vec{b}_i}{\partial q^j} \cdot \vec{b}_k = \frac{\partial g_{ik}}{\partial q^j}$$

and subtract

$$\vec{b}_i \cdot \frac{\partial \vec{b}_j}{\partial q^k} + \frac{\partial \vec{b}_i}{\partial q^k} \cdot \vec{b}_j = \frac{\partial g_{ij}}{\partial q^k}$$

(both of which are obtained from the first relation by a redesignation of indices), giving, upon use of $\partial \vec{b}_i / \partial q^j = \partial \vec{b}_j / \partial q^i$,

$$\frac{\partial \vec{b}_i}{\partial q^j} \cdot \vec{b}_k = \frac{1}{2}\left[\frac{\partial g_{jk}}{\partial q^i} + \frac{\partial g_{ik}}{\partial q^j} - \frac{\partial g_{ij}}{\partial q^k}\right] \equiv [ij, k], \tag{E.40}$$

where $[ij, k]$ is the **Christoffel symbol of the first kind**, also clearly symmetric in the indices i and j. Noting from (E.37) that

$$\frac{\partial \vec{b}_i}{\partial q^j} \cdot \vec{b}_k = \sum_l \Gamma^l_{ij} \vec{b}_l \cdot \vec{b}_k = \sum_l \Gamma^l_{ij} g_{lk},$$

Differentiation of Tensors 415

then
$$\sum_l \Gamma^l_{ij} g_{lk} = [ij, k]. \tag{E.41}$$

Multiplying by g^{mk} on both sides and summing on k,
$$\sum_{k,l} \Gamma^l_{ij} (g_{lk} g^{mk}) = \sum_k g^{mk} [ij, k].$$

By (E.20),
$$\sum_k g_{lk} g^{mk} = \sum_k g_{lk} g^{km} = \delta^m_l,$$

leaving
$$\Gamma^m_{ij} = \sum_k g^{mk} [ij, k] = \frac{1}{2} \sum_k g^{mk} \left[\frac{\partial g_{jk}}{\partial q^i} + \frac{\partial g_{ik}}{\partial q^j} - \frac{\partial g_{ij}}{\partial q^k} \right]. \tag{E.42}$$

Thus, the Christoffel symbols can be evaluated solely from knowledge of the metric of the space.

Returning to the acceleration (E.38) and lowering indices by means of (E.24), we have
$$A_k = \sum_l g_{kl} \ddot{q}^l + \sum_{i,j,l} \dot{q}^i \dot{q}^j \Gamma^l_{ij} g_{kl}$$
$$= \sum_i g_{ki} \ddot{q}^i + \sum_{i,j} \dot{q}^i \dot{q}^j [ij, k], \tag{E.43}$$

using (E.41). Newton's law for a single particle of mass m can then be written in covariant form as
$$Q_k = mA_k, \tag{E.44}$$

where Q_k is the generalized covariant force component; this is a relation valid in any curvilinear system.

Next, we want to look at partial derivatives of a covariant vector. From (E.3), we have upon differentiation
$$\frac{\partial V'_i}{\partial q^k} = \sum_j \frac{\partial V_j}{\partial q^k} \frac{\partial x^j}{\partial q^i} + \sum_j V_j \frac{\partial}{\partial q^k} \left(\frac{\partial x^j}{\partial q^i} \right)$$
$$= \sum_{j,l} \frac{\partial V_j}{\partial x^l} \frac{\partial x^l}{\partial q^k} \frac{\partial x^j}{\partial q^i} + \sum_{j,m} V'_m \frac{\partial q^m}{\partial x^j} \frac{\partial^2 x^j}{\partial q^i \partial q^k},$$

where in the second sum on the right, called the **affine term**, we have used the fact that

$$V_j = \sum_m V'_m \frac{\partial q^m}{\partial x^j}$$

from inversion of (E.3). It will be shown via the problem set that the affine derivative expression is just our old friend the Christoffel symbol. Hence we have

$$\Gamma^m_{ik} = \sum_j \frac{\partial^2 x^j}{\partial q^i \partial q^k} \frac{\partial q^m}{\partial x^j}$$

and

$$dV'(i) \equiv \sum_k \frac{\partial V'_i}{\partial q^k} dq^k = \sum_{j,k,l} \frac{\partial V_j}{\partial x^l} \frac{\partial x^l}{\partial q^k} \frac{\partial x^j}{\partial q^i} dq^k + \sum_{k,m} V'_m \Gamma^m_{ik} dq^k, \quad (\text{E.45})$$

where $dV'(i)$ depends on the index i but does not actually have covariant or contravariant properties of transformation, since $\partial V'_i/\partial q^k$ does not transform like a second-rank tensor as long as the affine term is present. Now, the change of the vector V'_i comes from two sources: the intrinsic physical variation and that due to the change of coordinate axes as we move from point to point in a curvilinear system. The latter part is clearly the affine term in (E.45), because for a constant vector $\partial V_j/\partial x^l = 0$ (all j and l), leaving only the affine term. The intrinsic physical variation of the vector is then given in general by

$$\sum_k \left(\frac{\partial V'_i}{\partial q^k} - \sum_m \Gamma^m_{ik} V'_m \right) dq^k,$$

so that

$$V'_{i;k} \equiv \frac{\partial V'_i}{\partial q^k} - \sum_m \Gamma^m_{ik} V'_m \quad (\text{E.46})$$

is called the **covariant derivative** (primes can now be omitted); it is the "true" derivative adjusted for the coordinate system, and it transforms like a second-rank tensor. In a similar way, the covariant derivative of the contravariant components of The vector is found to be

$$V^i_{;k} \equiv \frac{\partial V^i}{\partial q^k} + \sum_m \Gamma^i_{km} V^m. \quad (\text{E.46}')$$

There are many interesting properties of covariant differentiation (E.20) could be pursued if sufficient space were available. (For example, the covariant deriva-

Differentiation of Tensors   417

tive of the metric tensor is zero; it behaves as a constant with respect to this type of differentiation.) Instead, we turn our attention to a discussion of some familiar operators from vector calculus and how they appear in the tensor formalism.

Since we have shown that the gradient of any scalar Φ transforms covariantly, (E.24) and (E.22) yield

$$\nabla \Phi = \sum_j \frac{\partial \Phi}{\partial q^j} \vec{b}^j = \sum_{i,j} \frac{\partial \Phi}{\partial q^j} g^{ji} \vec{b}_i = \sum_i \left(\sum_j g^{ji} \frac{\partial \Phi}{\partial q^j} \right) \vec{b}_i, \qquad (E.47)$$

the parentheses giving the contravariant ith-component of the gradient. This expression will be needed shortly.

The divergence of a vector $\vec{V}$ in rectangular coordinates is simply $\sum_i (\partial V^i / \partial x^i)$. That definition is generalized in the tensor calculus to read

$$\operatorname{div} V \equiv \sum_i V^i{}_{;i} = \sum_i \frac{\nabla V^i}{\partial q^i} + \sum_{i,j} \Gamma^i_{ij} V^j,$$

where the quantity $V^i{}_{;i}$ is a covariant derivative acting on contravariant components of a vector, the contraction of this second-rank tensor producing a scalar in all coordinate systems. (In a Cartesian frame, Γ^i_{ij} is zero, and the second sum is not present.)

This expression, however, is not the most convenient one for practical usage. With the aid of interchange of dummy indices and the symmetry of the metric tensor, one can write that

$$\sum_{i,k} g^{ik} \frac{\partial g_{jk}}{\partial q^i} = \sum_{i,k} g^{ki} \frac{\partial g_{ji}}{\partial q^k} = \sum_{i,k} g^{ik} \frac{\partial g_{ij}}{\partial q^k},$$

so

$$\sum_i \Gamma^i_{ij} = \frac{1}{2} \sum_{i,k} g^{ik} \left[\frac{\partial g_{jk}}{\partial q^i} + \frac{\partial g_{ik}}{\partial q^j} - \frac{\partial g_{ij}}{\partial q^k} \right] = \frac{1}{2} \sum_{i,k} g^{ik} \frac{\partial g_{ik}}{\partial q^j}$$

from (E.42). To cast the last expression into a simpler form, note (from Appendix C) that any determinant, such as that of the metric tensor, can be written in terms of cofactors as

$$g = \sum_k g_{kl} \cdot \operatorname{cofactor}(g_{kl}).$$

Since the cofactor of g_{ij} does not contain g_{ij} (by definition), then

$$\frac{\partial g}{\partial g_{ij}} = \operatorname{cofactor}(g_{ij}) = g^{ij} g,$$

the final step following from (E.20) upon multiplying both sides of the last equation by g_{ij} and summing on i. From this last result, we find

$$\frac{\partial g}{\partial q^j} = \sum_{i,k} \frac{\partial g}{\partial g_{ik}} \frac{\partial g_{ik}}{\partial q^j} = g \sum_{i,k} g^{ik} \frac{\partial g_{ik}}{\partial q^j} = 2g \sum_i \Gamma^i_{ij},$$

or

$$\sum_i \Gamma^i_{ij} = \frac{1}{2g} \frac{\partial g}{\partial q^j} = \frac{1}{\sqrt{g}} \frac{\partial(\sqrt{g})}{\partial q^j},$$

whereupon

$$\operatorname{div} \vec{V} = \sum_i \frac{\partial V^i}{\partial q^i} + \sum_j \frac{1}{\sqrt{g}} \frac{\partial(\sqrt{g})}{\partial q^j} V^j = \frac{1}{\sqrt{g}} \sum_i \frac{\partial}{\partial q^i} (\sqrt{g}\, V^i). \quad (E.48)$$

The natural consequence of the last relation and (E.47) is that the Laplacian becomes

$$\nabla \cdot (\nabla \Phi) = \nabla^2 \Phi = \frac{1}{\sqrt{g}} \sum_{i,j} \frac{\partial}{\partial q^i} \left[\sqrt{g}\, g^{ji} \frac{\partial \Phi}{\partial q^j} \right]. \quad (E.49)$$

Finally, the curl in tensor calculus is defined by means of a (second-rank) difference of covariant derivatives: Let

$$V_{ij} \equiv V_{j;i} - V_{i;j} = \frac{\partial V_j}{\partial q^i} - \frac{\partial V_i}{\partial q^j},$$

so that we can now specify

$$\operatorname{curl} \vec{V} \equiv \frac{1}{\sqrt{g}} [V_{23}\vec{b}_1 + V_{31}\vec{b}_2 + V_{12}\vec{b}_3], \quad (E.50)$$

the other terms of (E.46) canceling. Having defined the common operators and demonstrated the usefulness of covariant differentiation, we shall conclude this discussion here.

PROBLEMS

E.1 Work out the relation between covariant and contravariant components of a vector in spherical coordinates, obtaining base and reciprocal base vectors as well.

E.2 Consider the perpendicular components r_x, r_y in the skewed coordinate frame.

(a) Find the matrices $\mathbf{T}$, $\mathbf{T}^{-1}$, and $\mathbf{G}$ starting from (E.47).

(b) Show that r^x and r^y now transform in unusual fashion.

(c) Rotate the coordinates through an angle θ, finding r'_x and r'_y in terms of r_x and r_y.

(d) Check your results in the previous part by means of a geometrical diagram.

(e) Find r'_x and r'_y in terms of r, θ, ϕ, and α.

E.3 (a) Show that in a transformation from rectangular to curvilinear coordinates q, the Christoffel symbols are given by

$$\left\{\begin{matrix} k \\ i\ j \end{matrix}\right\} = \sum_l \frac{\partial^2 x^l}{\partial q^j \partial q^i} \frac{\partial q^k}{\partial x^l}.$$

(b) Evaluate for spherical coordinates the quantities

$$\left\{\begin{matrix} 1 \\ 2\ 3 \end{matrix}\right\} \quad \text{and} \quad (x - y\cot\alpha)\ \left\{\begin{matrix} 3 \\ 2\ 1 \end{matrix}\right\}.$$

E.4 A vector field has a constant value C directed along the x-axis. The corresponding scalar potential is $-Cr\cos\phi$ in plane-polar coordinates r and ϕ. Show that the components of the potential gradient transform covariantly. Then, show that the covariant derivative of the vector field vanishes.

Appendix F
GROUPS, LIE ALGEBRAS, AND THE FORMAL STRUCTURE OF MECHANICS

F.1 THE THEORY OF GROUPS

Earlier in the text, we have been concerned with the symmetries and associated invariances of a physical system. There is a branch of mathematics which is extensively concerned with symmetries: group theory. Our attention will now turn to this area, with particular emphasis placed on some Lie (pronounced LEE) groups and algebras which bear a close relation to the formalism of dynamics.

In general, a **group** is a set of elements $A, B, C, \ldots$ which can be combined through a multiplication procedure established by definition. The number of distinct elements in the set is called the group **order** g, and it can be either finite or infinite. The elements may be abstract symbols, or they may correspond to explicit algebraic or geometrical operations. In all cases, however, the elements of a group must satisfy the four properties below:

1. **Closure:** The product of any two elements in the set must itself be a member of the set.
2. **Associativity:** Any product of set elements must satisfy the usual associative law of multiplication. Thus, for any three elements A, B, and C, we have
$$(AB)C = A(BC).$$
3. **Identity:** The set must contain an identity I as one of the elements, such that, for any element A,
$$IA = AI = A.$$
4. **Inverse:** For any element A in the set, there must exist in the set an inverse element (denoted A^{-1}) such that
$$AA^{-1} = A^{-1}A = I.$$
(The identity element, of course, will be its own inverse.)

The Theory of Groups 421

The crucial group operation is the multiplicative process. Note that (as for matrices) the commutative law is not required for groups in general. Whenever $AB = BA$ does hold for every pair of elements A and B, the group is said to be **Abelian**.

To illustrate the group properties as the theory is being developed, let us consider the group of geometrical operations that transform the simple figure of a square into itself. There are eight operations in the plane, rotations and reflections, which achieve this goal; they will be chosen as in Table F.1. Labeling the corners of the square with numbers 1 through 4, beginning at the upper right and going around clockwise, it is clear that the numbers will be shifted by each of the eight operators as indicated in the table. Other operators could just as well have been selected (such as a 90° *clockwise* rotation operator, which is equivalent to R_3), but would not have been independent of those given.

Next, consider the product of two successive operations obtained by applying first the operator on the right, then the one on the left of the product. Using $R_2 R_1$ as an example and consulting Table F.1 gives

$$\text{Effect of } R_1 \text{ only:} \quad {}^1_4\square{}^2_3$$

$$\text{Effect of } R_2 \text{ after } R_1 \text{ (product):} \quad {}^3_2\square{}^4_1$$

The combined effect of *both* operations is exactly the same as would have been obtained from the original square with R_3 alone, so we can write $R_2 R_1 = R_3$, which illustrates closure. Proceeding in this manner, the entire multiplication table for the square operators can be established, with each particular entry being

TABLE F.1 *Basic Geometrical Operators for Transformation of Square Into Itself*

Operator Symbol	Operation	Effect on Square
I	Identity	${}^4_3\square{}^1_2$
R_1	90° Counterclockwise rotation	${}^1_4\square{}^2_3$
R_2	180° Counterclockwise rotation	${}^2_1\square{}^3_4$
R_3	270° Counterclockwise rotation	${}^3_2\square{}^4_1$
H	Reflection through horizontal axis	${}^3_4\square{}^2_1$
V	Reflection through vertical axis	${}^1_2\square{}^4_3$
D_1	Reflection through diagonal joining corners 1 and 3	${}^2_3\square{}^1_4$
D_2	Reflection through diagonal joining corners 2 and 4	${}^4_1\square{}^3_2$

the product of the element in the left column (left factor) with that of the topmost row (right factor), as shown in Table F.2.

Note that each row (or column) of Table F.2 contains every element of the group once and only once, as is readily shown. The order of the two factors is sometimes relevant, so the group is *not* Abelian; that these eight operators do form a group is clear, however. The table exhibits the closure property, and the other properties are also checked easily with its aid. (Thus, the fact that $R_1 R_3 = I = R_3 R_1$ means that R_1 and R_3 are the inverses of one another, etc. For later convenience, it might be noted that each element is its own inverse, with the exception of R_1 and R_3.)

Any collection of elements in a group (usually a smaller subset) which in themselves satisfy the group postulates is called a **subgroup** of the initial group. Thus, I, R_2 form a subgroup of order 2, while I, R_1, R_2, R_3 form a subgroup of order 4, as is evident from Table F.2.

There are other ways of distinguishing smaller subsets within a group. For any elements A and B, $B^{-1}AB$ is called the **conjugate** of A with respect to B. All elements conjugate to a particular element A are said to be in the same **class**; the class formed from A must contain A itself, since $I^{-1}AI = A$ is such a conjugate. Suppose that B and C are two other elements in the class formed from (containing) A. The class could just as well have been formed from the conjugates of B (or C) instead, implying that B and C should be conjugate to each other if both are conjugate to A. The proof of this assertion is simple: If $B = D^{-1}AD$ and $C = E^{-1}AE$, then $A = ECE^{-1}$ and $B = D^{-1}(ECE^{-1})D = (E^{-1}D)^{-1}C(E^{-1}D)$, showing the conjugate nature of B with C through the product $E^{-1}D$. (We have used the fact that $(E^{-1}D)^{-1} = D^{-1}E$, as is clear.) Any group can now be divided into classes of elements, all the elements within a given class being conjugate to one another.

For the square, there are five classes, one containing I by itself, one containing R_2 alone, one with R_1 and R_3, one with H and V, and one with D_1 and D_2, showing that geometrically similar operators tend to go into the same class. In an

TABLE F.2 Multiplication Table for Square

	I	R_1	R_2	R_3	H	V	D_1	D_2
I	I	R_1	R_2	R_3	H	V	D_1	D_2
R_1	R_1	R_2	R_3	I	D_1	D_2	V	H
R_2	R_2	R_3	I	R_1	V	H	D_2	D_1
R_3	R_3	I	R_1	R_2	D_2	D_1	H	V
H	H	D_2	V	D_1	I	R_2	R_3	R_1
V	V	D_1	H	D_2	R_2	I	R_1	R_3
D_1	D_1	H	D_2	V	R_1	R_3	I	R_2
D_2	D_2	V	D_1	H	R_3	R_1	R_2	I

The Theory of Groups

Abelian group, $B^{-1}AB = B^{-1}BA = A$ for all A and B, so each element A is in a class by itself.

A group such as that of the square has a certain formal or abstract structure which is in no way dependent upon geometrical associations for its meaning. As an example, a set of eight matrices could be determined which would satisfy the product rules of Table F.2; these matrices would satisfy the group postulates just as the geometrical operators did. The elements of the group of matrices and those of the group of operators can be put into an exact one-to-one correspondence with each other, and the groups are then said to be **isomorphic** to one another. Suppose next that the only matrix available were the identity matrix; by itself it would form a rather trivial group, yet the multiplication rules could also be satisfied by having a many-to-one correspondence, or **homomorphism**, between the elements of the geometrical group and this solitary matrix. When an abstract group has a correspondence with a set of concrete elements that preserve the product rules like the matrices, we say that the latter form a **representation** of the group; the representation is called **faithful** when it refers to an isomorphism, and this is the situation in which we are usually most interested.

For each operator A of the group, we form a square nonsingular matrix $\Gamma(A)$ in the faithful representation. For the matrices, the product rule $A = BC$ would become $\Gamma(A) = \Gamma(B)\Gamma(C)$. Once one such matrix representation has been obtained, many others could be found by similarity transformation through a matrix **T**. Thus, for any A, if $\Gamma'(A) \equiv \mathbf{T}^{-1}\Gamma(A)\mathbf{T}$, then

$$\Gamma'(B)\Gamma'(C) = \mathbf{T}^{-1}\Gamma(B)\mathbf{T}\mathbf{T}^{-1}\Gamma(C)\mathbf{T}$$

$$= \mathbf{T}^{-1}\Gamma(B)\Gamma(C)\mathbf{T} = \mathbf{T}^{-1}\Gamma(A)\mathbf{T} = \Gamma'(A),$$

again satisfying the product rule. Representations related through a similarity transformation are called **equivalent**, in that either the set of Γ-matrices or the Γ' matrices could represent the underlying abstract group equally well.

One of the most important quantities in group theory is called the **character** of a given matrix. We have seen that corresponding matrices of equivalent representations are related through a similarity transformation. Two matrices of this type may appear quite different in form, yet they must have something in common: one such property is their trace, an invariant under the transformation. We define the character $\chi^{(j)}$ as

$$\chi^{(j)}(P) \equiv \sum_i \Gamma_{ii}^{(j)}(P) \equiv \text{Tr } \Gamma^{(j)}(P), \qquad (F.1)$$

for the operator P in the jth representation. The character is independent of which one of several equivalent representations is the one utilized. Furthermore, within a single representation, all elements of a given class are conjugate to one another, an operation which for matrices is merely a similarity transformation. Therefore, all matrices have the same character for elements in the same class. (This is readily verified for the group of the square, once its representations have been worked out.) Several important relationships of group theory involve the character in some way.

F.2 LIE GROUPS AND ALGEBRAS

Up to this point, nothing has been said about **continuous** groups, which have a nondenumerably infinite number of elements. The basic principles were more easily discussed with the aid of a finite group, yet Lie groups are of the continuous type and require a more detailed analysis. The rotation group is a familiar example from physics; every different rotation angle corresponds to a different group element, and this angle can be varied continuously.

We want to investigate the subject of Lie groups here so that the reader will gain some understanding of the underlying mathematical foundation on which classical mechanics is based; in addition, Lie group theory is currently being employed extensively in various aspects of research in modern physics. Our intention is to provide a readable introduction to the subject, but due to space and background limitations, it will certainly not be sufficiently detailed and rigorous so as to satisfy a mathematician.

As a basic definition, we shall state that a **Lie group** is a continuous group in which every element is associated with the n coordinates of a point in an n-dimensional real Euclidean space, and for which the composition law or multiplication rule of the elements is given in terms of analytic functions. The integer n is called the **dimension** of the group and is assumed finite.

Conventionally, the origin, or point with coordinates of zero, is taken as the location associated with the identity element I. Further, it will be understood that all points of interest will lie in the "neighborhood" of the origin, so that we are only going to deal with **local** Lie groups. (The word neighborhood is being used in an intuitive nonrigorous sense, for we don't want to go too far into topological concepts.) The **global** structure, which includes points far from the origin, will not usually be considered.

In accordance with these ideas, an element A of the group is assumed to be in a one to one correspondence with a point having coordinates $\alpha^1, \alpha^2, \ldots \alpha^n$, while the element B will correspond to a point of coordinates $\beta^1, \beta^2, \ldots \beta^n$. Denoting the coordinates of the product $C \equiv AB$ by $\gamma^1, \gamma^2, \ldots \gamma^n$, then

$$\gamma^j = f^j(\alpha, \beta), \quad \text{where } A = A(\alpha), \quad B = B(\beta), \quad C \equiv AB = C(\gamma).$$
(F.2)

Lie Groups and Algebras   **425**

The γ^j for the product C are specified analytic functions f^j of the coordinates of A and B. From the choice of the origin as the identity element, it follows at once that $f^j(\alpha, 0) = \alpha^j$ from $AI = A$, and $f^j(0, \beta) = \beta^j$ from $IB = B$.

As a simple one-dimensional example of a Lie group, consider the real numbers as represented by points on the x-axis, with the group "product" rule being just ordinary addition. If the element A corresponds to x-coordinate α, B to β, and C to γ, the relation (F.2) is merely

$$\gamma = \alpha + \beta,$$

a simple analytic function indeed. In a trivial fashion, the closure and associativity properties are satisfied, the identity element has the coordinate zero, and the inverse of any element D is the one which has for its x-coordinate the negative of the D coordinate. All the basic requirements for a Lie group have then been fulfilled.

However, the Lie groups with which we shall primarily be concerned are those of nth-order matrices. Recall that in mathematical terminology a **vector space** V of dimension n is a set of elements having the usual vector properties. A **linear operator** on this space is an operator P such that

$$P(C_1\vec{r}_1 + C_2\vec{r}_2) = C_1(P\vec{r}_1) + C_2(P\vec{r}_2),$$

where $\vec{r}_1$ and $\vec{r}_2$ are vectors and C_1 and C_2 are constant coefficients. The set of all invertible linear operators on the space is said to form the **general linear group** for that space. In addition, such operators can be given a matrix representation consisting of nonsingular square matrices. The group of operators will be isomorphic to the matrix group; the designation GL(n, R) for *G*eneral *L*inear transformation group on *R*eal numbers in n dimensions refers to both these groups and related others. Though we shall not show it, it turns out that most Lie groups are isomorphic to some subgroup of GL(n, R).

Of course, it is necessary to provide some rule by means of which these matrices can be assigned coordinates. In this regard, it is customary to choose what are essentially the n^2 real matrix elements themselves as the coordinates in an n^2-dimensional Euclidean space. The meaning behind this choice is that GL(n, R) is really of dimension n^2, though subgroups of smaller dimensionality are of frequent interest. To be strictly correct, in accord with the convention adopted for the identity, the unit matrix must be subtracted from the given matrix before the identification of coordinates is made; in that way, all coordinates will go to zero as the identity matrix is approached in a continuous fashion.

Let us investigate the composition functions f^j. Since the elements of the group are associated with points that are near the origin of coordinates, a Taylor expansion can be made about the origin in order to yield information on the coordinates of a product:

$$\gamma^j = f^j(\alpha, \beta) = \gamma^j(0,0) + \sum_{i=1}^{n} \left[\frac{\partial \gamma^j(0,0)}{\partial \alpha^i} \alpha^i + \frac{\partial \gamma^j(0,0)}{\partial \beta^i} \beta^i \right]$$

$$+ \frac{1}{2} \sum_{i,l} \left[\frac{\partial^2 \gamma^j(0,0)}{\partial \alpha^i \partial \alpha^l} \alpha^i \alpha^l + \frac{\partial^2 \gamma^j(0,0)}{\partial \beta^i \partial \beta^l} \beta^i \beta^l + 2 \frac{\partial^2 \gamma^j(0,0)}{\partial \alpha^i \partial \beta^l} \alpha^i \beta^l \right] + \cdots.$$

(Double-zero notation indicates that all α's and β's are to be set equal to zero in the function *after* any required differentiation.)

If both A and B are the identity element, then so is C, and $\alpha^i = 0 = \beta^i = \gamma^i$, whereupon $\gamma^j(0,0)$ must be zero. Using the fact that $f^j(\alpha, 0) = \alpha^j$ and $f^j(0, \beta) = \beta^j$, one can conclude that

$$\frac{\partial \gamma^j(0,0)}{\partial \alpha^i} = \delta_{i,j}, \qquad \frac{\partial^2 \gamma^j(0,0)}{\partial \alpha^i \partial \alpha^l} \equiv 0$$

from the first relation, and

$$\frac{\partial \gamma^j(0,0)}{\partial \beta^i} = \delta_{i,j}, \qquad \frac{\partial^2 \gamma^j(0,0)}{\partial \beta^i \partial \beta^l} \equiv 0$$

from the second. All of this means that, in general,

$$\gamma^j = f^j(\alpha, \beta) = \alpha^j + \beta^j + \sum_{i,l} \frac{\partial^2 \gamma^j(0,0)}{\partial \alpha^i \partial \beta^l} \alpha^i \beta^l + \cdots$$

$$= \alpha^j + \beta^j + \sum_{i,l} a^j_{il} \alpha^i \beta^l + \cdots, \quad a^j_{il} \equiv \frac{\partial^2 \gamma^j(0,0)}{\partial \alpha^i \partial \beta^l}. \tag{F.3}$$

Let us introduce a change of coordinate variables:

$$(\alpha^j)' \equiv \alpha^j - \frac{1}{2} \sum_{i,l} a^j_{il} \alpha^i \alpha^l,$$

$$(\beta^j)' \equiv \beta^j - \frac{1}{2} \sum_{i,l} a^j_{il} \beta^i \beta^l,$$

$$(\gamma^j)' \equiv \gamma^j - \frac{1}{2} \sum_{i,l} a^j_{il} \gamma^i \gamma^l.$$

Inversion of the first two expressions and use of (F.3) in the third show that to second order in the new primed coordinates

$$(\gamma^j)' = (\alpha^j)' + (\beta^j)' + \frac{1}{2}\sum_{i,l}(a_{il}^j - a_{li}^j)(\alpha^i)'(\beta^l)'.$$

This transformation demonstrates that the part of the a_{il}^j which is symmetric in the indices i and l can easily be eliminated, while the antisymmetric part, proportional to

$$C_{il}^j \equiv a_{il}^j - a_{li}^j$$

cannot be. The C_{il}^j are the local **structure constants** of the Lie group, and as such are of fundamental importance. More will be said about them shortly. With their use, eq. (F.3) becomes (omitting primes)

$$\gamma^j = \alpha^j + \beta^j + \frac{1}{2}\sum_{i,l} C_{il}^j \alpha^i \beta^l. \tag{F.3'}$$

Next, consider the inverse of any operator A with coordinates α^j. In view of the fact that $AA^{-1} = I$, with I having coordinates that are identically zero, it follows from the product rule that the coordinates of A^{-1} must be $-\alpha^j$, at least to first order. (This is reminiscent of the ideas involved in the discussion of Chapter 6 on the inverse rotation matrix for small rotations.) Actually, it is easy to show that the relation holds to second order as well.

Another type of operator which will be of interest later is the **commutator** of A and B, the definition familiar from quantum mechanics:

$$[A, B] \equiv AB - BA. \tag{F.4}$$

Note that in a general Lie group, only one operation is defined-multiplication. Sums or differences as implied in (F.4) would therefore be meaningless. But if the elements A and B are specifically taken as matrices from $GL(n, R)$, then the commutator is well defined. Returning to the expression (F.3') and substituting it in (F.4), the jth coordinate of the commutator is

$$([A, B])^j = \left(\alpha^j + \beta^j + \frac{1}{2}\sum_{i,l} C_{il}^j \alpha^i \beta^l + \cdots\right) - \left(\beta^j + \alpha^j + \frac{1}{2}\sum_{i,l} C_{il}^j \beta^i \alpha^l + \cdots\right)$$

$$= \frac{1}{2}\sum_{i,l}(C_{il}^j - C_{li}^j)\alpha^i\beta^l = \sum_{i,l} C_{il}^j \alpha^i \beta^l, \tag{F.5}$$

upon interchange of indices and employment of the antisymmetry property of C_{li}^j.

In general, it is possible to proceed by an alternative to the above expansions, by using the group associativity to obtain partial differential equations for the composition functions f_j (provided that the latter have well-behaved partial derivatives with respect to the α's and β's). These equations can only be solved if certain integrability conditions are satisfied, and the latter relations lead to the set of structure constants because of their relation to the group composition law. (Note that many Lie groups have the same set of structure constants; all of them share a common structure in the neighborhood of the identity element.)

We want to proceed in a less formal manner, however. Consider an element $A(\varepsilon)$ in the Lie (matrix) group under investigation, the coordinates $\alpha^j(\varepsilon)$ of which depend only on the small, real, and positive parameter ε, such that $A(0) = I$ for $\varepsilon = 0$. As ε is changed in value, the various elements $A(\varepsilon)$ for differing ε may in themselves form a group, called a **one-parameter subgroup** of the Lie group. Through proper choice of the parameter ε, it turns out that it is always possible to write the multiplication law in the **standard form**

$$A(\varepsilon_1)A(\varepsilon_2) = A(\varepsilon_1 + \varepsilon_2). \tag{F.6}$$

(Because of the additive nature of this composition rule, the subgroup must clearly be Abelian.)

As a consequence of (F.6), we have at once that

$$A(\varepsilon)A(-\varepsilon) = A(0) = I, \tag{F.6'}$$

hence,

$$A^{-1}(\varepsilon) = A(-\varepsilon). \tag{F.7}$$

Differentiating (F.6') gives, with use of (F.7),

$$\frac{dA(\varepsilon)}{d\varepsilon}A^{-1}(\varepsilon) + A(\varepsilon)\frac{d[A^{-1}(\varepsilon)]}{d\varepsilon} = 0,$$

or

$$\frac{dA}{d\varepsilon} = -A\frac{dA^{-1}}{d\varepsilon}A,$$

where the order of the factors must be handled with care inasmuch as quantities such as $A(\varepsilon)$ are matrices (or operators) of some type which in general do not commute. Now, we have

$$A(\varepsilon)\frac{d[A^{-1}(\varepsilon)]}{d\varepsilon} = A(\varepsilon)\frac{dA(-\varepsilon)}{d\varepsilon}$$

$$= A(\varepsilon)\lim_{\Delta\varepsilon \to 0}\left[\frac{A(-\varepsilon - \Delta\varepsilon) - A(-\varepsilon)}{\Delta\varepsilon}\right]$$

$$= \lim_{\Delta\varepsilon \to 0}\left[\frac{A(-\Delta\varepsilon) - A(0)}{\Delta\varepsilon}\right] \equiv -\left(\frac{dA(\varepsilon)}{d\varepsilon}\right)_{\varepsilon=0},$$

Lie Groups and Algebras 429

where the definition of a derivative has been employed along with (F.6). Let us set

$$t \equiv \left(\frac{dA}{d\varepsilon}\right)_{\varepsilon=0}, \tag{F.8}$$

so that $dA/d\varepsilon = tA$, a differential equation having solution

$$A(\varepsilon) = e^{\varepsilon t} \tag{F.9}$$

with $A(0) = I$ used to evaluate the constant of integration. In view of the matrix or operator t in the exponent, the expression for $A(\varepsilon)$ must be evaluated through a series expansion of the exponential:

$$A(\varepsilon) = I + \varepsilon t + \frac{\varepsilon^2}{2!}t^2 + \cdots \tag{F.10}$$

For obvious reasons, the quantity t is called the **infinitesimal generator** of the one-parameter subgroup. We represent it by t for "tangent vector" (though it is really an operator or matrix), since the coordinates $\alpha^j(\varepsilon)$ of $A(\varepsilon)$ are dependent on ε and a curve is traced out in n-space as ε is varied.

It turns out that every tangent vector to some "curve" that is differentiable in ε (or group element) is the infinitesimal generator of a group element and thereby satisfies (F.9). The set of all the infinitesimal generators t is called the **Lie algebra** $\mathscr{L}$ of the Lie group, and its elements form a linear vector space, although we shall not prove these assertions.

To complete the formal definition of the Lie algebra, we need to specify an important quantity known as a **Lie bracket**. For an understanding of this concept, consider two generators

$$t \equiv \left(\frac{dA}{d\varepsilon}\right)_{\varepsilon=0} \quad \text{and} \quad s \equiv \left(\frac{dB}{d\varepsilon}\right)_{\varepsilon=0}.$$

The product of the elements AB is in the Lie group by closure, but there is no requirement that the product ts of the generators will be in the Lie algebra; in general, the product of two generators does *not* produce a third one. The Lie bracket, on the other hand, will be defined in a way that is closely related to this product but so as always to be contained in $\mathscr{L}$. (The elements of $\mathscr{L}$, however, are not necessarily in the Lie group.)

For many situations in modern physics, the Lie bracket $[t, s]$ of two generators t and s is simply taken as their commutator, which can be shown to be a member of the Lie algebra. In general, the Lie bracket must satisfy certain fundamental properties, regardless of its specific form. (It is easy to show that the commutator does so.) These properties include:

1. Antisymmetry: $[t, s] = -[s, t]$. (F.11)
2. Linearity: $[C_1 t + C_2 s, r] = C_1[t, r] + C_2[s, r]$ (C_1, C_2 real). (F.12)
3. Jacobi Identity: $[[t, s], r] + [[s, r], t] + [[r, t], s] = 0$. (F.13)

As an element of the Lie algebra, the bracket depends heavily on these properties for its use in calculations.

As a vector space, a Lie algebra can be described in terms of a basis set. Suppose that the basis vectors are denoted by $\mu_1, \mu_2, \ldots \mu_n$. It is possible to write

$$t = \sum_i t^i \mu_i, \quad s = \sum_i s^i \mu_i, \quad r = [t, s] = \sum_i r^i \mu_i,$$

where the t^i, s^i, and r^i are the generator components in this basis. Note that

$$[t, s] = \left[\sum_j t^j \mu_j, \sum_k s^k \mu_k \right] = \sum_{j,k} t^j s^k [\mu_j, \mu_k].$$

But since the commutator bracket $[\mu_j, \mu_k]$ is in $\mathscr{L}$, we can also expand it in the μ-basis with coefficients C_{jk}^l:

$$[\mu_j, \mu_k] = \sum_l C_{jk}^l \mu_l. \tag{F.14}$$

Substitution of (F.14) yields

$$[t, s] = \sum_{j,k,l} t^j s^k C_{jk}^l \mu_l = r = \sum_i r^i \mu_i = \sum_l r^l \mu_l.$$

Equating coefficients, we have

$$r^l = \sum_{j,k} t^j s^k C_{jk}^l, \tag{F.15}$$

which gives the bracket components in terms of those of t and s. Inasmuch as this last relation is identical with (F.5) upon a change of indices, it becomes clear that the constants C_{jk}^l are merely the structure constants of the group.

Let us consider an explicit example. Consider the group of operators $G(l, m, n)$ which act on functions $f(q)$ of the variable q such that

$$G(l, m, n) f(q) = f(q + n) e^{l + mq},$$

where all quantities concerned are real. By an examination of the basic postulates, it is not difficult to show that the G-operators form a group. Next, form three one-parameter subgroups by taking one of the parameters or "coordinates" of the G-operator as ε, the others as zero:

$$G_1(\varepsilon,0,0)f(q) = f(q)e^\varepsilon,$$

$$G_2(0,\varepsilon,0)f(q) = f(q)e^{\varepsilon q},$$

$$G_3(0,0,\varepsilon)f(q) = f(q+\varepsilon).$$

With the aid of (F.8), we can find the tangent vectors for each of these subgroups:

$$t_1 f(q) \equiv \left[\frac{d(G_1 f(q))}{d\varepsilon}\right]_{\varepsilon=0} = [f(q)e^\varepsilon]_{\varepsilon=0} = f(q),$$

$$t_2 f(q) \equiv \left[\frac{d(G_2 f(q))}{d\varepsilon}\right]_{\varepsilon=0} = q f(q),$$

$$t_3 f(q) \equiv \left[\frac{d(G_3 f(q))}{d\varepsilon}\right]_{\varepsilon=0} = \left[\frac{df(q+\varepsilon)}{d\varepsilon}\right]_{\varepsilon=0} = \frac{df(q)}{dq}.$$

The actual expressions can only be worked out in detail by letting each G act on $f(q)$.

The tangent vectors correspond to the identity, multiplication by q, and differentiation with respect to q. If the Lie brackets are taken as commutators, the various possibilities are

$$[t_1, t_2] f(q) = (t_1 t_2 - t_2 t_1) f(q) = 0,$$

$$[t_1, t_3] f(q) = 0,$$

$$[t_2, t_3] f(q) = q\frac{df}{dq} - \frac{d}{dq}(qf) = -f(q), \quad \text{or} \quad [t_2, t_3] = -t_1.$$

The structure constants, most of which vanish, can be written down at once from (F.14).

F.3 THE ROTATION GROUP IN THREE-DIMENSIONAL SPACE AND THE GROUP SU(2)

A rotation matrix $\mathbf{A}$ in three-space is a member of the group $GL(3, R)$; it is not convenient, however, to follow the previous suggestion of taking the matrix elements themselves as the coordinates. The reason for this is that the nine elements of the matrix are not independent but are subject to the orthogonality conditions discussed earlier. As a consequence, only three parameters (such as the Euler angles) must be given to specify the elements of a rotation matrix; we are therefore dealing with a dimension three subgroup of GL(3, R) called the **special orthogonal group** SO(3, R). Even though this rotation group cannot be parameterized in a simple way, we know that any product of two rotations (corresponding to a single equivalent rotation) has elements which are determined by analytic trigonometric functions of the elements in the individual factors; the rotation group then must be a Lie group.

The analysis proceeds from the condition (5.12) which always holds for an orthogonal matrix:

$$\mathbf{A}\mathbf{A}^{-1} = \mathbf{A}\tilde{\mathbf{A}} = \mathbf{1}.$$

Regarding **A** as a function of the single parameter ε, differentiation of this condition produces

$$\frac{d\mathbf{A}}{d\varepsilon}\tilde{\mathbf{A}} + \mathbf{A}\frac{d\tilde{\mathbf{A}}}{d\varepsilon} = \mathbf{0}.$$

Taking the limit as $\varepsilon \to 0$ and noting that $\mathbf{A}(0) = \tilde{\mathbf{A}}(0) = \mathbf{1}$, we see from (F.8) that

$$\mathbf{t} + \tilde{\mathbf{t}} = \mathbf{0}, \quad \text{or} \quad \mathbf{t} = -\tilde{\mathbf{t}}.$$

In other words, the generator and hence the Lie algebra of the rotation group consist of matrices that are entirely antisymmetric. As a consequence, one can adopt a basis set

$$\boldsymbol{\mu}_1 \equiv \begin{pmatrix} 0 & 0 & 0 \\ 0 & 0 & -1 \\ 0 & 1 & 0 \end{pmatrix}, \quad \boldsymbol{\mu}_2 \equiv \begin{pmatrix} 0 & 0 & 1 \\ 0 & 0 & 0 \\ -1 & 0 & 0 \end{pmatrix}, \quad \boldsymbol{\mu}_3 \equiv \begin{pmatrix} 0 & -1 & 0 \\ 1 & 0 & 0 \\ 0 & 0 & 0 \end{pmatrix}$$

(F.16)

and write

$$\mathbf{t} = t^1\boldsymbol{\mu}_1 + t^2\boldsymbol{\mu}_2 + t^3\boldsymbol{\mu}_3; \tag{F.17}$$

any antisymmetric matrix is surely a linear combination of the matrices in the μ-basis. In view of the fact that

$$\boldsymbol{\mu}_1\boldsymbol{\mu}_2 = \begin{pmatrix} 0 & 0 & 0 \\ 0 & 0 & -1 \\ 0 & 1 & 0 \end{pmatrix}\begin{pmatrix} 0 & 0 & 1 \\ 0 & 0 & 0 \\ -1 & 0 & 0 \end{pmatrix} = \begin{pmatrix} 0 & 0 & 0 \\ 1 & 0 & 0 \\ 0 & 0 & 0 \end{pmatrix}$$

and

$$\mu_2\mu_1 = \begin{pmatrix} 0 & 1 & 0 \\ 0 & 0 & 0 \\ 0 & 0 & 0 \end{pmatrix},$$

then,

$$[\mu_1, \mu_2] = \begin{pmatrix} 0 & -1 & 0 \\ 1 & 0 & 0 \\ 0 & 0 & 0 \end{pmatrix} = \mu_3,$$

whereupon $C_{12}^3 = 1$ from (F.14). In general, the structure constants are given by

$$C_{jk}^l = \delta_{jkl},$$

where δ_{jkl} is the Levi–Civita density of Chapter 6.

Relations of this type have a vaguely familiar ring to them. The connection to a well-known operation can now be easily demonstrated. Employing (F.17) and the fact that

$$[\mu_j, \mu_k] = \sum_l \delta_{jkl}\mu_l, \qquad (F.18)$$

we have, for the Lie bracket of the tangent vectors **s** and **t**,

$$[\mathbf{s}, \mathbf{t}] = \left[\left(\sum_j s^j\mu_j\right), \left(\sum_k t^k\mu_k\right)\right]$$

$$= \sum_{j,k} s^j t^k [\mu_j, \mu_k] = \sum_{j,k,l} s^j t^k \delta_{jkl}\mu_l$$

$$= \sum_l (\vec{s} \times \vec{t})_l \mu_l. \qquad (F.19)$$

The Lie bracket of two vectors in the algebra of the rotation group is the same thing as the cross product of the vectors in three-space! The real three-dimensional vector space with cross product as bracket is probably the most common example of a Lie algebra in practice.

To understand the pure effect of the basis matrices, set $t^1 = t^2 = 0$, $t^3 = 1$, so that $\mathbf{t} = \mu_3$ for an example. From (F.10), we know that

$$\mathbf{A}(\varepsilon) = \mathbf{1} + \varepsilon\mu_3 + \frac{\varepsilon^2}{2!}(\mu_3)^2 + \cdots.$$

Since
$$(\mu_3)^2 = \begin{pmatrix} 0 & -1 & 0 \\ 1 & 0 & 0 \\ 0 & 0 & 0 \end{pmatrix} \begin{pmatrix} 0 & -1 & 0 \\ 1 & 0 & 0 \\ 0 & 0 & 0 \end{pmatrix} = \begin{pmatrix} -1 & 0 & 0 \\ 0 & -1 & 0 \\ 0 & 0 & 0 \end{pmatrix},$$

we have
$$(\mu_3)^3 = \mu_3(\mu_3)^2 = -\mu_3, \qquad (\mu_3)^4 = -(\mu_3)^2,$$

and so on, therefore

$$A(\varepsilon) = 1 + (\mu_3)^2 - (\mu_3)^2 \left(1 - \frac{\varepsilon^2}{2!} + \frac{\varepsilon^4}{4!} - \cdots \right) + \mu_3 \left(\varepsilon - \frac{\varepsilon^3}{3!} + \cdots \right)$$

$$= 1 + (\mu_3)^2 - (\mu_3)^2 \cos \varepsilon + \mu_3 \sin \varepsilon$$

$$= \begin{pmatrix} \cos \varepsilon & -\sin \varepsilon & 0 \\ \sin \varepsilon & \cos \varepsilon & 0 \\ 0 & 0 & 1 \end{pmatrix},$$

which is just the rotation matrix of Chapter 5 (in the active interpretation). The generalization to an arbitrary direction $\hat{n}$ for the axis of rotation is straightforward, giving

$$A(\varepsilon) = e^{\varepsilon \hat{n} \cdot \vec{\tau}}.$$

The commutation rule (F.18) is the basic ingredient of the Lie algebra for the rotation group. We have seen that the basis set (F.16) satisfies this rule and in fact other matrices of odd order can be found which will also do so. There is, however, a Lie group of even-order (2 × 2) matrices which bears a close relation to the rotation group; it is called SU(2) or SU(2, C), for *S*pecial *U*nitary group over the *C*omplex numbers, and the matrices of this group will similarly satisfy (F.18).

Nevertheless, the matrices of SU(2) do *not* form a representation of the rotation group because they do not obey the correct *global* properties. In particular, these entities do not reduce to the unit matrix when ε is set equal to 2π, but rather to its negative! For small ε, however, the situation is quite different—the matrices of SU(2) behave in accordance with the multiplication laws of the rotation group and so one can speak of a *local* representation.

The **Pauli spin matrices**, familiar from quantum mechanics, can be used to form a representation for the group SU(2):

$$\sigma_1 \equiv \begin{pmatrix} 0 & 1 \\ 1 & 0 \end{pmatrix}, \qquad \sigma_2 \equiv \begin{pmatrix} 0 & -i \\ i & 0 \end{pmatrix}, \qquad \sigma_3 \equiv \begin{pmatrix} 1 & 0 \\ 0 & -1 \end{pmatrix}. \qquad \text{(F.20)}$$

The spin matrices themselves do not satisfy (F.18), but upon multiplication of each by the constant $-i/2$ that requirement will be obeyed. For a rotation of amount ε about the direction specified by $\hat{n}$, we then have, by analogy with (F.9) and the previous work,

$$U(\varepsilon) = e^{\varepsilon \hat{n} \cdot (-i/2)\vec{\sigma}} = e^{-(i/2)\varepsilon \hat{n} \cdot \vec{\sigma}}. \tag{F.21}$$

$U(\varepsilon)$ is the exponential matrix operator for SU(2), while $\vec{\sigma}$ is a vector with the spin matrices as components. Expanding, since

$$\hat{n} \cdot \vec{\sigma} = n_1 \sigma_1 + n_2 \sigma_2 + n_3 \sigma_3 = \begin{pmatrix} n_3 & n_1 - in_2 \\ n_i + in_2 & -n_3 \end{pmatrix}$$

and

$$(\hat{n} \cdot \vec{\sigma})^2 = \begin{pmatrix} n_3 & n_1 - in_2 \\ n_1 + in_2 & -n_3 \end{pmatrix} \begin{pmatrix} n_3 & n_1 - in_2 \\ n_1 + in_2 & -n_3 \end{pmatrix} = \begin{pmatrix} 1 & 0 \\ 0 & 1 \end{pmatrix} = \mathbf{1},$$

we find that

$$U(\varepsilon) = \mathbf{1} - i\frac{\varepsilon}{2}(\hat{n} \cdot \vec{\sigma}) + \frac{1}{2}\left(-\frac{\varepsilon^2}{4}\right)(\hat{n} \cdot \vec{\sigma})^2 + \cdots$$

$$= \left(1 - \frac{1}{2}\left(\frac{\varepsilon}{2}\right)^2 + \cdots\right)\mathbf{1} - i(\hat{n} \cdot \vec{\sigma})\left(\frac{\varepsilon}{2} - \frac{1}{3!}\left(\frac{\varepsilon}{2}\right)^3 + \cdots\right)$$

$$= \cos\frac{\varepsilon}{2}\mathbf{1} - i\sin\frac{\varepsilon}{2}(\hat{n} \cdot \vec{\sigma}). \tag{F.22}$$

(Note that for rotation about a particular axis, only the corresponding component of $\vec{\sigma}$ is needed.) It is clear that

$$U^{-1}(\varepsilon) = \cos\frac{\varepsilon}{2}\mathbf{1} + i\sin\frac{\varepsilon}{2}(\hat{n} \cdot \vec{\sigma}) \tag{F.23}$$

and equally obvious on inspection is that

$$U^{-1}(\varepsilon) = U^{\dagger}(\varepsilon), \tag{F.24}$$

so the matrix $U(\varepsilon)$ is unitary. Furthermore, its determinant is unity:

$$\cos^2\left(\frac{\varepsilon}{2}\right) + n_3^2 \sin^2\left(\frac{\varepsilon}{2}\right) + (n_1 - in_2)(n_1 + in_2)\sin^2\left(\frac{\varepsilon}{2}\right)$$

$$= \cos^2\left(\frac{\varepsilon}{2}\right) + (n_1^2 + n_2^2 + n_3^2)\sin^2\left(\frac{\varepsilon}{2}\right) = \cos^2\left(\frac{\varepsilon}{2}\right) + \sin^2\left(\frac{\varepsilon}{2}\right)$$

$$= 1 = |U(\varepsilon)|. \tag{F.25}$$

The adjective "special" in the designation of SU(2) refers to the $+1$ determinant. For this same reason, $U(\varepsilon)$ is called a unitary **unimodular** matrix. Any such matrix corresponds to some value of the angle ε; note that $U(2\pi) = -1$, in agreement with previous statements.

The question remains as to how a transformation of coordinates in three-space (physically represented by a rotation) can be accomplished by means of the 2×2 matrix $U(\varepsilon)$. To answer that question, it is necessary to establish with a little algebra (see the problem set) that

$$U(\varepsilon)\sigma_i U^{-1}(\varepsilon) = \sum_{j=1}^{3} A_{ji}(\varepsilon)\sigma_j, \qquad (F.26)$$

a relation which follows at once from (F.22) and the properties of the spin matrices. In this expression,

$$A_{ji}(\varepsilon) = \delta_{j,i}\cos\varepsilon + n_j n_i(1 - \cos\varepsilon) - \left(\sum_k \delta_{jik} n_k\right)\sin\varepsilon, \qquad (F.27)$$

which is just eq. (5.39) for the full rotation matrix as a function of arbitrary ε and $\hat{n}$. Next, define a 2×2 matrix S such that

$$S \equiv x\sigma_1 + y\sigma_2 + z\sigma_3 = \sum_{i=1}^{3} x_i \sigma_i = \begin{pmatrix} z & x - iy \\ x + iy & -z \end{pmatrix}. \qquad (F.28)$$

Transforming S with a similarity transformation through $U(\varepsilon)$, one obtains

$$U(\varepsilon)SU^{-1}(\varepsilon) = \sum_i x_i (U\sigma_i U^{-1}) = \sum_{i,j} x_i A_{ji} \sigma_j$$

$$= \sum_j \left(\sum_i A_{ji} x_i\right)\sigma_j = \sum_j x'_j \sigma_j \equiv S' = \begin{pmatrix} z' & x' - iy' \\ x' + iy' & -z' \end{pmatrix},$$

$$(F.29)$$

where we have made use of (F.26) and (5.4). The determinant of S, which is here just $-(x^2 + y^2 + z^2)$, is, as usual, an invariant of the similarity transformation. The matrix U acting in a strange complex two-space is therefore directly related to the orthogonal transformation or rotation which occurs in the real three-space.

More generally, a transformation in a complex two-dimensional space requires a transforming matrix of the form

$$\begin{pmatrix} \alpha & \beta \\ \gamma & \delta \end{pmatrix},$$

where the matrix elements are all complex quantities, called by some authors the **Cayley–Klein parameters**.[1] (Other authors prefer to use the four quantities $\cos\varepsilon/2$

---

[1] E. Wigner, *Group Theory*, revised edition, Academic, New York, 1959, p. 159.

and the components of $\hat{n} \sin \varepsilon/2$, designating them by this name.) The eight quantities that are the real and imaginary parts of the Cayley–Klein parameters are subject to the conditions of unitarity and unimodularity which are imposed on the matrix, leaving a total of only three that are independent. Inasmuch as we have seen earlier that the orientation of a rigid body is always specified by only three parameters (the Euler angles or some other such set), this result is not surprising. Furthermore, if the coordinate transformation of (F.29) is actually carried out, using the matrix $\begin{pmatrix} \alpha & \beta \\ \gamma & \delta \end{pmatrix}$ for $\mathbf{U}(\varepsilon)$ and its conjugate transpose for $\mathbf{U}^{-1}(\varepsilon)$, the coefficients in the transformation will be given in terms of the Cayley–Klein parameters. One can then directly identify these coefficients with those from the real orthogonal matrix (5.18), the procedure furnishing a simple way of relating the Cayley–Klein parameters to the Euler angles. At this point, we shall not pursue such an approach, because in (F.22) we already have obtained a convenient explicit form for the transformation matrix.

Suppose that the product of two rotations is considered in the two-dimensional formalism. Let the first rotation be through angle α about the axial direction given by $\hat{n}$, the second by β about $\hat{m}$. From (F.22), one can write for the product

$$\left[\cos\frac{\beta}{2}\mathbf{1} - i\sin\frac{\beta}{2}(\hat{m}\cdot\vec{\sigma})\right]\left[\cos\frac{\alpha}{2}\mathbf{1} - i\sin\frac{\alpha}{2}(\hat{n}\cdot\vec{\sigma})\right]$$

$$= \left(\cos\frac{\alpha}{2}\cos\frac{\beta}{2}\right)\mathbf{1} - i\left[\left(\sin\frac{\alpha}{2}\cos\frac{\beta}{2}\right)\hat{n} + \left(\sin\frac{\beta}{2}\cos\frac{\alpha}{2}\right)\hat{m}\right]\cdot\vec{\sigma}$$

$$- \left(\sin\frac{\alpha}{2}\sin\frac{\beta}{2}\right)(\hat{m}\cdot\vec{\sigma})(\hat{n}\cdot\vec{\sigma}).$$

Noting that

$(\hat{m}\cdot\vec{\sigma})(\hat{n}\cdot\vec{\sigma})$

$$= \begin{pmatrix} m_3 & m_1 - im_2 \\ m_1 + im_2 & -m_3 \end{pmatrix}\begin{pmatrix} n_3 & n_1 - in_2 \\ n_1 + in_2 & -n_3 \end{pmatrix}$$

$$= \begin{pmatrix} (m_1 - im_2)(n_1 + in_2) + m_3 n_3 & m_3(n_1 - in_2) - (m_1 - im_2)n_3 \\ (m_1 + im_2)n_3 - m_3(n_1 + in_2) & (m_1 + im_2)(n_1 - in_2) + m_3 n_3 \end{pmatrix}$$

$$= (\hat{m}\cdot\hat{n})\mathbf{1}$$

$$+ i\begin{pmatrix} m_1 n_2 - m_2 n_1 & (m_2 n_3 - m_3 n_2) - i(m_3 n_1 - m_1 n_3) \\ (m_2 n_3 - m_3 n_2) + i(m_3 n_1 - m_1 n_3) & -(m_1 n_2 - m_2 n_1) \end{pmatrix}$$

$$= (\hat{m}\cdot\hat{n})\mathbf{1} + i(\hat{m}\times\hat{n})\cdot\vec{\sigma},$$

the product becomes
$$\left[\cos\frac{\alpha}{2}\cos\frac{\beta}{2} - \sin\frac{\alpha}{2}\sin\frac{\beta}{2}(\hat{m}\cdot\hat{n})\right]\mathbf{1}$$
$$\mathit{i}\left[\left(\sin\frac{\alpha}{2}\cos\frac{\beta}{2}\right)\hat{n} + \left(\sin\frac{\beta}{2}\cos\frac{\alpha}{2}\right)\hat{m} + \left(\sin\frac{\alpha}{2}\sin\frac{\beta}{2}\right)(\hat{m}\times\hat{n})\right]\cdot\vec{\sigma}.$$

The first bracket (coefficient of **1**) is the cosine of half the equivalent single rotation angle, as is seen from (F.22), while the second bracket is related to the direction of the equivalent rotation. The result is in perfect accord with (6.10) and has been obtained much more simply. (It may be worthwhile to point out that for exponential matrix operators of type (F.21), one cannot simply multiply e^A by e^B and get $e^{(A+B)}$; that familiar result follows only when A and B commute with one another, as is not difficult to prove. Consequently, we have followed a procedure which began with (F.22) instead.)

It is illuminating to discuss the theory a bit further. Applying (F.26) repeatedly, we find that

$$\mathbf{U}(\beta)[\mathbf{U}(\alpha)\sigma_i\mathbf{U}^{-1}(\alpha)]\mathbf{U}^{-1}(\beta) = \mathbf{U}(\beta)\left[\sum_j A_{ji}(\alpha)\sigma_j\right]\mathbf{U}^{-1}(\beta)$$

$$= \sum_j A_{ji}(\alpha)\left[\mathbf{U}(\beta)\sigma_j\mathbf{U}^{-1}(\beta)\right]$$

$$= \sum_{j,k} A_{ji}(\alpha)A_{kj}(\beta)\sigma_k = \sum_k [\mathbf{A}(\beta)\mathbf{A}(\alpha)]_{ki}\sigma_k,$$

or

$$[\mathbf{U}(\beta)\mathbf{U}(\alpha)]\sigma_i[\mathbf{U}(\beta)\mathbf{U}(\alpha)]^{-1} = \sum_k [\mathbf{A}(\beta)\mathbf{A}(\alpha)]_{ki}\sigma_k, \qquad (F.30)$$

a relation which has the form of (F.26) for this product operation. That is, there is a mapping from SU(2, C) to SO(3, R) which preserves the group multiplicative operation. But it turns out that there are two elements in SU(2) for each one in SO(3), so the correspondence between the two groups is a homomorphism, not an isomorphism. This is easily seen from the fact that $\mathbf{A}(\varepsilon + 2\pi) = \mathbf{A}(\varepsilon)$ for the 3×3 matrices, while, from (F.22),

$$\mathbf{U}(\varepsilon + 2\pi) = \cos\left(\frac{\varepsilon + 2\pi}{2}\right)\mathbf{1} - i\sin\left(\frac{\varepsilon + 2\pi}{2}\right)(\hat{n}\cdot\vec{\sigma})$$

$$= -\cos\frac{\varepsilon}{2}\mathbf{1} + i\sin\frac{\varepsilon}{2}(\hat{n}\cdot\vec{\sigma})$$

$$= -\mathbf{U}(\varepsilon),$$

so both $\mathbf{U}(\varepsilon)$ and $-\mathbf{U}(\varepsilon)$ are mapped into $\mathbf{A}(\varepsilon)$.

For the Lie algebras of the two groups, however, the situation is quite different; they are isomorphic. The basic commutation rule (F.18) is satisfied in both cases, and, by exponentiation, we obtained expressions for $\mathbf{A}(\varepsilon)$ and $\mathbf{U}(\varepsilon)$. There is then a one-to-one correspondence, an isomorphism, between the elements of the Lie algebras; the local structure of both groups is the same. (Any group representation faithful in the vicinity of the identity element leads one to the structure properties of the Lie algebra, and there is a mathematical theorem known as **Ado's theorem**, which states that every finite-dimensional Lie algebra has such a faithful representation in terms of matrices.)

F.4 OTHER GROUPS OF INTEREST IN PHYSICS

It is desirable to digress from our analysis in order to introduce a few other groups. There are several groups in mathematics which have important applications in physics. We have already mentioned the general linear group $GL(n, R)$, which has for its elements all the invertible linear transformations of an n-dimensional vector space, and is isomorphic to the group of nonsingular $n \times n$ matrices; $SO(3, R)$ is one significant subgroup.

More generally, a linear transformation can be accompanied by a translation, giving what is called an **affine transformation**

$$\mathbf{X}' = \mathbf{R}\mathbf{X} + \mathbf{k}. \tag{F.31}$$

Here, $\mathbf{X}'$, $\mathbf{X}$, and $\mathbf{k}$ are column vectors, and $\mathbf{R}$ is a matrix. Usually, $\mathbf{R}$ provides the details of a rotation, while $\mathbf{k}$ expresses the translation for each component. A second such relation

$$\mathbf{X}'' = \mathbf{R}'\mathbf{X}' + \mathbf{k}'$$

in combination with the first yields a product

$$\mathbf{X}'' = \mathbf{R}'(\mathbf{R}\mathbf{X} + \mathbf{k}) + \mathbf{k}' = (\mathbf{R}'\mathbf{R})\mathbf{X} + (\mathbf{R}'\mathbf{k} + \mathbf{k}'),$$

which demonstrates the group closure. If desired, a group of nonsingular $(n + 1) \times (n + 1)$ block matrices can be constructed, of type

$$\begin{pmatrix} \mathbf{R} & \mathbf{k} \\ \mathbf{0} & 1 \end{pmatrix},$$

where $\mathbf{k}$ is an $n \times 1$ column vector, while $\mathbf{0}$ is a row vector that is $1 \times n$. The group of affine transformations is isomorphic to the matrix set.

Inasmuch as any translation preserves vector differences, hence vector lengths, it is a possible rigid body motion. But in that case, the associated matrix $\mathbf{R}$ must necessarily be orthogonal. Those affine transformations which are related to rigid body motions form a subgroup of the affine group known as the **Euclidean group**. The latter is a six-parameter group in three-space that carries our usual notions of geometry; it contains both the translation and rotation groups as subgroups.

Next we discuss the **Galilei group**, which is the group of transformations between Cartesian coordinate systems that serve as the inertial frames of the Newtonian theory. Although they are usually seen in abbreviated form, the most general Galilean transformation is actually of the type

$$\mathbf{X}' = \mathbf{R}\mathbf{X} + \mathbf{v}t + \mathbf{X}_0, \qquad t' = t + t_0. \tag{F.32}$$

The second equation is a time transformation which assumes that the clocks in the two frames are not synchronized, even though they run at the same rate. The first equation is a vector relation which allows for the possibilities that the primed and unprimed axes are rotated at some angle with respect to one another, that there is relative motion at velocity $\mathbf{v}$, and that the two system origins are not coincident at time zero (differing by $\mathbf{X}_0$). One could think of one-parameter subgroups of the associated Lie algebra as corresponding to three rotation axis directions, three velocity directions, three directions for the origin displacement, and one time displacement. The Galilei group is therefore ten-dimensional in its most general form.

Extending the Galilean concepts into the relativistic domain, one meets the **Lorentz group**, the terminology of which is complicated by the fact that it comes in two varieties. First, there is the **homogeneous Lorentz group**, which is concerned with transformations between two frames in uniform relative motion, or frames rotated in space relative to one another, or both. The origins, however, do coincide in space-time, so one then has a six-parameter group. Secondly, there is the **inhomogeneous Lorentz group** (or **Poincaré group**), which allows for space-time translations as well and gives a ten-parameter group by analogy with the Galilei group.

There are other groups of great importance in modern physics. The special unitary group SU(2) has already been discussed. There is a similar group of order three, called SU(3). It is a three-dimensional entity with eight parameters and a set of generators consisting of eight matrices that are 3×3 but otherwise much like the spin matrices. The **Eight-Fold Way** developed by Gell-Mann and Néeman,[2] which has been widely heralded in the elementary particle physics of the past two decades, is based on the symmetries of this group. The principal idea of the theory is that the eight fundamental baryons have masses which are taken as equal in zeroth approximation, thus forming a set of eight degenerate quantum

---

[2] M. Gell-Mann et al., *The Eight-Fold Way*, W. A. Benjamin, New York, 1964.

states of a strong interaction Hamiltonian; these are in a sense identified with the irreducible representations of SU(3). The actual baryon mass spectrum is then obtained by splitting the degeneracies through the introduction of a perturbation potential that is not invariant under the symmetries of SU(3). The consequences of this theory have been rewarding, but we cannot delve into the matter in greater detail here.

F.5 LIE GROUPS AND MECHANICS

We shall now explain the relationship of the Lie group theory to classical mechanics. Consider any functions $f(q, p, t)$ of the appropriate dynamical variables for some problem, the behavior of f being sufficiently reasonable so that repeated differentiation is possible. Such functions form a real Lie algebra, considered simply as a vector space with certain properties that will be explored in this section. In classical mechanics, the Lie bracket for this algebra is defined to be that of Poisson!

As justification for this choice, note that the requirements which must be satisfied by any Lie bracket (antisymmetry, linearity, Jacobi identity) have been established for the Poisson bracket in Chapter 8. Until now, the operator-matrix treatment presented has always taken the Lie bracket to be a commutator, but the latter is well recognized in the literature as the quantum-mechanical analogue of the classical Poisson bracket. Therefore, the identification does seem reasonable, and we shall use Poisson brackets when constructing realizations of Lie algebras for mechanics. (A **realization** is simply a generalization of the concept of representation; the latter term is usually restricted in application to a set of matrices or operators, while the former can also include sets of functions or other entities with appropriate group or algebraic properties.)

Also, we note that the set of all possible canonical transformations forms a group. Thus, suppose there is a canonical transformation

$$Q = Q(q, p, t), \quad P = P(q, p, t),$$

followed by another one,

$$Q' = Q'(Q, P, t), \quad P' = P'(Q, P, t).$$

Poisson brackets will be preserved for any two dynamical quantities f and g:

$$[f, g]_{q,p} = [f, g]_{Q,P} = [f, g]_{Q',P'}. \tag{F.33}$$

The equality of the first and third members implies that a single canonical transformation can be found from the (q, p) to the (Q', P') set, which demonstrates closure. The other group properties can be established rather trivially.

If in a given situation the canonical transformations are such as to involve analytic functions of continuous parameters, then we will be dealing with a particular Lie group. The rotation group $SO(3, R)$ is an example.

To illustrate the relevant ideas without great difficulty, consider once more eqs. (8.54) of the infinitesimal canonical transformation:

$$dq_i = \varepsilon \frac{\partial G}{\partial p_i}, \qquad dp_i = -\varepsilon \frac{\partial G}{\partial q_i}.$$

If the characterizing parameter ε is taken as the angle $d\phi$ for an infinitesimal rotation around the z-axis, the coordinate changes of some particular particle in a system are given by

$$dx = -y\,d\phi, \qquad dy = x\,d\phi, \qquad dz = 0,$$

and similarly

$$dp_x = -p_y\,d\phi, \qquad dp_y = p_x\,d\phi, \qquad dp_z = 0.$$

Equations (8.54) then tell us that the generator has the form

$$G = xp_y - yp_x \equiv L_z$$

in terms of the angular momentum. As the system rotates through $d\phi$, a dynamical function $f(q, p)$ will change its numerical value slightly because its coordinates have undergone small changes. But we have

$$\frac{df}{d\phi} = \sum_i \left(\frac{\partial f}{\partial q_i} \frac{dq_i}{d\phi} + \frac{\partial f}{\partial p_i} \frac{dp_i}{d\phi} \right) = \sum_i \left(\frac{\partial f}{\partial q_i} \frac{\partial G}{\partial p_i} - \frac{\partial f}{\partial p_i} \frac{\partial G}{\partial q_i} \right) \equiv [f, G],$$

the last step following from (8.6). The formal solution to this differential equation is

$$f(\phi) = f_{\phi=0} + \phi \left(\frac{df}{d\phi} \right)_{\phi=0} + \frac{\phi^2}{2!} \left(\frac{d^2 f}{d\phi^2} \right)_{\phi=0} + \cdots.$$

The derivatives

$$\frac{df}{d\phi} = [f, G], \qquad \frac{d^2 f}{d\phi^2} = \left[\frac{df}{d\phi}, G \right] = [[f, G], G],$$

and so on, are all to be worked out and evaluated at $\phi = 0$. In more succinct notation,

$$f(\phi) = e^{\phi \Omega_0} f_0, \qquad \Omega \equiv [\ , G]$$

where the zero subscripts indicate that the bracket operator acts on f in each term of the expansion of $e^{\phi\Omega_0}$, the result then being evaluated at $\phi = 0$. The meaning of this result is that one can obtain the value of a function f after a *finite* rotation through angle ϕ by means of Poisson brackets involving the generator (the angular momentum component L_z).

As another useful example of the procedure suggested, let us instead take G as the Hamiltonian H of some system, which generates a time transformation according to eq. (8.55). With no explicit time dependence present, a development similar to that above gives

$$\frac{df}{dt} = [f, H]$$

and

$$f(t) = f_{t=0} + t\left(\frac{df}{dt}\right)_{t=0} + \cdots = e^{t\Omega_0}f_0, \quad \text{where} \quad \Omega \equiv [\ , H]$$

is just the time development operator of Chapter 8.

Thus, consider the familiar problem of a mass falling freely under gravity. The Hamiltonian is

$$H = \frac{p^2}{2m} + mgy,$$

with solution

$$y(t) = y(0) + t[y(0), H] + \frac{t^2}{2!}[[y(0), H], H] + \cdots.$$

The necessary Poisson brackets are obtained from eq. (8.6) by direct calculation:

$$[y(t), H] = \frac{\partial y}{\partial y}\frac{\partial H}{\partial p} - \frac{\partial y}{\partial p}\frac{\partial H}{\partial y} = \frac{\partial H}{\partial p} = \frac{p(t)}{m}, \qquad [y(0), H] = \frac{p(0)}{m},$$

$$[[y(t), H], H] = \left[\frac{p}{m}, H\right] = \frac{\partial(p/m)}{\partial y}\frac{\partial H}{\partial p} - \frac{\partial(p/m)}{\partial p}\frac{\partial H}{\partial y} = -\frac{1}{m}\frac{\partial H}{\partial y} = -g.$$

Higher brackets will vanish, so the solution is

$$y(t) = y(0) + \frac{p(0)}{m}t - \frac{1}{2}gt^2,$$

in terms of the usual constants representing initial values of position and momentum. This discussion shows how infinitesimal canonical transformations can be used to develop the solution for a finite time t. The Lie algebraic formalism with the Poisson bracket is the theoretical framework underlying many of the relations in Chapter 8.

With either the classical or the quantum-mechanical analysis of a physical system, there may be symmetry operations which leave the Hamiltonian invariant and form a group. In the case of the infinitesimal canonical transformations just discussed, both the Hamiltonian and the form of the canonical equations are left invariant under the operations of a group of dynamical transformations generated by a constant of the motion. While the group corresponds to the symmetries of the system, one may actually be more interested in the Lie algebra of the group, inasmuch as it often has simpler properties and corresponds to the conserved quantities so dear to the heart of a physicist.

As an example, we have just seen that rotations around the z-axis had the constant L_z for a generator. Consider the Poisson bracket for two components of the angular momentum of a particle:

$$\left[L_x, L_y\right] = \left[(yp_z - zp_y), (zp_x - xp_z)\right] = xp_y - yp_x = L_z.$$

In general, for arbitrary indices,

$$\left[L_j, L_k\right] = \sum_l \delta_{jkl} L_l. \tag{F.34}$$

This is the classical analogue of the famous quantum commutation relation for angular momentum; note that it is in agreement with (F.18) for the rotation group. If L_x and L_y are constants of the motion, Poisson's theorem (Chapter 8) ensures that L_z and hence the entire vector $\vec{L}$ are also constants. But we have shown in eq. (8.17) that the fundamental bracket of two canonical momenta should always vanish; the meaning of (F.34) must then be that two Cartesian components of $\vec{L}$ can never be chosen simultaneously as the canonical momenta of a given problem.

Functions of the dynamical variables may have additional structure other than that imposed by Lie group or algebraic considerations. Often, there are auxiliary conditions which must be satisfied as well, the group-algebraic structure providing an overall framework.

Hopefully, this appendix has given the reader some insight into the basic Lie concepts. Many of the detailed complexities have been omitted; the reader should consult more specialized texts for further information.

PROBLEMS

F.1 The fact that a group multiplication table (such as Table F.2) contains every element of the group once and only once in each row or column is known as the **rearrangement theorem**. Prove this theorem in general.

F.2 Consider the group of six rotation-reflection operators which transform an equilateral triangle into itself. Work out the multiplication table for this group and find its classes.

F.3 Consider the four matrices

$$A_1(\varepsilon) = \begin{pmatrix} 1+\varepsilon & 0 \\ 0 & 1 \end{pmatrix}, \quad A_2(\varepsilon) = \begin{pmatrix} 1 & 0 \\ 0 & 1+\varepsilon \end{pmatrix},$$

$$A_3(\varepsilon) = \begin{pmatrix} 1 & \varepsilon \\ 0 & 1 \end{pmatrix}, \quad A_4(\varepsilon) = \begin{pmatrix} 1 & 0 \\ \varepsilon & 1 \end{pmatrix},$$

all of which reduce to the identity in the limit as $\varepsilon \to 0$. Can each of these represent a one-parameter subgroup in standard form? Calculate the tangent vectors for all four matrices. Using these tangents as a basis, find the Lie brackets and hence the structure constants of the Lie algebra.

F.4 Show by explicit calculation that the relation (F.26) is correct.

F.5 Find the composition law (multiplication rule) for the Galilei group, and show that its elements satisfy group postulates.

F.6 Consider again the group of operators $G(l, m, n)$ of Section F.2. Let $\mu_1 \equiv 1$, $\mu_2 \equiv q$, $\mu_3 \equiv -p$ be operators which act on functions $f(q, p)$ of the dynamical variables.

(a) Setting

$$e^{+\varepsilon \mu_i} = 1 + \varepsilon[\ , \mu_i] + \frac{\varepsilon^2}{2!}[[\ , \mu_i], \mu_i] + \cdots,$$

work out $e^{+\varepsilon \mu_i} f(q, p)$ for each μ_i, using Poisson brackets in the evaluation of the exponentials.

(b) How do you interpret the results?

References

During the preparation of the manuscript for this text, numerous sources were consulted. It would neither be possible nor in fact desirable to list all of these books and articles. Instead, we give below a brief selection of books which can be endorsed as generally useful and well written. They are arranged chronologically within each section, with the latest date usually given first.

TEXTBOOKS

General Mechanics for Physicists—Graduate Level

E. Desloge, *Classical Mechanics* (two volumes), Wiley–Interscience, New York, 1982.

A. Fetter and J. Walecka, *Theoretical Mechanics of Particles and Continua*, McGraw-Hill, New York, 1980.

H. Goldstein, *Classical Mechanics*, 2nd edition, Addison-Wesley, Reading Mass., 1980.

> For almost three decades, this has been the dominant text at the graduate level. What better testimonial can be given?

R. Rosenberg, *Analytical Dynamics of Discrete Systems*, Plenum, New York, 1977.

J. Bartlett, *Classical and Modern Mechanics*, Univ. of Alabama Press, University, Alabama, 1975.

E. Saletan and A. Cromer, *Theoretical Mechanics*, John Wiley, New York, 1971.

S. Groesberg, *Advanced Mechanics*, John Wiley, New York, 1968.

D. Ter Haar, *Elements of Hamiltonian Mechanics*, North-Holland, Amsterdam, 1961.

H. Corben and P. Stehle, *Classical Mechanics*, 2nd edition, John Wiley, New York, 1960.

L. Landau and E. Lifshitz, *Mechanics*, Addison-Wesley, Reading, Mass., 1960.

L. Page, *Introduction to Theoretical Physics*, 3rd edition, Van Nostrand, Princeton, 1952.

J. Slater and N. Frank, *Mechanics*, McGraw-Hill, New York, 1947.

E. Whittaker, *A Treatise on the Analytical Dynamics of Particles and Rigid Bodies*, 4th edition, Cambridge Univ. Press, Cambridge, England, 1937.

Mechanics—Undergraduate Level

V. Barger and M. Olsson, *Classical Mechanics—A Modern Perspective*, McGraw-Hill, New York, 1973.

T. Bradbury, *Theoretical Mechanics*, John Wiley, New York, 1968.

K. Symon, *Mechanics*, 2nd edition, Addison-Wesley, Reading, Mass., 1960.

S. McCuskey, *An Introduction to Advanced Dynamics*, Addison-Wesley, Reading, Mass., 1959.

R. Becker, *Introduction to Theoretical Mechanics*, McGraw-Hill, New York, 1954.

R. Lindsay, *Physical Mechanics*, Van Nostrand, Princeton, 1950.

Textbooks on Mathematics Relevant to Mechanics

E. Butkov, *Mathematical Physics*, Addison-Wesley, Reading, Mass., 1968.

G. Arfken, *Mathematical Methods for Physicists*, Academic, New York, 1966.

H. Margenau and G. Murphy, *The Mathematics of Physics and Chemistry*, Vol. 1, 2nd edition, Van Nostrand, Princeton, 1956.

J. Rosenbach and E. Whitman, *Essentials of College Algebra*, Ginn and Co., Boston, 1951.

Textbooks on Group Theory in Physics

E. Sudarshan and N. Mukunda, *Classical Dynamics—A Modern Perspective*, John Wiley, New York, 1974. This is now the standard text on the group theoretical approach to mechanics.

J. Belinfonte and B. Kolman, *A Survey of Lie Groups and Lie Algebras with Applications and Computational Methods*, Society for Industrial and Applied Mathematics, 1972.

M. Tinkham, *Group Theory and Quantum Mechanics*, McGraw-Hill, New York, 1964.

M. Hamermesh, *Group Theory*, Addison-Wesley, Reading, Mass., 1962.

Textbooks on the Theory of Relativity

R. Adler, M. Bazin, and M. Schiffer, *Introduction to General Relativity*, 2nd edition, McGraw-Hill, New York, 1975.

R. Skinner, *Relativity*, Blaisdell, Waltham, 1969.
W. Rindler, *Essential Relativity*, Van Nostrand Reinhold, Princeton, 1969.
F. Sears and R. Brehme, *Introduction to the Theory of Relativity*, Addison-Wesley, Reading, Mass., 1968.
A. French, *Special Relativity*, W. W. Norton, New York, 1968.

Textbooks on the Theory of Oscillations

C. Hayashi, *Nonlinear Oscillations in Physical Systems*, McGraw-Hill, New York, 1964.
J. Stoker, *Nonlinear Vibrations*, Interscience, New York, 1950.
A. Andronow and C. Chaikin, *Theory of Oscillations*, Princeton U. Press, Princeton, 1949.
P. Morse, *Vibration and Sound*, second edition, McGraw-Hill, New York, 1948.

Other General Textbooks of Interest

V. Arnold, *Mathematical Methods of Classical Mechanics*, Springer-Verlag, Berlin, 1978.
W. Thirring, *Classical Dynamical Systems*, Springer-Verlag, Berlin, 1978.
The last two references describe classical mechanics in the language of topology, a useful approach in recent years.
S. McCuskey, *Introduction to Celestial Mechanics*, Addison-Wesley, Reading, Mass., 1963.
R. Lindsay and H. Margenau, *Foundations of Physics*, Dover Reprint, New York, 1957.
L. Landau and E. Lifshitz, *The Classical Theory of Fields*, Addison-Wesley, Reading, Mass., 1951.

JOURNALS

The articles are listed in the order of the sequence of the associated material within each chapter. (AJP designates the *American Journal of Physics*.)

Chapter 1

L. Eisenbud, "On the Classical Laws of Motion," *AJP* **26**, 144 (1958).
R. Weinstock, "Laws of Classical Motion: What's F? What's m? What's a?" *AJP* **29**, 698 (1961).
C. O'Sullivan, "Newton's Laws of Motion: Some Interpretations of the Formalism," *AJP* **48**, 131 (1980).

J. Satterly, "Moments of Inertia of Skeleton Figures and Shells," *AJP* **27**, 172 (1959).
G. Martin, "Unwinding a Spool," *AJP* **26**, 194 (1958).
C. Collinson, "Definition of an Elastic Collision," *AJP* **45**, 579 (1977).
C. Adler, "Connection between Conservation of Energy and Conservation of Momentum," *AJP* **44**, 483 (1976).

Chapter 2
E. Desloge and R. Karch, "Noether's Theorem in Classical Mechanics," *AJP* **45**, 336 (1977).

Chapter 3
J. Ray, "Nonholonomic Constraints and Gauss' Principle of Least Constraint," *AJP* **40**, 179 (1972).

Chapter 4
I. Freeman, "An Interesting Property of the Kepler Ellipse," *AJP* **45**, 585 (1977).

Chapter 5
R. Hilborn, "A Note on Euler Angle Rotations," *AJP* **40**, 1036 (1972).
C. Leubner, "Coordinate-Free Rotation Operator," *AJP* **47**, 727 (1979).
R. Romer, "Matrix Description of Collisions on an Air Track," *AJP* **35**, 862 (1967).
J. de Felicio and D. Redondo, "Linear Collisions Revisited," *AJP* **49**, 147 (1981).
E. Tuttle, "Mohr's Circle and the Determination of Moments of Inertia," *AJP* **45**, 396 (1977).

Chapter 6
G. Rego, "On Euler's Theorem on Rigid Body Motion," *AJP* **47**, 466 (1979).
D. Parker, "An Analytic Derivation of the Relationship Between the Angular Velocity Vector and the Euler Angles and Their Time Derivatives," *AJP* **37**, 925 (1969).
J. Wilkes, "Rotations as Solutions of a Matrix Differential Equation," *AJP* **46**, 685 (1978).
C. Leubner, "Correcting a Widespread Error Concerning the Angular Velocity of a Rotating Rigid Body," *AJP* **49**, 232 (1981).
D. Mott, "Torque on a Rigid Body in Circular Orbit," *AJP* **34**, 562 (1966).

Chapter 7

L. Chavda, "Matrix Theory of Small Oscillations," *AJP* **46**, 550 (1978).

Chapter 8

N. Lemos, "Canonical Approach to the Damped Harmonic Oscillator," *AJP* **47**, 857 (1979).

P. Campbell, "Classical Transformation Operators and the S-Matrix," *AJP* **36**, 931 (1968).

Chapter 9

R. Brehme, "A Geometric Representation of Galilean and Lorentz Transformations," *AJP* **30**, 489 (1962).

J. Ray, "Principle of Equivalence," *AJP* **45**, 401 (1977).

H. Ohanian, "What is the Principle of Equivalence?" *AJP* **45**, 903 (1977).

D. Fahnline, "A Covariant Four-Dimensional Expression for Lorentz Transformations," *AJP* **50**, 818 (1982).

Appendix F

G. Koster, *Notes on Group Theory*, Technical Report No. 8, Solid State and Molecular Theory Group, M.I.T., Cambridge, Mass., 1956.

B. Ram, "Physics of the SU(3) Symmetry Model," *AJP* **35**, 16 (1967).

T. Jordan and E. Sudarshan, "Lie Group Dynamical Formalism and the Relation between Quantum Mechanics and Classical Mechanics," *Rev. Mod. Phys.* **33**, 515 (1961).

Index

Abelian group, 421
Absolute elsewhere, future, past, 331
Acceleration, 2
Action and reaction, 3
Action-angle variables, 303
Action function, 83, 297
Active view of rotation, 132
Ado's theorem, 439
Advance of Mercury's perihelion, 361
Affine space, 349
Affine term, 416
Affine transformation, 439
Analytic function, 394
Angular momentum, 7, 15, 142
Angular velocity, 10, 179, 183
Anholonomic constraint, 36
Antiproton production, 345
Antisymmetric tensor, 169
Apparent velocity, 183
Appearance of rapidly moving object, 336
Apsidal points of orbit, 96
Associativity, 420
Asteroid rotation, 123
Atwood machine, 59
Autonomous system, 263
Axial vectors, 182

Base vectors, 403
Bead:
  on helix, 44, 60
  on hoop, 58
  on ring, 243
Bernoulli's solution for string, 229

Bertrand's theorem, 111
Birkhoff's theorem, 360
Black hole, 363
Body components of angular velocity, 189
Body cone, 199
Boost in relativity, 326
Brachistochrone, 74
Brehme diagram, 327
Brown, Ernest William, 31
Bruns' theorem, 118
Bulk modulus of solid, 162

Calculus of variations, 70
Canonical equations, 52, 277
Canonical form, 263
Canonical transformation, 286, 289
Cartesian tensor, 136
Cauchy-Riemann equations, 394
Causality, 332
Cayley-Hamilton theorem, 390
Cayley-Klein parameters, 436
Center of mass:
  calculations, 8
  definition, 8
  theorems, 13, 116
Central force, 14, 92
Centrifugal force, 185
Character of group element, 423
Characteristic equation, 145, 389
Characteristic function of Hamilton, 298
Chasles' theorem, 173
Chebyshev polynomials, 168

452   *Index*

Christoffel symbol of first kind, 414
Christoffel symbol of second kind, 413
Class of group, 422
Clausius, 109
Closure, 420
Cofactor, 381
Cofactor matrix, 174
Combination tones, 268
Comet trajectory, 123
Commutator, 427
Computer, 31
Configuration space, 79
Congruent transformation, 240
Conjugate of group element, 422
Conjugate momentum, 48
Conservative force, 17, 93
Constants of the motion, 50
Constraint, 35
Contact transformation, 289
Continuous group, 424
Continuous media, 253
Contraction of tensor, 137
Contravariant vector, 399
Convolution integral, 236, 270
Coordinate force, 339
Coordinate systems (illustrated), 2
Coordinate velocity, 337
Coriolis force, 186
Couple (torque), 31
Coupled springs, 45, 246, 259
Covariance in relativity, 341
Covariant derivative, 416
Covariant vector, 399
Cramer's rule, 385
Critically damped motion, 222
Cross product of vectors, 367
Cross-section, 112
Curl of vector, 368
Cyclic coordinate, 48
Cycloid, 76

D'Alembert's principle, 37
D'Alembert's solution for string, 227
Damped oscillator, 221
Deflection:
 of air mass, 188
 of light ray, 362
Degeneracy of frequencies, 306
Degrees of freedom, 37

Determinantal properties, 382
Dirac delta-function, 9, 374
Direction cosines, 128
Directrix for conic, 97
Dispersion relation, 250
Divergence of vector, 368
Dot product of vectors, 366
Double pendulum, 69
Dyadic or dyad, 138
Dynamic balance of tire, 151
Dynamics, 11

Earth's force-free motion, 200
Earth's interior composition, 165
Earth's precession and nutation, 213
Eccentric anomaly, 100
Eccentricity, 97
Eccentricity vector, 99
Eigenfunctions or eigenvectors, 144
Eigenvalues, 144
Eight-fold way, 440
Einstein summation convention, 137
Elastic collision, 19, 165
Electrical circuits, 46
Elementary row operation, 387
Ellipsoid of inertia, 196
Elongation, 158
Energy conservation theorem, 17, 117
Equilibrium point, 263
Equivalence principle, 347
Equivalent one-dimensional model, 106
Equivalent representation, 423
Euclidean group, 440
Euler angles, 134
Euler-Lagrange equation, 72
Euler's equations for rotation, 194
Euler's theorem on rotation, 173
Event (in relativity), 321
Event horizon of black hole, 363
Exponential form of rotation operator, 156
External force, 13

Faithful representation, 423
Fictitious forces, 185

Index

Finite rotation formula, 178
Focus of orbit, 97
Force-driven oscillator, 223
Force-free motion of rigid body, 195
Forces of constraint, 55
Foucault pendulum, 188
Fourier series for string, 228
Four-vector, 337
Friction, 12
Full rotation matrix, 136
Fundamental brackets, 281

Galaxies in astronomy, 109
Galilean transformation, 4, 319
Galilei group, 440
Gauss' theorem, 368
Generalized force, 39
Generalized momentum, 48
Generating circles, 269, 272
Generating function, 289
Geodesic, 90
Geometry, 349
Gibbs, J. W., 153
Gibbs-Appell function, 90
Gradient vector, 367
Gravitation, Law of, 26
Gravitational field intensity, 26
Group, 420
 of square, 421
Gyroscopic motion, 215

Hamiltonian density, 276
Hamiltonian function, 51
Hamiltonian system, 261
Hamilton-Jacobi equation, 297
Hamilton's canonical equations, 52, 277
Hamilton's principle, 78
Harmonic law, 102
Heavy symmetrical top, 205
Helmholtz' theorem, 164
Hermitian matrix, 145
Herpolhode, 197, 218
Holonomic constraint, 35
Homogeneous Lorentz group, 440
Homogeneous system of equations, 385
Homomorphism, 423

Ignorable coordinate, 48

Impact parameter, 111
Inertia, concept, 2
Inertia ellipsoid, 196
Inertial frame, 4
Inertia tensor, 141
Infinitesimal transformation, 309
Inhomogeneous Lorentz group, 440
Intermediate moment theorem, 201
Internal force, 13
Invariable plane, 197
Inverse cube force, 113
Inverse of matrix, 130, 386
Inverse square law, 96
Inversion of order, 380
Inversion operator, 175
Irrotational vector, 164
Isomorphism, 423
Isoperimetric problems, 76

Jacobian, 378
Jacobi identity, 281
Jupiter orbit, 104

Kepler's equation, 103
Kepler's harmonic law, 102
Kepler's law of areas, 94, 116
Kerr solution, 363
Kinematics, 2
Kinetic energy, 16
Kronecker delta-function, 129

Lagrange's equations, 40
Lagrangian density, 255
Lagrangian multipliers, 56
Lagrangian points, 122
Least action principle, 83
Least constraint principle, 85
Lenz vector, 99
Levi-Civita density, 180
Lie algebra, 429
Lie bracket, 429
Lie group, 424
Limit cycle, 266
Linear collisions, 19, 165
Linear transformation, 126
Line of nodes, 134
Locally inertial frame, 349

Loedel diagram, 327
Longitudinal wave, 164
Lorentz-Fitzgerald contraction, 325
Lorentz group, 440
Lorentz metric, 347, 352
Lorentz transformation, 321

Mach's principle, 5
Many-body problem, 115
Matrices, 129, 386
Maupertuis, 83
Mean anomaly, 103
Metric space, 349
Metric tensor, 402
Michelson-Morley experiment, 320
Minkowski force, 339
Minkowski space, 322
Minor in determinant, 381
Mixed tensor, 401
Moduli of elasticity, 161
Mohr's circle, 171
Moment, 7
Moment of inertia, 10, 143
Momentum, 3
Momentum-energy invariant relation, 341
Multi-valued function, 393

Newton's Laws of motion, 2
Newton's Law of universal gravitation, 26
Newton's Principia, 4
Noether's theorem, 64
Non-conservative force, 17
Non-holonomic constraint, 36
Nonlinear equations, 267
Non-singular matrix, 387
Normal coordinates, 237
Nutation of top, 208

One-dimensional crystal, 248
One-parameter subgroup, 428
Orthogonality conditions, 129, 132
Orthogonal transformation, 129
Overdamped motion, 222

Parallel axis theorem generalized, 144
Parity operator, 176
Particle, 1

Passive view of rotation, 132
Pauli spin matrices, 434
Perturbation theory, 311
Pfaffian constraint, 36
Phase plane, space, 261
Photon rocket, 364
Physical components of vector, 401
Pitch of satellite, 204
Poincaré group, 440
Poinsot construction, 195
Point transformation, 279
Poisson bracket, 280
Poisson bracket theorem, 283
Poisson's theorem, 286
Polar form of complex variable, 392
Polar vector, 182
Pole in barn paradox, 334
Pole of function, 395
Polhode, 197, 218
Position vector, 1
Potential energy, 17
Potential in gravitational field, 27
Potential well, 125
Pound-Rebka experiment, 363
Power laws of force, 110
Precession:
  of the equinoxes, 210
  of top, 208
Principal axes, 147
Principal axis theorems, 150
Principal diagonal (of determinant), 381
Principal function of Hamilton, 299
Principal moments of inertia, 147
Principal value of complex function, 393
Principle of equivalence, 347
Products of inertia, 143
Projectile motion, 300
Projection technique, 189
Proper acceleration, 338
Proper time interval, 330
Proper velocity, 337
Pseudoscalar, pseudovector, 182
Pure-strain dyadic, 158

Realization of group, 441
Rearrangement theorem, 444
Reciprocal base vectors, 404
Red-shift of spectrum, 363

Index

Reduced mass, 93
Relativistic Lagrangian, 342
Relativity:
 general, 346
 special, 319
Representation of group, 423
Residue of pole, 395
Response:
 to delta-function, 235
 of step function, 235
Restricted three-body problem, 118
Rheonomous constraint, 35
Ricci tensor, 354
Riemann curvature tensor, 353
Riemannian geometry, 350
Rigid body, 1, 173
Role of satellite, 205
Rotation dyanic, 152, 158
Rotation group, 431
Rotation matrices, 132, 135
Rotation-reversal theorem, 155, 192
Routh-Hurwitz criterion, 256
Runge-Lenz vector, 99
Rutherford scattering, 124

Saddle point, 265
Scalar product, 366
Scattering cross-section, 111
Schwarzschild radius, 363
Schwarzschild solution, 355
Scleronomous constraint, 35
Secular equation, 145
Self-excited oscillations, 268
Shear, 158
Shear modulus, 162
Shortest distance between two points, 72
Sifting property, 374
Similarity transformation, 140
Simple harmonic oscillator, 221
Singular point, 263
Skeleton figures, 10
Skewed coordinates, 408
Skydiver, 6, 31
Sliding hoop, 23
Solenoidal vector, 164
Space components of angular velocity, 191

Space cone, 199
Space-like interval, 330
Space shuttle, 364
Special orthogonal group, 432
Spherical pendulum, 69
Spin matrices, 434
Spool, 12, 32
Stable equilibrium, 220
Stable node, 264
Stability index, 273
Static balance of tires, 150
Statics, 11
Stokes' theorem, 369
Strain dyadic, 157
Stress tensor, 160
Structure constants, 427
Subgroup, 422
Superposition principle, 225
Symmetric matrix, 145
Symplectic approach, 278

"Tennis racket" theorem, 201
Tensor, Cartesian, 136
Tensor components, 136
Tensor, general, 398
Tensor of inertia, 141
Terminal velocity, 6
Time-dependent constraint, 62
Time-dependent perturbation theory, 311
Time-development operator, 311
Time dilatation, 325
Time-like interval, 330
Tippie-top, 209
Top, symmetrical, 205
Torque, 7
 on satellite, 202
Trace of matrix, 390
Transformation matrix, 129
Transform methods, 230
Transpose of matrix, 131
Transverse wave, 164
True anomaly, 100
Twin paradox, 333

Underdamped motion, 222
Uniform string, 226
Unimodular matrix, 435
Unitary matrix, 435
Unstable node, 264

Van der Pol equation, 276

Variation of constants, 313
Vector product, 367
Vector space, 425
Velocity, 2
Velocity addition formula, 364
Virial theorem, 109
Virtual displacement, 18

Virtual work principle, 18, 24

Work, 16
World line, 321

Yaw of satellite, 205
Young's modulus, 162